THE INCOMPARABLE MONSIGNOR

J. L. HEILBRON

the INCOMPARABLE MONSIGNOR

Francesco Bianchini's
world of science, history and court intrigue

OXFORD
UNIVERSITY PRESS

OXFORD
UNIVERSITY PRESS

CONTENTS

ACKNOWLEDGEMENTS

Owing to restrictions of movement and access caused by the pandemic, I have had to approach colleagues I did not know for copies of their work that, in the normal course, I would have consulted in a library. My requests were met with a spirit of generosity in accordance with the ideals of the old Republic of Letters championed by my protagonist Francesco Bianchini. I am particularly indebted to Riccardo Balestrieri, Joseph Connors, Edward Corp, Moti Feingold, Richard Maher, Brigitte Sölch, and Luis Tirapicos. At the outset of my studies of Bianchini I had the invaluable help of Ivano Dal Prete, like Bianchini a gentleman from Verona, and work of the contributors to international meetings about Bianchini in Verona and Augsburg. Like all scholars, I am indebted to the staffs, present and past, of research libraries. The institutions on whose resources I have relied and whose staffs I thank include the Biblioteca Civica and the Biblioteca Capitolare in Verona, the Bodleian (Oxford), the British Library (London), the Huntington (San Marino), and the Vallicelliana (Rome).

I must also thank many friends and colleagues for their patience in listening to my stories about The Incomparable Monsignor, among them Dan Kevles, Kanwal Misri, and Eileen Reeves. Alison Browning has heard them all, often, and has helped me refine them. I am indebted to Jeff Hodges for his inspired proofreading and owe special thanks to Stefano Gattei, who procured odd bits for me from Italian archives and helped in other ways to bring this work to press.

Knowledgeable editors are important members of the Republic of Letters. I am lucky to have one in Latha Menon at OUP. I thank her and her assistant, Jenny Nugee; Amanda Brown, who secured reproducible versions of the images needed; Hilary Walford for correct and charitable copyediting; and Roopa Vineetha Nelson for managing the production.

PROLOGUE

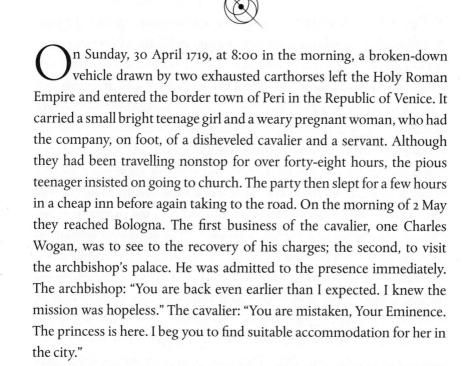

On Sunday, 30 April 1719, at 8:00 in the morning, a broken-down vehicle drawn by two exhausted carthorses left the Holy Roman Empire and entered the border town of Peri in the Republic of Venice. It carried a small bright teenage girl and a weary pregnant woman, who had the company, on foot, of a disheveled cavalier and a servant. Although they had been travelling nonstop for over forty-eight hours, the pious teenager insisted on going to church. The party then slept for a few hours in a cheap inn before again taking to the road. On the morning of 2 May they reached Bologna. The first business of the cavalier, one Charles Wogan, was to see to the recovery of his charges; the second, to visit the archbishop's palace. He was admitted to the presence immediately. The archbishop: "You are back even earlier than I expected. I knew the mission was hopeless." The cavalier: "You are mistaken, Your Eminence. The princess is here. I beg you to find suitable accommodation for her in the city."

After a few days in Bologna the company set off for Rome. They arrived on 15 May, slowed by the press of people welcoming the princess along the way. Among the first to bring her the pope's greetings after her arrival at the convent in which she would spend the next few months was the protagonist of our history, the incomparable Monsignor Francesco Bianchini. Why he? The answer fills most of this book. And who were Wogan and his princess and why were they in such a hurry? That too will be revealed.

Let us begin with "the never sufficiently to be praised" Monsignor, held by many of his contemporaries to have been the greatest Italian of his time, and by spokespersons from the generations that followed as "the greatest man that Italy produced in the [eighteenth] century," or, if judged by his depth and breadth of mind, the greatest Italian ever.[1] Among his accomplishments was writing a universal history from the creation to the fall of Assyria; recovering ancient calendars; discovering, excavating, and interpreting ancient buildings; designing a papal collection of antiquities later partially realized in the Vatican museums; undertaking a geodetic mapping of the papal states; confirming and publicizing Newton's theories of light and color; discovering several comets and a few variable stars; building the most beautiful astronomical instrument, and the most exact solar observatory in the world, in the basilica of Santa Maria degli Angeli in Rome; detecting the slow decline in the obliquity of the ecliptic and almost discovering the aberration of starlight (two key astronomical findings of the eighteenth century); and creating a map of nonexistent features on the invisible surface of Venus. His international reputation earned him election as a foreign associate (one of only eight) of the Académie royale des sciences of Paris and as a fellow (chosen on Newton's nomination) of the Royal Society of London.

All that was by no means all. As a trusted servant of Pope Clement XI, who reigned from 1700 to 1721, Bianchini helped execute the delicate balancing the papacy practiced during the War of the Spanish Succession, which pitted Britain, the Dutch Republic, and the Habsburg Empire against France and Spain. One of his assignments resulted in attachment to the cause and person of the Old Pretender, James III, the Stuart claimant to the thrones of England, Scotland, and Ireland. Bianchini embedded memorials of his involvement in events of European significance in the pavement of Santa Maria degli Angeli at locations corresponding to the sun's position on the dates on which they happened. This hard-copy history was a unique and, for those who can read it, extraordinary invention. Bianchini was subtle as well as clever. Although he accepted the Newtonian universe and the humanity of heretics, he managed to avoid the snares of conservative prelates and oppressive censors. He did not push against closed doors. No Galileo, he served his church and preserved himself, "a little saint and every inch a courtier."[2]

The story begins by explaining how its hero came to master the mathematical and historical sciences (and law and theology), the arts of diplomacy and dissimulation, and the patronage system that brought him into the company of popes and princesses. Chapter 1 takes him from his birth in Verona through his education in Bologna, Padua, and Rome; he emerged well trained in the Jesuit humanistic curriculum and, what sat uncomfortably with it, Galilean science. Chapter 2 finds him working up universal history and chronology from the records of all nations and reducing them to a single volume and two decks of playing cards. Chapter 3 brings together his astronomy and history at the solar observatory Clement commissioned at Santa Maria degli Angeli. It is still one of the best sights in the Eternal City. These interests took Bianchini to archaeology and early church history, and the task of superintendent of all ancient Latin inscriptions found in Rome. At the same time, his technical competence earned him the charge of overseer of two of the three main aqueducts in Rome's water supply.

Chapter 4 describes Bianchini's rise to chief intellectual of Rome and the main instrument in executing Clement's policy of using Roman monuments, art, science, and erudition to offset the loss of pontifical military power.[3] That demanded an agile performance by Bianchini because his increasing domestic importance competed with his cosmopolitan values. He decided to be a senator in the international Republic of Letters rather than the dictator of Italian culture. His even-handed internationalism aligned with papal policy during the Spanish war and involved him in the events that propelled Wogan's princess to Rome. These events, the subject of Chapter 5, included a diplomatic mission to France, which Bianchini extended to England ostensibly to see such marvels as the Parian marbles in Oxford and Sir Isaac Newton in London. Newton kindly overlooked Bianchini's position of domestic prelate to the Whore of Babylon and welcomed him as a fellow chronologist and astronomer. The cordial hospitality Bianchini received from well-placed English infidels enabled him to estimate for himself and the pope the chances that the successor to the reigning sovereign of England, Scotland, and Ireland, the Protestant Queen Anne, might be her nephew, the Pretender James. Chapter 6 turns at last to Wogan and the princess and Bianchini's recruitment to the Jacobite cause.

During the 1720s, Bianchini returned with released energy to his intellectual pursuits. Chapter 7 describes his wide-ranging productions in archaeology and early church history during the years between Clement's death (1721) and his own (1729). Although he was then in his sixties, he managed almost single-handedly to inaugurate a trigonometrical survey of the Papal States from Rome to Rimini and to map the suppositious features of Venus deploying a telescope 20 meters, yes, 65 feet, long. These are the subjects of Chapter 8. To be sure, the survey was a little out and the features spurious; but such was Bianchini's stature that redoing the Rome meridian anchored the famous survey by Roger Boscovich in the mid-eighteenth century and seeking spots on Venus kept astronomers busy for 150 years.

The final chapter disposes of the actors and describes several monuments to them that still reward inspection. Consideration of these monuments, which include part of the Vatican museums, makes a fitting conclusion, for, as Bianchini rightly insisted, historians ought not to omit the proofs and delights afforded by depiction of artifacts illustrative of the times they write about.

The conditions that made possible the life and work of Bianchini are hardly reproduceable today. He was sustained as a member of a collegial community of high-ranking ecclesiastics who competed in learning and spent their evenings listening to lectures in academies devoted to the physical sciences and early church history. Many of them were wealthy and none had the immediate care of a family. Church emoluments paid for their servants, lodgings, food, and learned leisure. They knew about fine art, ancient medals, Latin inscriptions, scientific instruments. They talked about the early Roman Catholic Church in the language it spoke. They lived in a perpetual seminar. Bianchini (see Fig. 1) was the exemplar of their culture.

Fig. 1. Francesco Bianchini (1662–1729) as he appeared during his excavations on the Palatine in the 1720s. FB, *Palazzo de' cesari* (1738), p. 1.

Material documenting his life is dauntingly extensive. The chief deposit, in Verona, contains more than ninety folio volumes, mostly in his hand. Other archives, in Verona, Rome, and London, have significant holdings. Over six thousand items of his correspondence are known, most of it unpublished.[4] His major printed works occupy two dozen volumes in Latin and Italian; his shorter writings many more. The number of relevant historical works ranging from the seventeenth century to the day before this volume went to press is enormous. Another visit to Italian archives to review this material would have been desirable. The pandemic ruled out trips that would have enabled me to fill a hole or two in the narrative, and check quotations and citations. In confronting these problems I have had the invaluable help of Professor Stefano Gattei of the University of Trent. And so I have reason to hope that I have not neglected any material that, if known to me, would have changed my characterization of the Incomparable Monsignor substantially.

1

A GALILEAN EDUCATION

Francesco Bianchini was a diplomat by nature and necessity. His ambitions were moderate, as were his means; his parents, Gaspare and Cornelia, were Venetian subjects of the merchant class, probably comfortably off but not wealthy enough to provide plentifully for their six children. Francesco's grandfather had settled in Verona around 1630 to escape the plague in his native Brescia; which he did, more by luck than design, for the plague hit Verona too, and proved more deadly there than in Brescia. This is a plague widely remembered; the beautiful church of Santa Maria della Salute in Venice visibly celebrates its termination, and a classic of world literature, Alessandro Manzoni's *The Betrothed*, takes place during it.[1] Although Verona did not prove a refuge from the plague, the city, lying at the intersection of trade routes from Venice to Milan and from Italy through the Brenner Pass to northern Europe, offered major incentives to energetic merchants who survived it. Verona also contained many ancient and medieval monuments for the inspiration of people who did not need or want to trade. Grandfather Bianchini became a middling middleman; his son Gaspare advanced the business; and his grandson Francesco became a scholar.

Francesco Bianchini entered this world on the twelfth day before Christmas in 1662. He would be the eldest male of six siblings. Perhaps birth order played a part in fixing the gentleness and helpfulness that contemporaries remarked in his character; he had to comfort his brothers and sisters after their mother's early death, when he himself was not quite ten, and he continued to care for them. There was not much disposable income: the family could not afford dowries for all the girls, and Francesco could not expect to advance significantly in substance or influence without a patron. The Lord provided one in good time, in the person of Pietro Ottoboni, then Bishop of Brescia, and a friend of one of Francesco's uncles. The bishop was the head of a Venetian family whose distinguished service to the city had enabled it to buy into the nobility at ruinous expense. Continuing to gamble, the Ottoboni invested their hopes and resources in the career of the Bishop of Brescia. They hit the jackpot four decades later, in 1689, when the aging bishop became pope.[2] Alexander VIII, as Ottoboni named himself, hastened to reward his relatives and their friends. Francesco would share in this outrageous nepotism. He had much in his favor besides his uncle's recommendation: loyalty, complaisance, and, what counted much among the Ottoboni, learning without affectation.

The Bianchini family lived at the Ponte delle Navi, a bridge over the Adige close to the church of San Fermo Maggiore[3]—a perfect spot for a bright boy interested in everything! Just to the west ran the wide river, implicit with the technological problems of canalization and flooding that occupied many of the mathematicians of Italy; to the east, the two-layered San Fermo, a Gothic structure built over a Romanesque one, a standing lesson in ecclesiastical architecture and history. A few hundred meters away stood the great relic of Roman times, the amphitheater, not much smaller, but much better preserved, than Rome's Coliseum. In the guidebook to Verona by Francesco's younger contemporary, fellow citizen, equal antiquarian, and frequent admirer, Scipione Maffei, we learn that in his time, the theater could seat 22,000 spectators; perhaps the ancients crowded in 30,000; today it admits only 15,000 for its famous operatic performances thanks to safety arrangements unknown when gladiators slaughtered one another there. Maffei dwelt on numbers: numbers of entrances, arches, steps, stairways, corridors, tiers, and drains, and measurements of everything measurable by line and ruler. How did so many spectators protect themselves from sun and rain? How large and how many were the conduits for rainwater and less pleasant liquid wastes?[4] Mathematics and archaeology are not enemies.

Maffei described many more items that excited the polymath in his city of 40,000 souls. He mentioned remains of ancient bridges and baths, walls, and a great gate in such good shape that he could fault only the abundance of its ornament. There were churches almost beyond numbering, many with medieval origins; collections of ancient coins and sculptures open to the public; the "noble and magnificent monuments" of the Scaligeri, from whom the learned and opinionated Joseph Scaliger, with whose chronology Bianchini would do battle, claimed descent; and old and new forms of fortification, on which he would become an expert. Ancient inscriptions abounded in private collections to which young Bianchini had access; Maffei eventually brought many of them together with other artifacts in a public *Museo lapidario* still to be found close to the amphitheater. In the churches of Verona, Francesco could contemplate many fine paintings by Venetian masters and, in the cathedral, a bit of bare wall where, in good time, his fellow citizens would place a monument to him (see Plate 16).[5] He had a taste for art and a talent for drawing. To finish with his serious boyhood amusements, he liked listening to music and, perhaps, making it, if it is true that he was a passable performer on the zither.[6]

With the Jesuits in Bologna

After he had finished his preliminary studies of humanities in Verona, the next obvious place for Francesco was a Jesuit college. Verona did not have one. Nor did any other city in the Veneto. The Jesuits had not recovered from their ouster from Venetian territory in 1606–7 in reprisal for an interdict declared against the Republic by the bellicose Pope Paul V, to whom Francesco was indebted for two more initiatives that would affect his career. Paul endorsed the Inquisition's prohibition of the Copernican system as contrary to scripture and absurd in philosophy and, more usefully, commissioned a high-pressure aqueduct, the Aqua Paola; Bianchini would have to sidestep the prohibition and supervise the aqueduct. But first he must have a Jesuit education. His parents sent him to Bologna, where, like Shakespeare's Veronese gentleman, he would begin "to see the wonders of the world abroad."[7] He was then eleven years of age. To help him enlarge his views, an uncle left him a legacy that not every boy of his age would have welcomed: a fund that he had to expend on good books between the ages of sixteen and thirty. Thus Francesco at the brink of adolescence began to acquire the heavy burden of scholarship.[8]

The distance from Verona to Bologna is about 145 km by road, a journey of two or three days by coach. Entering the northern districts of the city from the west and heading toward his college, Francesco would have passed the basilica of San Petronio, "the head and patron of the city," as a contemporary guidebook has it. Among the many items the book extolled in the basilica was one especially noteworthy, something unique in the world, an instrument for measuring the sun's noon height, "much more exact than any previously made," installed in 1655 by Giovanni Domenico Cassini, the most celebrated observational astronomer of his time. Thirty years later Bianchini would improve on this pioneering instrument with his grand *meridiana* in Rome. The guidebook mentioned other material connected with the basilica: a "perpetual" Easter calendar and what may mystify a modern, a table of times of local noon. Why a table? In Italy in Bianchini's time the day began half an hour after sunset and the interval between then and the following nightbreak was divided into twenty-four equal hours; consequently, noon occurred at 19:40 at midwinter and 16:21 at midsummer. Italian timekeeping reliably foxed foreigners.[9] The twin towers of Trinità dei Monti in Rome each had a clock, one (the only one in Rome) telling north European time, the other Italian time. It was unlikely (the opinion is Bianchini's) that the sacristan who regulated them understood them. Later Bianchini would try to teach the Jacobite court in Rome how to tell time.[10]

Continuing east past the basilica on Bologna's main road, now the Via Rizzoli, Francesco saw two medieval towers looming in the distance, one of which, the Torre degli Asinelli, had been used by a group of Jesuits led by Giovanni Battista Riccioli to check Galileo's law of free fall. Riccioli was the author of the standard handbook of astronomy of the time, a reliable compendium of hard fact about the universe and soft arguments against the Copernican system. He died two years before Francesco arrived at the college at which he had taught and where his shadow lingered. Having made proper obeisance at the Torre degli Asinelli, Francesco's party would have turned right into the via Castiglione, proceeding south another 500 meters to the "Jesuit Island," a large triangle bordered by the *vie* Castiglione, Chiari, and Cartoleria. The island accommodated the Jesuits' church of Santa Lucia, then recently remodeled in the style of their main church in Rome, the Gesù; their college for intending Jesuits and staff; a residence for the sons of nobles and another, the Collegio del Beato Luigi Gonzaga, for the sons of citizens and merchants.[11] Francesco was to reside at Beato Luigi for most of the seven years 1673 to 1680.

Beato Luigi opened in 1645 and had sixty or seventy boarders during Bianchini's time. Each paid 27.5 lire a month, about 5 scudi, which, converted to a yearly salary, would have been just enough for a single person to live on.[12] The younger Beato boys attended the lower school in grammar and rhetoric at the Jesuit College. The older boys could sit in on courses on logic, philosophy, and mathematics, a right then only recently acquired. Thirty years earlier, in 1641, Pope Urban VIII bowed to the university's demands that the Jesuits not compete with it in course content, and access to Jesuit philosophy and mathematics was restricted to Jesuit novices. Compliance was sporadic. When the papal legate tried to crack down, the college responded by suspending instruction in its lower school, until the loss of this essential service forced the city fathers and the university to surrender. It may surprise our era of quantitative illiteracy that the struggle centered on the right to offer courses in mathematics. The victory of the Jesuits in 1672 enabled Francesco to study with a maverick mathematician and natural philosopher, Giuseppe Ferroni, who had honed his skills with disciples of Galileo; Ferroni was "a rarity, the sort of thing found in an arcade," but a knowledgeable mathematician and effective pedagogue.[13]

It is to relax your minds, Ferroni would say to his students before showing them an experiment and developing its geometrical explanation. He enjoyed referring to principles quite contrary to the Aristotelian physics Jesuits taught using scary concepts like the horror of the vacuum. Francesco had to defend no fewer than eighty "physical theses" from Aristotle in his last year at college.[14] Ferroni did not publish his counter physics, nor his good advice to confess ignorance rather than profess philosophy when cornered; for he expected that the censors would not approve his proving, "with many apt experiments, the truth of the teachings of our admirable mutual master Galileo." That did not stop him from insinuating his views in print: using the old dodge of a dialogue, he put them in the mouths of two of his clever students, one a nascent Jesuit, the other Francesco.[15] Adimanto (as Ferroni disguised Francesco, probably in reference to a character in Plato's *Republic* notable for his discretion), speaks for the establishment.[16] When the other interlocutor, Silvio, declares a taste for romances and a preference for Copernicus, "although I know [his system] is considered false," prudent Adimanto replies that, although Copernicus had spoken openly with the encouragement of several bishops and a pope, severe laws had constrained speculation after the great discovery that "divine scripture speaks only too clearly about the stability of the earth and the movement of the sun." Silvio confesses that he had commended the dangerous doctrine only to pave the way to a clinching argument he had devised against it. The two novice astronomers then join minds to describe in detail, and so to teach, the system they propose to demolish.[17]

Adimanto offers the curious argument that, according to Aristotelian physics, a stone let loose from a tower on an earth moving in the Copernican manner would travel faster toward the east at midnight than at midday. (This supposes that the tower's daily and yearly motions align when it is opposite the sun at night and oppose when it is under the sun at noon; imagine the tower at the equator.) That does not happen: "Therefore, the Copernican system is false."[18] Very clever, novel, subtle, Silvio replies, and also (what he does not say) an attack on Galileo's favorite but flawed argument for Copernicus, which applied the same discrepancy in velocity to create the tides. Clever Silvio offers another argument, an ironical variant of a popular physical objection to a rotating earth. Do you recall the story in 2 Kings 20:11, in which Hezekiah asks for a sign that he will be cured of an illness, as the Lord had promised? The prophet Isaiah asks whether a sudden jump in the shadow on a sundial would do as a sign. Indeed, says the sick king, provided the shadow goes backward. And so it came to pass. How could such a thing happen on a Copernican earth? Stopping its spinning would have knocked down all the buildings and people in the world, which we know did not happen, because no report of it has come down to us. The argument from the sundial was an old chestnut in the Jesuits' anti-Copernican cupboard.[19]

Gentle Adimanto accepts Silvio's argument as erudite, robust, plausible, and unanswerable. The only way out for Copernicans, the two agree, is to appeal to God, for "a new miracle [that] kept the men and buildings standing."[20] Their argument was neither erudite nor unanswerable. It was a caricature of one pushed by Riccioli, which even he came to regard as indecisive.[21] Thus did Ferroni teach Francesco not only the elements of modern astronomy but also the merits and mechanisms of dissimulation in dealing with people in power unwilling to concede a lost game. The situation required tact. "I regret [wrote one of Ferroni's teachers to an imprudent student] that you have mentioned Copernicus unnecessarily, and the sentence against him; I exhort you to be more cautious in the future."[22] The seeds of Bianchini's diplomacy were planted early.

Like many bright boys impressed by the learning and discipline of their Jesuit teachers, and, in his case, by the sophistication of their teaching, Francesco wanted to join their order. His father objected: as the eldest son, Francesco had family duties incompatible with the vow of chastity. Although he did not fulfill those duties, the objection was apt: the boy had no business becoming a Jesuit. He lacked the calling. He did not want the cure of souls, missionary work, or responsibility for delivering the sacraments. He did take minor orders, eventually; but he ventured no further into the priesthood, though promised rapid advancement if he did.[23] Believing that sacrifice expressed the height of adoration of God's power and authority, he wrestled hard to win the everlasting contest "between the lower appetite of the senses and the dictates of reason."[24] Whether he ever lost the contest is not recorded.

Bianchini's nephew and first biographer, Alessandro Mazzoleni, mentioned "singular piety" as the first among his uncle's character traits. This singularity, owed intellectually if not emotionally to his many years with the Jesuits, was an uncomplicated compartmentalization. Bianchini deployed the rigorous test of reason in worldly matters while fully accepting the incomprehensible dogma of his church. "He believed firmly and was quite content to discover that the high mysteries of our Holy Faith are impenetrable." He saw no disabling inconsistency in defending Galileo and Newton and collecting historical material subversive of scripture while acting as a censor.[25] He could practice all the conventional religious virtues while discharging doubtful political missions. Though firm in faith, he was unblemished by bigotry; and, though he practiced every detail of liturgical and procedural protocol when he could, he thought it impertinent to suppose heroic observance pleasing to God. Perfect yourself by all means, but do not copy the mistake of astronomers who carry astronomical calculations to unobservable exactness. Meditate upon the cross with apt (but not rapt!) attention; it will straighten out your life just as (the analogy is Bianchini's) the right tool can prop up a ruinous tower. Remember always that blessedness lies in an inexplicable conformity of the "peaceful spread and operation in us of the Creator" with the delights of the exercise of reason.[26] Do not vex yourself: "The Lord calls us friends not servants." It is enough to "love to know and know to love."[27] There was no melancholy in Bianchini's religious makeup.

When he left the Jesuits at the age of eighteen, Francesco had the tools to master the learning of the time: fluency in Latin and some knowledge of Greek; a grasp of mathematics and initiation into Galilean science, both its content and its epistemology; and, what proved enabling for much of his work, his gift for drawing, which his teachers helped him develop into a skill. With these acquisitions, Francesco returned to the Veneto to attend the famous university in Padua. He enrolled as a student of theology. Father Ferroni had urged him also to listen to the lectures of Geminiano Montanari, who held a chair of mathematics in the faculty where Galileo had taught. Obedient Bianchini listened and soon acknowledged Montanari as his master.

With a Galilean in Padua

Montanari became a Galilean owing to a cause outside Francesco's psychological reach. It was a fight over a woman. The escapade obliged young Montanari to flee from Florence, where he was studying law, to Vienna.[28] There he ran into Paolo del Buono, a disciple of Galileo, who was simultaneously advising the Holy Roman Emperor about the drainage of mines and the Grand Duke of Tuscany about the program of a select academy devoted to Galileo's legacy. When in 1658 Montanari deemed it safe to return to Florence, the Grand Duke's Accademia del Cimento ("Academy for Experiment") had been in existence for a year. With del Buono's introduction, Montanari was able to attend its meetings, where he saw phenomena, like the perplexing rise of liquids in narrow tubes, that would direct his own experimental work.[29] To become a complete Galilean he had also to master astronomy. This he did in his hometown, Modena, by helping a practiced observer draw up a new ephemeris of the planets. After two years of this work, in 1664, by then an expert calculator, practical observer, and instrument designer, Montanari was deemed fit by Riccioli and other authorities to teach mathematics at the University of Bologna.[30]

Montanari threw himself into the good work of spreading Galilean science, or *fisicomatematica* as its practitioners called it. He attracted many students by repeating the demonstrations he had seen performed in Florence by "the first academy of philosophy promoted by experiments."[31] He followed up the old experiments on capillary rise with new accurate measurements and an atomistic theory of the phenomena.[32] Luckily, rival academicians challenged his results. Montanari responded by bringing the dispute to the judgment of the Royal Society of London. This characteristically aggressive move unexpectedly softened his behavior. The society found that his wrangling diminished the effectiveness of his "uncommon acuteness of mind" and advised him to stop. "Wrangling is for pettifogging attorneys, not for philosophers."[33]

Montanari took the advice and through the society became a principal informant about natural knowledge in Italy to the international brotherhood of experimental philosophers.[34] That was an unusual honor. In general the society was suspicious of Italian science because of such extravagances as an analogy developed by one of Montanari's rivals, Luc'Antonio Porzio of Positano, between the capillary rise of water particles and the clumping of maggots in rotten cheese. Anyone can repeat the experiment and understand the analogy by holding the cheese aloft: the creatures avoid falling from its moldy part by clinging to their mates fixed in its sounder part.[35] Intolerant of such speculations and eager to stay in touch with the society and its science, Montanari tried to learn English, "so as to know perhaps before anyone else some small things from the experiments of [Robert] Boyle and others published in that blessed language."[36] By working with Montanari on the astronomical observations forwarded to the society, Francesco learned something of the ethos and honors of the international Republic of Letters. Later he would be a main vehicle for the importation of Newton's science into Italy and try in his turn to master English.

The sort of physics Montanari taught Bianchini appears from his investigation of a bizarre phenomenon he read about in the Royal Society's *Transactions*. It concerned Dutch tears or Prince Rupert's drops, small tadpole-shaped pieces of glass that can be hit on the head with a hammer without breaking but that pulverize when their tails are snapped.[37] Montanari spent many weeks at the glassworks in Murano and Bologna examining minutely how methods of annealing, temperature regime, glass types, impurities, shapes, and sizes affected the performance of the tears. From this information he offered a plausible cause (not "naked truth," for good Galileans did not deal in that) of the explosive power: the straining of the internal particles of the glass by the quick tempering of the surface. As he showed by careful measurement, the strain left the drops between 1.5 percent and 3.5 percent less dense than untempered glass. The sudden release of internal strains at the tails caused the splintering.[38] It was, and was considered, a model work. Montanari wrote it up for the Royal Society.[39]

As an astronomer, Montanari could claim a more notable discovery. He seized on a comet that appeared in 1664, measured its progress by an instrument he had invented, corrected the measurements for refraction, estimated the comet's daily proper motion, and determined that its parallax placed it well beyond the moon, within the space that Aristotle had reserved for the unchanging heavens.[40] That was not his notable discovery. What then? Something incredible: the stars themselves are subject to fits. In strains worthy of Galileo, Montanari announced to the Royal Society the discovery of "many novelties concerning the heavens unheard of through the ages." Some stars varied in brightness, waxed and waned; no wonder that the ancients disagreed about the number of the Pleiades, since one of its seven stars sometimes disappeared. Montanari uncovered many other irregularities by comparing Galileo's depiction of the Pleiades with his own, drawn with "every possible accuracy," through a 63-power telescope he had made himself.[41] What caused the variability? His answer, in words "abhorred by many [philosophers], I DO NOT KNOW."[42]

The Royal Society circulated news of the stellar disappearances, and Montanari's stock rose on the international market. He was now recognized as "outstanding," "most ingenious," and "very skillful" in astronomy and mathematics, "most diligent" as an observer, and an experimenter of "uncommon learning and inventiveness." His growing reputation moved Bologna up a notch on the short list of places where *fisicomatematica* flourished.[43] It flourished not only in the city's university but also, unexpectedly, in its basilica. There, at Cassini's *meridiana*, Montanari observed many things, notably, on the evening of 31 March 1676, a great fleeting awesome fiery meteor. Combining his own measurements with others', he reckoned that the fireball sped by Bologna at a height of 150,000 braccia (say 70 km); had the same angular size as the moon but much greater brilliance; hissed as it flew; and exploded tens of seconds later as it passed near Livorno.

Montanari's account of the meteor accurately portrayed an apparition frightful where light pollution does not dim it. "It is almost impossible to have a more careful description of a bolide." So judged an authority who knew that the fireball was the incandescent envelope of extraterrestrial matter burning up in the atmosphere.[44] This capital fact Montanari did not know, and, like his contemporaries, he thought that the fireball emanated from earth and might fall back on it. Was that a reason for fear? Observation indicated that the *fiamma* (the incandescent envelope) had a diameter of a little under a kilometer. If it fell on Bologna, would it consume the city? No! Montanari figured that its great size easily accounted for its great luminosity without preternatural heating. He found by experiment that the number of candles needed to provide a given illumination increased with the square of the distance from them; hence the *fiamma*, standing at 150,000 braccia, had the intensity of $(150,000)^2/(1.5)^2 =$ 10 billion candles. Say sixteen candles per square meter, less intense than the lights on a birthday cake.[45]

Montanari's estimate of the *fiamma*'s height made the atmosphere at least ten times higher than the accepted estimate inferred from the refraction of starlight. It will not be necessary to follow his testing of Boyle's law, his measurements of refraction in partial vacuums, or his attempts to calculate a reliable hypsometric formula.[46] Nothing decisive resulted from these labors.[47] In the end Montanari issued the warning that *fisicomatematica* could not reliably reach the heavens. Stick with what you can touch and trust. "I am used to philosophizing from first-hand experiments . . . I fear that we all mislead ourselves when we want to discuss things that take place far from us, applying to them the same concepts we use for terrestrial things that we have in our hands."[48] Bianchini made good use of this doctrine in his historical studies.

Although the city fathers were proud of their mathematicians (they doubled Montanari's salary twice) and its *meridiana* (the only item in town that some visitors thought worthy of remark), Montanari came to think that his efforts on behalf of *fisicomatematica* were insufficiently appreciated in Bologna.[49] In 1678 he moved to Padua (see Fig. 2). He made himself useful there and in Venice as a practical man as well as a pedagogue. In Venice he furnished an observatory, reputed in Italy as the best in Europe

Fig. 2. Bianchini's *maestro* Geminiano Montanari when professor of astronomy at the University of Padua. Charles Patin, *Lyceum patavinum* (1682), after p. 108.

after the Royal Observatory in Paris, for the rich and influential Venetian patrician Girolamo Correr.[50] The new Paduan professor fired off a set of practical artillery tables deduced from experience and Galileo's parabolic trajectories. He reasoned "with order and clarity" (an echo of Descartes's method) about the ills of the "circulation" (an echo of Harvey's cardiology) of the Venetian currency.[51] He became an expert on the navigability of the rivers of the Veneto and the silting of the delta of the Po. And he mounted a prolonged physico-mathematical attack on astrology, which, as a professor of mathematics at Bologna, he had been obliged to teach, "without prejudice," largely for the use of physicians, although he had not been able to believe in it or in medicine since boyhood (see Fig. 3).[52] Above all, in Padua he breathed a freer air. He remarked to Cassini that while observing comets in Bologna he had seen evidence of the earth's motion, "of which I can now speak more boldly since I am out of the claws of the priests."[53]

Like Ferroni, Montanari sugarcoated epistemology in dialogues whose appeal may be inferred from one on inanity, or the nature of vacuum, which he composed a year or two before he left Bologna. The colloquists are himself, the architect and *fisicomatematico* Guarino Guarini, and a few dead philosophers. Aristotle and Descartes come forward to deny the existence of vacuum and dispute the nature of matter. "I gathered that they were almost as much in the dark as I." Guarini could not account for emptiness or inanity either, but, in the manner of Simplicio, the blinkered philosopher in Galileo's famous dialogue on world systems, stuck with received opinion until he had a better one. "I have no reason to believe that vacuum does not exist ... except that I have not seen completely convincing proofs; and without them I have seldom left my Aristotle."[54] Good advice but no solution. How to advance? Montanari appealed to the highest authority he knew.

"O great Galileo!" What am I to believe? I cannot conceive of a void space. Yet, if space were filled with very subtle particles, would these particles not have to be composed of still subtler ones, and so on, regressing infinitely, "as Renato des Cartes has done with very little credit?"[55]

Fig. 3. Bianchini's playful representation of Montanari's lectures on astronomy. Under Horace's slogan, "he threw away his paint pots and big words," the blindfolded figure of astrology, with her books and symbols, indicates Montanari's attitude to her vain and verbose subject. Biblioteca Civica, Verona, MS 2833, fo. 206d.

Galileo: "Tell me, Prof. Montanari, whether you understand the infinite."
Answer: "I understand only that I do not have the intellect to understand
it." Galileo: "Excellent . . . You have learned everything that can be learned
about infinity."[56] Inanity is infinite dilution of matter. It is incomprehensible, even to a philosopher.[57] Francesco took this point to heart. In his
notes on his teacher's lectures, he records that most of what we want to
know about nature is beyond our grasp.[58] Jesuits might "seek a middle
way whatever that might be between truth and falsity" in the matter of the
vacuum; but, even if we suppose its existence, that is to grant very little,
since the concepts of non-being and incorporeity involve metaphysical
questions "too far from experimental teaching" to investigate further.[59]

The lesson of moderation, even renunciation, in seeking the ultimate
principles of natural philosophy perfectly matched Bianchini's principles
and piety. It reduced the main questions of life and science to placing
the border between the rational and the incomprehensible, the know-
able and the ungraspable. Limitations to the reach of reason had to be
acknowledged; but premature surrender to impotence or authority was as
harmful as arrogance and overreach. Francesco's admiring emulation of
his guru extended beyond epistemology to ethics, to Montanari's dedication to students and public welfare. The master reciprocated. The disciple
became his collaborator and the heir of his books and instruments. Since
Montanari died in 1687, Francesco had the use of this inheritance, which
stuffed his already crowded library, for his entire career.[60]

The disciple wrote out several of the master's teachings at length. One
ingeniously approximates an apparently unknowable quantity, the size of
atoms, by mixing mathematics with "natural experiments, or the true history of the effects made by everyday operations of nature."[61] Take a piece
of copper wire an inch in diameter covered with a layer of silver of imperceptible depth; when drawn out through successive holes to a diameter
1/300 of an inch or less, the wire will still retain a silver lining. Then do as
Montanari did: rub a piece of the drawn wire with a bit of card 1,000 times
until the card comes away with traces of copper. If each stroke removes
a layer one particle deep, none can have a diameter greater than 1/300,000

of the imperceptible thickness of the original silver layer.[62] Say the thickness of the original layer was a tenth of a millimeter; Montanari's estimate of the size of a silver atom would then be 3×10^{-8} cm; quite a good result.

Further information about the range of subjects Francesco learned at Padua may be inferred from his notes on Montanari's lectures. They begin with trigonometry and problems worked with logarithms of the trigonometric functions. Then come military architecture, mechanics, experimental physics including pneumatics, hydrostatics with much on Archimedes, motion (collision, impulse, center of gravity, free fall, resistance), atomic theory, and praise of Galileo.[63] Mechanics, especially simple machines, is treated in detail, as is fortification, down to tables of angles of sight from the various bastions. Francesco's illustrations, as exemplified by his depictions of capillary rise, pulleys, and a capstan, are full of realistic embellishments (see Fig. 4 (a–c)). Galileo appears often as a master of the workings of abstruse things like screws, an authority on motion, and a marvelous contriver of "speculations . . . on physical-mathematical matters."[64] In short, a course still desirable.

Montanari may have enrolled Francesco in his pursuit of the comet of 1680–1, for soon the student had learned enough to be able to spot and plot a comet on his own, in particular the comet that disclosed itself in July 1683.[65] Francesco's unusually keen eyesight gave him an advantage in hunting for vanishing stars as well as for comets. Examining the constellation Leo with particular attention, he and Montanari recorded many discrepancies between their observations and stellar intensities given in the standard star atlas of the time, Johannes Bayer's *Uranometria* (1603); a typical record by Francesco reads, "My observations made at Padua 2 April 1683: the star o in the [lion's] foot equal to the star η, star ν hardly visible . . . star o in the tail near star β did not appear at all, φ two new [stars] but very small . . ." (see Fig. 5). Apparently stars had a "genius for instability."[66] That was a wishful overestimate. Of the many stars he accused of fickleness, perhaps only two were true variables. No matter. For young Francesco, participating in observations that nightly brought

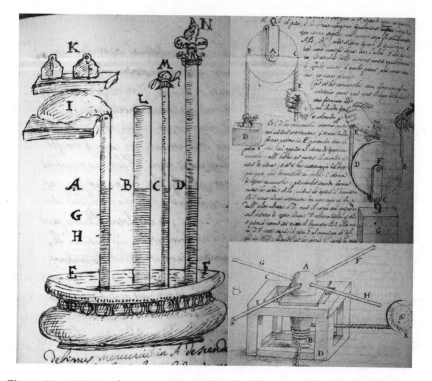

Fig. 4. Montanari's physics as recorded in Bianchini's notes and drawings, 1682 and 1683: capillary rise to equal heights regardless of tube bore, pulleys at work, and a capstan. Biblioteca Civica, Verona, MS 2833, fos 183^r, 136^v, 133^v, respectively.

new evidence against received views about the heavens was a heady experience. He discovered that he could make discoveries! And discoveries all the more valuable for agreeing with the then ultramodern cosmology of the "bizarro genio Francese" Descartes, in which a star can easily convert into a comet or planet, and the only ideas clear and distinct enough to design a cosmos are the concepts of mechanics.[67]

Francesco discovered several people important for his future in Montanari's entourage. Among them were Gianantonio Davia and Luigi Ferdinando Marsigli, both from Bologna, who would commence their careers as *fisicomatematici* and rise respectively to Roman cardinal and imperial general. Another was Correr. Observing the young man's talents and malleability, Correr thought to broaden his interests to include

Fig. 5. Variable stars in the constellation Leo sketched by Bianchini during hunts for them with Montanari. Francesco Porro, *Schizzi di carte celeste* (1902), tab. XII.

money. Or, rather, ancient coins and their application to history, which fascinated Correr more even than stars. Francesco resisted: he was an astronomer, not an antiquarian. Correr: "Why should a mathematician not be an antiquarian? Who better?" A mathematician finds truth in demonstrations, the historian by observation of ancient monuments. Both methods are secure. "Medals or stones are dependable if contemporary with the facts [they record] and exposed to public view." In this dictum Francesco glimpsed the contours of his later universal history.[68] He learned a similar lesson from a professor of medicine at Padua, Charles Patin, another man of science addicted to numismatics. Patin pointed out that many physicians collected old coins with the serious purpose of supporting historical study. "There is no study more delightful and useful, and . . . no way to learn it more quickly and enjoyably than by familiarity with medals."[69] Francesco began to cultivate the necessary familiarity.

Through another Paduan professor, Carlo Rinaldini, a former associate of the Accademia del Cimento, Francesco may have met Elena Lucrezia

Corner (not Correr!), the first woman to receive a doctorate in philosophy in Italy. She could defeat an entire faculty in disputations in any language they chose, knew astronomy and mathematics and physics as well as Galileo, and could twist theology better than her teachers. Need it be added that she was beautiful, modest, and pious (see Fig. 6)? Although a Benedictine oblate, she was a public figure during Bianchini's time at Padua and well within his range of possible acquaintance through her preceptor Rinaldini.[70] When the body that nourished her mind gave out in 1684, at the age of 38, Francesco composed an ode to her memory. She must be among the gods, he wrote, but where? No single place in

Fig. 6. A woman young Bianchini admired, Elena Lucrezia Corner. *Pompe funebri* (1686), after p. 12.

the Pantheon could do justice to her strength and striving. Where then? "Una tibi domus omnis Olympus."[71] Francesco liked smart pious well-bred women. Half a lifetime later he would become devoted to a young woman Elena's equal in fortitude, piety, and physical charms, lower in learning no doubt but higher in rank: the runaway princess we have yet to meet.

Physico-Mathematics in Rome

After graduating from Padua with a degree in theology, a lot of *fisico-matematica*, and a whiff of numismatics, Bianchini (as we shall now call him) went to Rome to study law at the undemanding civic university, the *Sapienza*.[72] The reigning pope was the now Blessed Innocent XI Odescalchi, the architect of the Holy League against the Turks, which a year earlier, 1683, had won its great victory before the gates of Vienna.[73] Several main actors in our story had ties to the commander credited with the victory, the Polish king Jan Sobieski. Highest on the list was the western counterpart to Sobieski, the stalwart Catholic James II Stuart, who succeeded to the thrones of England, Scotland, and Ireland in 1685. He did not enjoy them for long, however, for James's Protestant subjects threw him out for his efforts to return them to their old faith. James might have expected Innocent's support in improving the plight of Catholics in England. But the holy work of restoring the Stuarts could be accomplished only with the help of Louis XIV of France, and Innocent would do nothing that might strengthen Louis, whose irrepressible expansionism had caused him to wage war with Catholics as well as with Protestants.

While conquering his neighbors, Louis had helped the Turks keep the Austrians at bay. Worse yet, he had his tame clergy issue, in 1682, a set of "Gallican" propositions that made him head of the church in France with the right of appointment of bishops and control of income from unoccupied benefices. It therefore behooved the Holy Father to do what he could to weaken His Most Christian Majesty. There would be no papal help for Louis's ally James. That did not please a capital figure in our story, Giovanni Francesco Albani, the future Pope Clement XI, admired by many for his "perfect erudition in profane matters . . . and the profundity [of his knowledge] of sacred things." Albani advertised his support of James in the environment he preferred above all others, an academy of letters and science, and one particularly apt for his political purpose. This was the famous academy sponsored by the former Queen of Sweden, Christina, who had abdicated freely and emigrated in state to Rome, in marked contrast to James, who had retained his religion and fled ignominiously to France. Albani's advertisement contains the kernel of the policy that, as pope, he would pursue with Bianchini's diligent help.[74]

Albani told Christina's academy that James, Christina, and Jan Sobieski were royal witnesses to the Catholic faith from the western, northern, and eastern edges of Christian Europe. After appropriate compliments to his hostess, he dwelt on the asymmetry of Stuart and Sobieski in Innocent's foreign policy. Everyone knew that the reigning pope had helped to finance Sobieski's armies. And, in so far as James strove to return Britain "to the sweet obedience of the venerated laws of the Vatican," Innocent should have helped him too. Albani was too good a politician to remark that James orbited around the Sun King and that Innocent did not approve of the itinerary. The orator then attributed to Innocent his own strong desire to see "the clearest light of the faith spread widely and firmly over England." Events in the East supported hope for the West. "We already see with great satisfaction the squalid sullen Ottoman moon sinking swiftly towards its most deserved setting."[75] Pursuing the hope, Albani would bring about the physical union of a Stuart and a Sobieska to produce a family of claimants to the inheritance of James II.

Innocent's domestic programs had mixed results that did not make Rome a happy place during his reign. He inherited a huge debt, some 50 million scudi, about twenty times the Vatican's reliable annual income. Being frugal and austere by nature and a banker by inheritance, he was able to clamp down on expenses and cut the debt by ten percent. He was generous to the poor and uncharitable to prelates whose lifestyles brought scandal to the church. An authoritarian prude, he centralized power, controlled the theater, cancelled carnivals, fought luxury, and forbade French fashions; luckily, he was not interested in ideas. The learned circles to which Bianchini had access through Montanari continued to meet, although the pope was at war with Turks, at daggers drawn with Louis and eventually also with Christina, and at odds with fun-loving cardinals.[76]

The center of these unperturbed circles was Giovanni Giustino Ciampini, a senior official in the papal chancery who drafted papal bulls and consistorial decrees. He owed his decisive pull up the Vatican ladder, and the consequent lucrative opportunities to help petitioners frame pleas to the pope, to the same Pietro Ottoboni who patronized Bianchini.[77] Monsignor Ciampini had two great projects outside his day job: knowing everything and promoting the careers of people who could help him to omniscience. He had the requisite drive, self-confidence, acuity, and perseverance, and their side effects of haste, obstinacy, conceit, and oversensitivity. He generously shared his means and his learning, his collections of books, instruments, and artifacts that flowed into the corridors and stairways of his house, with experienced prelates and promising beginners alike. He was particularly attentive to the needs of younger scholars like Bianchini, to whom he gave the freedom of his overstocked house just behind the church of Sant'Agnese in Piazza Navona.[78]

Ciampini established two long-lived academies, one for ecclesiastical history, the other for physical science and applied mathematics. Since the historical at first did not engage the interests of Montanari's student, we shall leave it aside until it did. But Bianchini immediately felt at home in the *Accademia fisicomatematica* headed by Montanari's surrogate Ciampini.[79] It had come into existence in 1677 with the participation of members of Queen Christina's academy, where astronomy, astrology, and up-and-coming prelates held privileged positions. Among these prelates was Enrico Noris, a Veronese schooled by Montanari, on the way to the cardinalate, and the future popes Ottoboni and Albani. Among the astronomers they heard was young Bianchini, who told them about comets and stayed to hear the theories of Descartes and Boyle discussed favorably.[80] Like Christina's academy, Ciampini's was a center of social life for its celibate members. It met on Sundays at his house until the pressure of numbers required a larger venue. To enhance the pursuit of knowledge and sociability, Ciampini added to his offerings evening gatherings on every weekday but Wednesday and Saturday and corresponded with enclaves of the republic of letters throughout Europe.[81] "The lynx-eyed Abbot Bianchini" was soon responsible for carrying on correspondence about astronomy with such eminences as England's astronomer royal, John Flamsteed.[82]

The lynx was not the academy's leading man of science. That role belonged to a Jesuit, Francesco Eschinardi, professor of mathematics at the Collegio Romano, who did not share the principles Montanari taught.[83] Eschinardi's main problem in natural philosophy was how to handle Galileo. "It often happens that the works of great writers are read with a judgment disturbed by their reputations, which does not allow for any suspicion of error." And so Eschinardi quibbled over little mistakes in Galileo's writings.[84] Yet, when required to discuss moving bodies, Eschinardi followed Galileo's lead and so produced a most original demonstration of Galilean ballistics.[85] An arquebus ball had come suddenly through the window of a room in which Eschinardi was sitting. He could see no window opposite from which a rival theorist could have fired. Reasoning then from the geometry of the situation, he deduced that the parabolic path of the ball had begun in a nearby tower. With the courage of a missionary, Eschinardi climbed the tower to discover a young man sitting there cheerfully shooting at pigeons. News of his feat traveled. Murder suspects tried to enlist him to prove their innocence by showing that they could not have been in position to fire the fatal shot. He declined the opportunity to set a precedent in jurisprudence.[86]

Heaven being a dangerous place for Jesuits, Eschinardi kept his astronomy close to observation and lobbied for the construction of a *meridiana* in Rome to improve on the values obtained in Bologna's San Petronio.[87] In mundane affairs he stayed with practical topics such as carriage design, the operation of rudders, and the improvement of clocks.[88] As a *fisico-matematico* he took on all things calculable without doctrinal entanglements: the shapes of optical images, the operation of burning mirrors, the amplification of sound, the floods of the Tiber, thermometers, and much else.[89] Among the much else was the barometer, but, as it involved concepts of the vacuum, atomism, and infinity, Eschinardi was content to observe that they had their difficulties. In that he agreed with Montanari; but, whereas the Paduan professor exploited them despite their obscurity, the Roman Jesuit rejected them because of it.[90]

Bianchini probably heard some of the five discourses Eschinardi presented to the *Accademia fisicomatematica* around the time of his arrival in Rome. One proposed digging a canal between the Mediterranean and the Red Sea; another analyzed (and rejected) the alleged power of the remora (a small fish) to stop large ships; a third speculated that comets consist of bits left over from the fashioning of the planets and so move like them—that is, "by a certain intrinsic force inclining [them] to circular or spiral motion, etc., analogous to the force that ordinary bodies have to go to the center of the earth." Eschinardi offered this combination of Galileo and Aristotle, which would underlie Newton's gravitational theory, as the common opinion of modern philosophers.[91] Yet, he said, correctly, the real motions are complex spirals; we make fictions for calculation; and the best available fiction is one that leaves the earth as we experience it, at rest.[92]

Ciampini's interests in *fisicomatematica* concerned such useful matters as magnetism, the movement of vehicles, the flight of birds, and artificial wings.[93] He therefore rated the Dutch engineer Cornelis Meijer, who had arrived in Rome in 1675 and worked on the levees of the Tiber under several popes, very highly.[94] The feeling was mutual. Meijer thought that Ciampini's academy deserved much of the credit for improvements in Rome because of its promoting "useful discoveries continually drawn from the inexhaustible mines of mathematics."[95] He published his first collection of gems from these mines in 1681. It coupled items in the pure Galilean experimental tradition with two proposals for the enhancement of Rome.[96] One, promoting restoration of navigation on the Tiber, amounted to a business venture; the other, transforming the main obelisks of the city and their *piazze* into sundials, carried the gnomonic art to unreachable heights. Meijer proposed an obelisk for Piazza Monte Cavallo on the Quirinal with two monstrous globes at its base, one terrestrial, the other celestial, both turned by clockwork, and for St Peter's square, which already had an obelisk, six planispheres, one each describing the systems of Ptolemy and Tycho, the other four the celestial and terrestrial globes (see Fig. 7). On its steps would be inscribed the dates of all the comets that had appeared since the time of Christ.[97] Meijer's subsequent collections of specimens from the mines of mathematics describe comets, lodestones, eyeglasses, and, in collaboration with Montanari, eclipses of the satellites of Jupiter.[98]

One of Meijer's designs still has some shelf life. It is a one-room flat for scholars complete with all work-at-home accessories and an advanced security system. Through a telescope and camera obscura the owner could detect would-be visitors without getting out of bed; through a periscope, happenings in the street; through acoustic tubes, plots and gossip throughout the building housing his cell. The well-protected hermit had pistols and daggers ready to hand to ward off attacks on his gold coins and scientific instruments. There were ample provisions for company, a corner for a dog, cages for birds, and a contrivance to allow chickens to enter to lay eggs. A fireplace kept the flat warm, provided running hot water, and the possibility of a distillery. Naturally there was a wine cellar. The place would have suited Bianchini. Among the necessities with which Meijer provided his one-room flat were a thermometer, microscope, and celestial sphere, cabinets for books, and, on the ceiling, receiving sunbeams from a mirror in a window, a meridian line.[99]

The upshot of Bianchini's first years in Rome was introduction to a key academy, integration into the wider intellectual milieu, practice in being

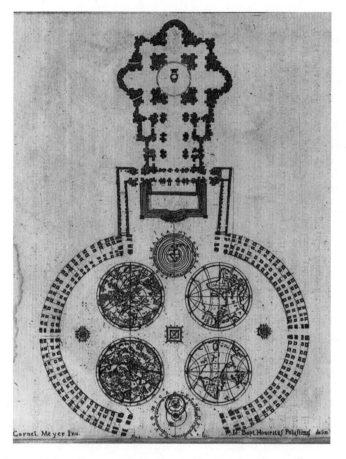

Fig. 7. Proposal for a huge sundial in front of St Peter's. The small square at the center of the colonnades is the obelisk then (and still) existing; the planispheres on the nave axis show the Tychonic and Ptolemaic systems; the others represent the celestial and terrestrial spheres. Cornelis Meijer, *L'arte* (1685), pl. XV.

a Galilean where Galileo was not entirely approved, and exposure to big projects by *fisicomatematici*. He was to have a hand in realizing Meijer's scheme to improve waterways and transform obelisks, and Eschinardi's proposal of a large Roman *meridiana*. These achievements would not come about soon, however, because, among other reasons, in 1686 Bianchini had to interrupt his advancement in Rome to take care of a family matter in Verona. A little matter of murder.

Back Home

The import–export business in Verona had its excitement. A transaction that involved a murder touched Gaspare Bianchini closely enough that it took his eldest son two years to persuade the authorities to drop the charges. While waiting, Francesco instructed local savants on the methods of philosophizing practiced in Padua and fished for a job in Rome.[100] He also took his first steps into espionage. That was around 1687, when Louis XIV, planning as usual to cause trouble for the Habsburgs, was building a fortress at Guastalla close to the Po river in the territory of his ally the duke of Mantua. Bianchini's skill in drawing, study of fortification under Montanari, knowledge of Verona's forts as described by Maffei, and ability to travel as a savant without raising suspicion, well qualified him for spying out the doings at Guastalla. On the recommendation of Correr, who was then serving as Venice's representative in Verona and had come to know Bianchini well, the Venetian Senate commissioned our scholar to record the plan of the fortress, estimate its dimensions, and identify its weak spots. All this he did (the weak spots were insufficient provision for artillery and awkward placement near a river) while hiding in a ditch (Fig. 8). He did so well that he was asked to suggest improvements in Venetian forts and ordered not to share his drawings with anyone.[101]

The episode provided two examples of the great truth that power is transient. For one, although experts rated the fortress of Guastalla able to withstand a three-month siege by 50,000 men if properly supplied, Spanish troops razed it without resistance in 1689, before it could be garrisoned. For another, the Duke of Modena, Mantua's enduring rival for the duchy of Guastalla, had not been able to obtain it despite the support of his sister Maria, the Queen of England. As the last ramparts at Guastalla took shape, Maria followed her husband James into exile.[102] Bianchini thus became aware of the Jacobite cause and its Italian connections long before he began to commemorate it in the pavement of Santa Maria degli Angeli.

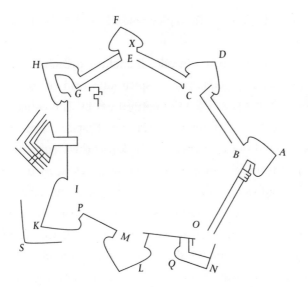

Fig. 8. First fruit of Bianchini's espionage, a plan of the fortress at Guastalla, 1686. Redrawn from FB, *Relazione sulla fortezza di Guastalla* (1885), end plate.

In his hometown Bianchini helped to establish an institution that would last longer than the fortifications of Guastalla. Collaborating with young physicians discontented with medical bosses and Galenic theories, he set up an academy of experimental science, the *Aletofili* ("truth-seekers"), to whom he taught the *fisicomatematica* he had learned from Montanari. Perhaps more to their interests, he showed how to apply the mechanistic physiology he had picked up in Rome to a current problem of medical concern.[103] A woman from Parma had claimed to be able to see the furnishings in her room in pitch darkness when awakened suddenly from a troubled sleep. Bianchini traced the symptoms, some of which he had experienced, to a superabundance of vital fluids in the nerves suddenly released through the eyes. The fluids cause the surfaces of objects they strike to vibrate, and the vibration, propagated back to the eye, causes the sensation of light and colors. The explanation, perhaps even then not entirely persuasive, was pregnant with the future. The "light of modern philosophy" dispels mystery, Bianchini said, and *fisicomatematica* can "exorcize ... superstition," making marvels no less marvelous by attaching them to plausible causes.[104] Had he Montanari's *fiamma* in mind?

Continuing in missionary mode, Bianchini preached that modern philosophy aimed to "construct a spiritual world of knowledge and understanding" that paralleled the natural world. The construction had to follow certain rules, "on the understanding that it may be improved by every just demand of experience."[105] We know these rules or principles, "the few, clear principles of quantity and motion that nature presents to everyone as unquestionable and that mathematicians suppose without needing proof."[106] And yet we also know that physical concepts are not truer than the experience on which they rest; we must not follow misguided philosophers who feign the shapes and sizes of the ultimate particles; we are just guessing in assigning causes.[107] A probable mechanical cause may be all that we need, however, for it is enough to make science useful. "There may be many experiments to do: but for this purpose we have eyes and hands given us by a Providence that wants us to know its contrivances."[108] And the guidance of Galileo, who showed us how to erect "the edifice of sound philosophy." Those who walk with him march in the van of history. "The philosophy brought to Italy by Pythagoras from Egypt and Greece, and buried there with the Roman republic, came forth again with the birth of Galileo, and is now close to delivering immortality to the Italian name."[109] As his international reputation grew, Bianchini would become an eloquent opponent of this sort of academic jingoism.

Most of what Bianchini transmitted to the Aletofili Montanari had not published. As his master's literary executor, Bianchini printed the lecture on atoms and a long unpublished "physical mathematical dialogue" on whirlwinds. To them he prefaced an account of Montanari's life. He emphasized his master's religiosity and insistence that science serve the public good. From another manuscript he added a snippet that showed to perfection Montanari's "clarity of intellect, quickness and correctness in application, facility and common sense in ordering and developing his principles, energy in finding the best ways to ascertain the truth, circumspection in discussing complex experiments, [and] the roaming (so to say) of his mind over the entire subject when contemplating its parts and consequences." The snippet measures the limit of resolution of the ordinary eye. Montanari found that his students could just distinguish two distant marks separated by one or two minutes of arc. Then, "roaming his mind over the entire subject," he estimated the size the fibers of the retina must have to achieve this discrimination. Since the distance from lens to retina averaged half an inch Venetian measure, it followed that the minimum separation of the fibers in the eyes of his students was 0.0004 cm, which revealed something of "the greatest and most admirable subtlety with which [nature] conducts its operations."[110] Montanari's result is of the right order of magnitude.

Bianchini's help in establishing the *Aletofili* had at least one important practical outcome: the physico-medical academicians won their battle against the local college of physicians.[111] That happened in 1700, a dozen years after Bianchini had returned to Rome. He had left the doctors with instructions for further work in experimental science and a fine slogan, *Aut docet aut discit*, "[the Academy] either teaches or learns," signifying that much remained to be added to what others had discovered. Follow the method, and enhance the glory, of the Galilean school, observe and experiment and report with befitting modesty. A true philosopher, literally a lover of truth, must not be arrogant. Bianchini's want of arrogance was widely remarked as unusual in "men of letters."[112] He shied away from disputes, avoided controversy, and spoke easily to everyone; in short, he behaved as if he knew neither mathematics nor Greek.[113]

Bianchini kept Ciampini informed about the prospering of experimental philosophy in Verona and asked his help in leaving the city as soon as possible. He wanted a position in Rome but no church benefice, for the unusual reason that, since he had not taken orders, he would violate the sacred canons by accepting one. He planned to live on very little, 120 or 150 scudi a year, and wanted nothing better than a place to read; he would settle for any employment in the Vatican Library. Ciampini referred this modest request to Emmanuel Schelestrate, the first *custos* of the library— its effective head under the Cardinal Librarian—who thought there might be an opening as a *scriptor* (cataloguer and copyist) of Greek works. When that fell through, a similar possibility in Hebrew surfaced; Bianchini threw himself into the language, of which he then knew scarcely a Masoretic dot; but the Hebrew post went the way of the Greek. Schelestrate sugared the pill: both he and Pope Innocent (still the 11th of that ilk) knew that Bianchini could do much better than ill-paying slavery over hard Hebrew texts.[114]

In the late fall of 1688 Bianchini could at last return to Rome. He planned to stay and so required a separate coach for his scholarly gear, which included, besides his books, a large clock and a larger telescope. (Since the city discouraged importation of mathematical instruments by taxing them, Bianchini also asked for help to store them outside the Piazza del Popolo.[115]) The Ottoboni immediately took him on as their librarian.[116] It was a much better job than Vatican *scriptor*, and a luckier one. The year after Bianchini had taken up his post, Cardinal Pietro Ottoboni acquired two thousand manuscripts and many printed books from the rich library of Queen Christina, who died in 1689. To this booty—literally booty, as Christina's magnificent holdings consisted largely of items stolen by her father, Gustavus Adolphus, from German collections during the Thirty Years War—Ottoboni added the splendid enrichments available to a nepotistic pope with a needy family. He became Alexander VIII the year Christina died and, being old himself, lost no time in distributing spoils. He did not forget Bianchini. "We are now Pope; what do you want, what can we give you?" "Your blessing." Alexander complied and added two pensions and a canonry aptly located in Santa Maria della Rotonda, the church established within the Pantheon of ancient Rome.[117]

The greatest gainer from Alexander's generosity was his great-nephew, another Pietro Ottoboni, whom he made a cardinal, placed in charge of the family library, allowed to read prohibited books, gave offices and sinecures worth 50,000 scudi a year, and established as Vice Chancellor of the Church. This post oversaw a warren of offices concerned with the issuance of papal bulls and other money matters; it brought a fine income (there was no office of Chancellor) and residence in the huge

Palazzo della Cancelleria just north of the Campo de' Fiori. Huge indeed: the palace housed a minor basilica (San Lorenzo in Damaso) and, in five rooms, the seventeen-thousand-volume Ottoboni library. And hugely in need of repair. Innocent XI had not filled the office of Vice Chancellor or maintained the building; Ottoboni, then twenty-two, had to see to its furnishing and decoration, for which he employed a team of artists for several years.[118] That suited him. He liked the arts. He had not been destined for the church and was enjoying a serious flirtation when named cardinal. He did not bother to become a priest, or even a subdeacon, for forty years, and before taking that drastic step continued in correspondence with his superannuated inamorata, much as Jonathan Swift was then unconvincingly romancing his Stella.[119] Ottoboni's true love was opera. He built a theater within the Cancelleria and commissioned works by Händel, Vivaldi, Scarlatti, and Corelli for performance there and elsewhere. He patronized all other arts, plastic and literary, and devoted his residual energy to increasing his collections of artifacts and offices. He spent so generously and lavishly and conspicuously that he died heavily in debt.[120] In the net, the Ottoboni may have kept nothing from the windfall of Alexander VIII.

Bianchini took greater pains to retain what he had. Soon after beginning as librarian to the newly prosperous Ottoboni, he thanked Alexander for the pleasure he anticipated in serving the pope, the scholar, and the public, and for the emolument the service brought. And he pledged faithfulness to Queen Christina's policy of open access: lovers of learning must be encouraged, whatever their status, "for those most attracted to serious studies will be most useful to the state." Perhaps he had himself in mind. He added that the size of a library measured the merit of its owner. Here he had the Ottoboni in mind. "He who builds libraries for himself and others . . . excels in the arts of a Prince."[121] Bianchini held his post as Ottoboni librarian for a dozen years and retained access to it and his good friend the Cardinal Vice Chancellor, with whom he was immediately on a basis of mutual trust.[122]

In combining scholarship with service as head of a princely library, Bianchini had many distinguished models.[123] He singled out his fellow Veronese and *fisicomatematico*, Enrico Noris, "one of the most illustrious persons of our age," who succeeded Schelestrate as first *custos* at the Vatican in 1692. Noris was a formidable scholar and, although, we are told, exceedingly modest and gentle, accepted no one as a colleague who could not study for more than eight hours straight; a requirement that must have reduced demand for desk space at the Vatican Library. Noris directed his compulsive studies toward history and chronology. We may justly see autobiography in Bianchini's description of Noris's reactions to his first exposure to Rome, "overwhelmed by the ancient monuments ... [and] the number of famous libraries crammed with printed books and manuscripts."[124] The relationship between the two Veronese savants became particularly close a few years before Noris's death in 1704, when, owing to his unexcelled mastery of ecclesiastical history and calendrical calculations, Noris was put in charge of the committee that oversaw the installation of Bianchini's *meridiana* at Santa Maria degli Angeli.[125]

Meanwhile Bianchini was studying history and chronology for more than eight hours a day in Ottoboni's library to produce a work that would win him a place in Noris's. The resultant minor masterpiece, completed

Fig. 9. Bianchini presenting his *Universal History* to Cardinal Pietro Ottoboni in the Cancelleria. FB, *IU*, after title page.

in 1697, celebrated its origin in a frontispiece showing the Cardinal Vice Chancellor receiving the work from Bianchini (in competition with a musician presenting a score) while two artists work nearby, one on items from Ciampini's collections (see Fig. 9). Through windows behind Ottoboni appear a Christian church and a classical building, representative of the ancient archaeology and ecclesiastical history to which Bianchini would devote as much time as he did to astronomy.

2

UNIVERSAL HISTORY

Alexander VIII's replacement, Innocent XII, an austere administrator, labored to simplify Rome's judicial system, improve the operation of its charities, discipline and educate its clergy, and stamp out nepotism. He did not require Bianchini's help in these endeavors. Since rapid turnover in papal policies and families made reliance on their patronage risky, Bianchini sought greater job security and scholarly opportunity than he had as a functionary in the household of a spendthrift cardinal who might lose his place in Roman musical chairs.

One option was teaching *fisicomatematica*. Bianchini kept it alive by editing more of Montanari's unpublished writings. One of these, on a hurricane that hit the Veneto in the summer of 1686, is arranged as a dialogue featuring Montanari, Davia, and another of Montanari's students, Giovanni Antonio Gozzadini. We are interested in Davia, by then advanced to bishop. He had switched from military to spiritual engineering under Innocent XI, who sent him as an emissary to Brussels.[1] Alexander VIII raised him to the episcopate, and Innocent XII returned him to diplomacy as nuncio to Cologne and then to Poland, where we shall meet him again. Meanwhile, he appeared in the hurricane dialogue, in which he flirts with such dangerous topics as Cartesian cosmology and the vacuum.[2] That was not helpful to him or Bianchini, because Innocent was then cracking down on "atheistic" intellectuals who followed modern physics too closely. The affair faded but left a legacy of discouragement that helped convince Bianchini to set aside natural philosophy and raise his game as librarian.[3]

He now aimed at the Vatican Library, not, as before, as a scribe, but somewhere near its head. There was space. Noris had resigned the office of first *custos* in 1695 on his promotion to cardinal. He had enjoyed the pomp of office that irritated his great enemies the Jesuits, but otherwise he did not much like the job. "[It] does not give me time for my studies . . . I prefer to be a writer than a custodian of books."[4] Five years later, in 1700, he returned as Cardinal Librarian. Meanwhile his former post of first *custos* remained open. It went to the long-serving second *custos*, Lorenzo Zaccagni, whose position was offered to Bianchini. He declined: something better had turned up.[5] The way to the new opportunity passed through universal history and Ciampini's historical academy, the *Accademia dei concili*.[6]

The Principles

An account of the world from Creation to the time of writing was a tired format by 1700. Nonetheless Bianchini saw a future in it. The key was to eschew biblical sources and to illustrate each period with a montage of appropriate objects—monuments, coins, gems, sculptures, pottery, sarcophagi, inscriptions—that forced its character on the memory. The project mobilized Bianchini's artistic skill in the montages and astronomical know-how in chronological calculations, and an ability to interpret archaeological artifacts as symbols of historical periods. To perfect this scheme he had to read up the histories and study the relics of all the ancient peoples represented in the libraries and collections he could exploit. Although reliance on physical artifacts and pagan sources greatly complicated the project, it brought one important simplification. It circumvented the most difficult historiographical problems of the day: the antiquity of man and the relationship between gentile and Jewish history.

When Bianchini planned his *Istoria universale*, the irksome question of the date of Creation deducible from Scripture was again in the news. St Augustine had declared it insoluble, and he had been right; but that did not stop Renaissance savants armed with precise chronology and the three languages from disputing whether the Hebrew or the Greek text of the Old Testament counted more reliably. Since the Greek version made the world a millennium or more older than the Hebrew, the question of reliability meant something to savants who calculated the likely span of human history at 6,000 or 7,000 years. By the middle of the seventeenth century the historian could choose among fifty competing dates for Creation; by the early eighteenth, 200, the difference between the earliest and latest amounting to 3,400 years and millions of words.[7] The lengthy dispute that began around 1690 between the Cistercian Paul Pezron and the Benedictine Jean Martianay, who defended the Greek and the Hebrew account respectively, suggests what was at stake.[8]

Pezron chose the longer Greek count to meet a challenge to the Bible as a general guide to human history. The challenger was a librarian, Isaac de la Peyrère, a French Calvinist, previously perhaps a Jew, and soon to be a Catholic, who published anonymously in 1655 his demonstration that men existed before Adam. Pre-Adamites provided parents for biblical people mysteriously unrelated to Adam, like those for whose instruction God branded Cain; put back Creation far enough to allow the Egyptians, Babylonians, and Chinese the seniority they claimed over the Jews; explained the high state of Chaldean culture in the time of Abraham; reduced the events of Jewish history, including its miracles and even the Flood, to local events; and canceled the problem of deriving the Chinese and the Americans from the line of Noah. The book aroused an ecumenical spirit: Jews, Protestants, and Catholics all condemned it. The Roman Inquisition, which eventually got hold of its author, allowed him to expiate his sin by confessing to the pope and converting to Catholicism. The pope, Alexander VII, offered Peyrère a benefice if he would stay in Rome as a pendant to his most spectacular convert, Queen Christina.[9]

Peyrère preferred to return to Paris and to lobby for Pre-Adamites, but now in the way the Jesuits argued for Copernican astronomy: it was a useful hypothesis, did not change the observed facts, and was known to be wrong only because a pope had said so. And, just as Copernican astronomy, though untrue, showed the Ptolemaic to be false, Peyrère's system revealed that the Old Testament could not be the history of the entire human race. Peyrère retired to an Oratorian convent outside Paris to refine his idea that the extant versions of the Old Testament are merely different abridgments of a lost larger history. His brother Oratorian Richard Simon publicized the unsettling conclusion: although scripture is infallible, no infallible date for Creation or unambiguous human history can be obtained from its available synopses.[10]

Besides offering some shelter for pre-Adamites, the Greek testament accommodated the tradition that the Messiah was to arrive during the sixth millennium. Here the Hebrew count fell short. So what? Martianay, the champion of the Hebrew text, argued briefly and unanswerably that, being inspired, it trumped arithmetic.[11] How could our would-be universal historian find firm ground? The writings of Peyrère and Simon (and, of course, Spinoza) cast doubt on the most authoritative source for the history of mankind before the rise of the Greeks. And, even if accepted as a reliable account of the Jews, the Old Testament did not agree at all with what the gentile nations said about themselves. This was the great problem faced by Bianchini. His resourceful solution made his reputation.

Everyone knew that gentile historians grossly exaggerated the age of their nations. To take a notorious example, Berossus the Babyonian, who lived at the time of Alexander the Great, taught that 432,000 years of recorded history, or perhaps 472,000, had run their course before the Flood. Against this boast, the claims of the Chinese and the Egyptians to an antiquity of 40,000 years, or perhaps 80,000, lacked imagination. Most Christian historians summarily dismissed these numbers as lies or conceits. Others interpreted the 432,000 years of the Babylonians as so many days.[12] Bianchini took a more principled course. From Marcus Terentius Varro, an underachieving contemporary of Cicero credited with writing 620 "books" covering all respectable subjects, Bianchini learned that the Greeks had required around a millennium to build up their arts and sciences. That made a very important yardstick. The Chinese, Egyptians, and Babylonians would have been imbeciles to have made so little progress during the vast stretches of time their annals boasted. In fact, according to Bianchini, their boasts were a very useful datum: they confirmed that the Flood, the universal deluge recorded in most gentile histories, had wiped out virtually all memory of the pre-diluvian past.[13] Most of what remains was transmitted in myths, which can function as a telescope for peering into the most remote periods of human history.

Critical Dates

The great question then was the date of the Flood. To fix it from pagan history alone, Bianchini invoked the yardstick he had deduced from Varro and the plausible assumption that the Greeks started from a higher state of civilization than Adam. Hence the time from the Creation of the world, of which all societies preserved a memory, to the Flood had to be more than a thousand years. By how much? Let us say half again as long. All respectable chronologists (that is, all who followed the Hebrew Bible) put the Flood halfway between Creation and the first Olympiad, which occurred around 800 years before the time of Caesar Augustus. Taking Creation to have preceded the Flood by 1,500 or, so as not to rush the patriarchs, 1,600 years, we have the beginning of human history about 800 + 1,600 + 1,600 = 4,000 years before Augustus. What could be freer from arbitrary assumptions? The calculation is as reliable as Bianchini's plausible premise of progress: "men of equal capacity for the sciences can in equal times create arts and learning that, if not completely equal, are at least proportional to the time and the paucity of means and workers."[14]

Pettifoggers might object that the premise of equal progress in equal times did not justify placing the Flood halfway between Creation and the first Olympiad. By the time he came to write up the history of the seventeenth century after Creation, Bianchini had found a uniformitarian argument apparently immune to cavil. Recent excavations at the base of Vesuvius had disclosed layers of soil interrupted at several depths by detritus from volcanic eruptions. The latest major eruption occurred about eight hundred years after the quake that destroyed Pompeii. If "ordinarily equal intervals produce corresponding effects," we might expect to find that a big eruption had occurred around 2400 BC. If only it could be connected with the Flood! It is scarcely credible that around 1690 an architect–antiquarian had detected a damp layer at just the expected level. To check the conclusion, Bianchini recommended a hunt for dampness left over from the Flood that destroyed humankind. "It is worthy of the curiosity of philosophers to try similar experiments in other mountains of Asia that throw out fire."[15] The suggestion did not produce an international geophysical year.

The date of Creation deduced assumption-free from the progress of mud and mankind agreed perfectly with that recoverable from the Hebrew Bible. Indeed, its coincidence to one part in a thousand with the value computed by the precise and learned Anglican Archbishop James Ussher, and to one part in a hundred with Creation according to two chronologists Bianchini admired, the Lutheran Sethus Calvisius and the Jesuit Dionysius Petavius (Denis Pétau), might seem an ecumenical miracle.[16] Not to Bianchini. He did not suppose that his 4,000 years was anything closer than a gross approximation; and he no doubt held, with Martianay, that it made no sense to be punctilious about dates to which the historical actors themselves attached no importance.[17] But even the gross approximation of 4,000 years secured a very important objective. As the censor who recommended publication of the *Istoria* remarked, it compressed pagan annals within the minimum span conventionally allotted to the history of the Jews. Hence Bianchini's answer to the pyrrhonism of Peyrère: considerations drawn from gentile historians indicate that, when properly deflated, their annals will agree well enough with the Old Testament. Polygenetic Creation, an indefinitely old world, and men before Adam were not required.

In calculating Creation, Bianchini had an eye to devising an organizing system for the largely unknown and undatable history of humanity (apart from the Jews) before the Olympiads. Varro had worked with three ages: the unknown and uncertain, from Creation to Flood; the mythical and fabulous, from Flood to first Olympiad; and, thenceforward, the historical. Bianchini split the first age into two and symmetrized the whole to produce four ages of ten centuries each. The removal of the Flood to the middle of the second "uncertain" age had the advantage of emphasizing again how little the human race remembered of the times immediately after almost all of it had drowned.[18]

The symmetrizing into four equal ages made explicit the symbolic character of Bianchini's world history. As he put it, he aimed to inculcate comprehension of human development, not apprehension of disconnected facts. Disconnection had been the hallmark of universal histories, even those Bianchini admired. Calvisius's *Opus chronologicum* is an unrelenting sequence of factoids. Its last, posthumous edition of 1685 begins with the separation of the waters on 27 October in year 1 and proceeds through 900 double-column pages down to December 5633 and the destruction under Louis XIV of Huguenot churches in France. (Calvisius's *Chronologicum*, rated by the grand chronologist Joseph Scaliger as "most reliable, useful, and packed with learning," followed the Hebrew count.) Petavius's abridged chronology of 1682 has several features in common with the history Bianchini would write. It emphasizes pagan history, provides genealogies of the heroes of myth, treats them in a euhemerist manner, periodizes events rather than lists them, and occasionally introduces a chapter on the arts and sciences.[19] Still, Petavius remained too concerned with chronology to provide a big picture. Or, to change to Bianchini's metaphor, Calvisius and Petavius played upon single instruments, whereas history requires a harmonious orchestra. "History without chronology is music without a beat; [but] annals without history are beats without music."[20]

An astronomical metaphor might fit the work and the author even better. The assumption of equal progress made in equal periods, more or less, corresponds to a sun-centered view of the planets; seen from the earth, however, they do not advance steadily, and sometimes even retrogress, much more in keeping with the stuttering progress of humankind. Over periods long enough, however, and here a century will do, the net motion is always forward and roughly proportional to time elapsed. The parochial historian focuses on the vicissitudes of a country or people, their triumphs, failures, rises, and falls; the universal historian studies aggregates and averages, humankind in mean motion, and might readily accept the implausible proportion of equal progress in equal time. Or, to return to Bianchini's analogy, the forty centuries of average constant progress provided the steady beat. It was up to him to write the music.

At the head of the description of each century he put a montage that caught its essence. As the subtitle to the *Istoria* makes clear—"proved with monuments and illustrated with symbols of the ancients"—Bianchini regarded the montages as demonstrations, as well as representations, of historical developments. He conceived that the ancients had two ways of preserving their history, one suited to fixing every particular circumstance, the other to comprehending a series of interconnected subjects. "The former they called the art of letters, the latter the understanding of symbols. Each invention is nothing other than a way of communicating thoughts: the first more general but more difficult to learn ... the second more restricted but faster because more directly connected to the immediate impression that the imagination receives from the feelings." Symbols cannot express the operations of the mind, such as affirming, denying, comparing, deducing, "in which the substance and variety of thinking consists;" they convey their meaning directly, and can be understood even if the language of their inventors should perish, "as we see in the insignia of magistrates, in paintings, marbles, shields, gems, and seals, of our own people, and of barbarians." Altogether, Bianchini chose and drew over 140 objects, mostly from printed sources, to give his readers direct access to ancient history.[21]

He planned to continue into modern history, from Augustus to 1600. To accommodate the more abundant material, he subdivided the centuries into scores, each with its own image, amounting to eighty in all. The two volumes would contain 120 carefully constructed images. In the event, he did not push his history beyond the thirty-second century after Creation, because, he explained, his book was becoming large and he had reached historical time; and also, perhaps, because he was paying the costs of publication. He did complete all 120 images, which he issued before the *Istoria* as sugar-coating for historical study. They were to be cut out for a card game from which players could learn simultaneously the main lines of human development and the role of chance in history. Forty cards divided into four suits constituted a deck. After a little practice following suit and sequence, players would know not only the order but also the meaning of the images. Bianchini's academic friends were not above playing it.[22]

In this manner, Bianchini would form Ciceronian cosmopolitans at home everywhere because they grasped the essential elements of history and science. A person so formed transcends time as well as space. "[He is] one of the republic of all men, born to extend himself to, and converse virtually with, every age, although obliged to restrict himself to a certain place and time."[23] Images abet cosmopolitan communication. Their sequence shows at a glance the life cycles of kingdoms and principalities, their growth, maturity, decay, ruin. Their power appears also in the sciences, which have been transmitted primarily by symbols and representations, like geometrical figures and armillary spheres. "Wherefore we must declare our great obligation to the ancients, who not being content with discovering the sciences and arts, also multiplied ways of transmitting them; and even after inventing a method of writing down sounds with letters, cultivated the practice of summarizing and illustrating thoughts with symbols."[24] Bianchini not only designed the montages of symbols (the images), but probably also engraved most of those illustrating the centuries in *Istoria universale*. For special items, like the frontispiece, he commissioned established artists, the most accomplished of whom was Rome's *Commisario delle Antiquità*, Piero Santi Bartoli, famous as an illustrator of archaeological finds and classical edifices for the lavish folios then prized by connoisseurs.[25]

Avoiding scriptural problems by pretending to deal only with pagan histories would not have worked if the church had not acceded already to the sort of erudition the project required. It was a dangerous accommodation. The tools that Bianchini and other scholars developed to analyze gentile texts were turned easily, and eventually, against the Bible. In the last years of the reign of Innocent XII, however, Bianchini could capitalize on the fleeting equilibrium of accommodation and traditional belief to gain a reputation. A generation ago a prominent professor of Italian literature at the University of Rome, Riccardo Merolla, ranked Bianchini as the most important exponent, and the *Istoria universale* the best example, of Roman erudition around 1700 for the sensibility of his historiography, the richness of his documentation, and his ability to keep his results within the expanding envelope of the allowable. His astronomy offers an exact parallel. He never advocated heliocentrism, but insinuated it quietly, "for convenience," as restrictions eased.[26] He was very accommodating.

The claims and counterclaims of Protestant and Catholic authorities locked in erudite combat over church history, though productive of books, had demonstrated to bystanders that the study of the past could no more uncover the truth than it could turn scholars into gentlemen. As Bianchini developed his forty-century format, sceptics like Pierre Bayle were illustrating the principle enunciated by Felix de la Mothe le Vayer that, since historians must rely either on the notoriously unreliable testimony of others or on direct observations distorted by their interests, "there is almost no certainty in anything that the most famous historians have spread abroad and future historians will produce." The wise man discounts historical reconstructions as he does portraits of beautiful women.[27] To find the true features of the woman falsely beautified—that is, the kernel of truth in historical documents—the anti-pyrrhonist looked to the direct testimony of coins and inscriptions rather than to copies of texts. This was the lesson Bianchini received from Correr and Patin during his Paduan years, and from Ciampini and Raffaello Fabretti, an accomplished antiquarian close to the Ottoboni, whom Bianchini sometimes helped with inscriptions in Rome.[28] To be sure, the objects in Bianchini's images seldom date from the times he took them to illustrate. There was no hope for a coin issued on the day after Creation.

To employ artifacts for his purpose, Bianchini assumed that symbolic representations existed in the minds of human beings long before the inventions of the arts needed to instantiate them and that the instantiations faithfully preserved the intent of the symbolic representations. A hazardous assumption to be sure. In favor of it he argued that both the symbolic representations and their physical realizations were public property. Although artists might manifest a personal style in their designs, the result could not differ much from public expectation if it were to accomplish its purpose. "Whence if the facts [recorded by Roman inscriptions and medals] were not true, men living at the time would not have preserved them in memorials corroborated by many signs of public authority." Depend upon it. "It is incredible how quick our mind is to extract the signs of truth from objects: and the force with which truth insinuates itself firmly into the mind is equally admirable and efficacious."[29]

In practice Bianchini took much from the ancient historians despite their dubiety. He tried to apply where he could the sort of criticism practiced by the Benedictine monk Jean Mabillon and the scholars who took his *De re diplomatica* (1681) as their guide.[30] Bianchini knew the man personally. While on a trip to Italy in 1685–6 to hunt for manuscripts, Mabillon attended a meeting of Ciampini's *Accademia fisicomatematica*. He was then at his prime as leader of the "Maurist" school of historians, composed initially of the Benedictines of Saint-Maur headquartered at Saint-Germain in Paris; an imposing and influential figure, he has enjoyed such honorifica as "the greatest historical scholar of the seventeenth century," or, better, "the Galileo of scholarly history."[31] By grit and scholarship he had cut out much that was fabulous from the medieval history of France, including most (fifty-five of eighty) Benedictine saints. Mabillon heard Bianchini give a talk at Ciampini's academy on a mathematical subject and was impressed.[32] And no doubt Bianchini gave ear to Mabillon's methods for testing the reliability of documents by consideration of their ink, paper, parchment, handwriting, form, seals, every physical characteristic accessible to touch or sight. Armed with authentic sources and unflinching erudition, Mabillon could slaughter saints with impunity.[33] Bianchini would practice Maurist methods in scrutinizing the lives of early popes.

Another visitor to Ciampini's academy who thought hard about history was Leibniz. He and Bianchini had much to discuss, including steps to obtain a revocation of the Vatican's prohibition of the Copernican system, and they corresponded for some time after Leibniz had left Italy in 1690. Many of Leibniz's attitudes toward history have exact counterparts in Bianchini: admiration of Calvisius, Petavius, and Mabillon; interest in universal history, including its extension to China; attention to pedagogy and mnemonic images; insistence on the use of "monuments" as well as literary sources; and, above all, persistent attention to the place of history in the body of science—that is, to the problem of historical truth.[34] We may depend on ancient historians for information, but we must depend on ourselves for "the refinement of judgment we call criticism, [which] investigates marbles and manuscripts of every age and country in order to separate even in matters of little moment the certain from the doubtful." Euhemerist analysis of fables and criticism of artifacts can reconstruct reliable history without recourse to what Bianchini accepted as the "true depiction" in the Bible.[35]

He had powerful tools for his impossible task. His knowledge of materials, mathematics, and languages; his doctrine of symbols as an answer to historical pyrrhonism; his association with accomplished scholars such as Ciampini, Leibniz, and Mabillon; his skill in drawing; and his position as keeper of the Ottoboni library, "this wonderful library, this excellent museum, which supplied the historical proofs described in [my] book," and which Mabillon's disciple, Bernard de Montfaucon, rated as second only to the Vatican in Hebrew, Greek, and Latin sources (see Fig. 10). As librarian, Bianchini kept current with books produced and imported into Italy and profited from conversations with scholars resident in or visiting Rome.[36] Moreover, he had access to collections of ancient coins and artifacts owned by Ottoboni, Ciampini, and other wealthy men. Bianchini put all Rome under contribution for proofs for his *Istoria universale*.[37] He also

Fig. 10. The entrance to Bianchini's domain, the Ottoboni library in the Cancelleria. The globes signify the old and the new worlds of geography and philosophy. FB, *IU*, fo. b^r.

taxed himself, and not only by study. Very likely he had to borrow from friends and family in Verona to meet the considerable expense of paper, typesetting, and woodblocks.

The History

"In the beginning . . ." (Bianchini's image for Creation (see Fig. 11)) gives priority to symbols recalling the foreplay of the games of the Roman circus. He remarked that the preliminary seven circuits made by the chariots, and the twelve *carceri* (the "prisons" in the arena from which the chariots entered the races) represented the order of the world: the number of planets, zodiacal signs, months. Fig. 11.1, taken from the decoration on a lamp in Bianchini's collection, shows a four-horse chariot and a symbol of the seven circuits; Fig. 11.3, from a medal then recently unearthed in Rome, boasts an Apollo figure crowned with solar rays, carrying Mercury's caduceus, and driving the quadriga of the sun around the zodiac. Bianchini's montage also includes the goalposts of the circus topped with cosmic eggs (11.2, taken from a bas-relief and a coin). Fig. 11.4 represents the ancient Eleusinian mysteries: the goddess of agriculture Demeter

Fig. 11. Century 1: Creation. FB, *IU* 67.

searching by torchlight for her daughter Persephone as she is just about to emerge from her winter in Hades to renew the earth. Bianchini placed the two women directly under the first point of Aries, the beginning of spring. Fig. 11.6, from a lamp, alludes to Neptune, the ocean, water, and the separation of the waters at Creation. Other ancient peoples' commemorations of creation and renewal left such bric-a-brac as Egyptian obelisks with their undecipherable hieroglyphics and (11.5) the stag sacrificed every spring by the natives of Florida. Aristotle was right: common people everywhere believed in a divine creation, "before the errors of those who called themselves philosophers threw it into doubt."[38]

A slight, unacknowledged hint from scripture suggested that the first step into the Age of the Unknown was the golden age of innocence. The image (see Fig. 12), taken from a bas-relief, features a Prometheus figure (12.2) whom Bianchini identified, with much scholarly fireworks, with

Fig. 12. Century 2: The Golden Age. FB, *IU* 77.

Saturn. The identification makes use of the figure's sandal (its removal) and several references to the saturnalia, like the festooning of the building (12.4) and the food on the table (12.5). An orgy is not in preparation, but rather an annual reversal of roles in which the well-to-do wait on their servants (the trio under the table). The Saturnalia thus recalls the joyful days when Saturn reigned before the introduction of slavery and other misfortunes. This happy recollection is compromised, however, by a hint, preserved in Fig. 12.1, at our parents' error in Eden. Looking far afield as usual, Bianchini found that fall stories, like tales of a golden age under a benign Saturn, occurred in all the cultures he studied. Note that this Saturn, Saturn I, was not the bad-tempered Saturn II who lived after the Flood; mythographers must be on guard against confusing different fabulous characters with the same name, as Cicero had warned while pointing out six different Hercules, three Jupiters, five Mercurys, as many Minervas, and an indefinite number of Vulcans. Yet it would be wrong to use this multiplicity to deny the existence of real people and true episodes behind the fables. With Plato, Livy, Tertullian, and Augustine, we must recognize that "truth has a certain odor of eternity, impervious to time: and its perennial fragrance filters through the densest tissue of lies." We can choose to sniff the fables or remain in ignorance. "But woe to learning if it dismisses mythology as terra incognita!"[39] We would have no way of knowing about much of human history.

According to the poets, the Age of Silver saw the beginnings of astronomy, arithmetic, and agriculture. Bianchini assigned the age's commencement to the third century after Creation. Saturn I was still around, as he lived for a thousand years. So did most of his family, as if they were Hebrew patriarchs. That suggested a correspondence between the inventors of the arts, like Saturn II and Uranus, and the house of Adam down to Noah. Bianchini drew up partial family trees, ending with Noah and Uranus, who was known to have drowned in the Ocean. Apparently (Bianchini concluded) a long-lived character named Uranus invented mathematics and astronomy (whence the name of its muse), no doubt in the service of agriculture: not real astronomy, the astronomer Bianchini added, "but the study of the seasons and meteorological phenomena, not the theory of celestial motions." So much we can infer from myth and poetry. We can recover Uranus's knowledge in its most developed form from surviving fragments of the astronomy and arithmetic taught by the Egyptian Toth, Cicero's Mercury V (are you following, dear reader?) just before the Flood. Chinese records, collected by Jesuit missionaries, indicated an equally early enlightenment in exact science in the Far East. No matter that writing had not been invented yet. The calendar stones of the Americans show that determined astronomers can do without it.[40]

To illustrate this last point, Bianchini included an abacus-like calculator in the image for Century III (see Fig. 13). A bas-relief of the god Annui, to whom the Egyptians ascribed the invention of science in general, also appears (13.1), carrying a celestial globe (13.2); under which a perplexed

Fig. 13. Century 3: The Silver Age. FB, *IU* 93.

Chinese, leaning on an armillary sphere and studying his "hieroglyphs," confirms that his countrymen knew as much as the Egyptians, and at the same time. They must have had their own versions of Annui, Toth, and Mercury V. The centerpiece, an ancient plough, symbolizes the art that began civilization. To the left, a Minerva examines boundary stones of the sort enlarged in Fig. 13.4; agriculture required surveying fields, thus more arithmetic and even geometry. There can be no doubt that before the Flood people knew agriculture and the limited mathematics and astronomy it required.

> The fabulous names of Uranus and Jupiter and Minerva do not detract in the least from this fact when the most ancient nations, like the Chinese, Chaldeans, and Egyptians, attest to the same thing: to having learned from their ancestors who lived close to the time of the Flood that people then existed who practiced these arts, and compared parts of quantities by means of numbers.[41]

The Age of Copper commencing in the fourth century continued the rise of the arts and the decline of virtue. We infer the decline from the creation of weights and measures to combat fraud. Is it not written, "divers weights, and divers measures, both of them are alike abominations to the Lord" (Proverbs 20:10)? The Copper Age also saw the invention of the first musical instruments. In general, the arts most necessary to human survival, like agriculture, came first; music belonged among them for its role in sacrifices and ceremonies. Arts necessary only to procure luxury goods from afar, like navigation, came last. The mnemonic image for the Copper Age features the raising of a boundary stone and some musical instruments. Bianchini explained that the stone and the instruments both referred to the law, "infused in the soul with the reason, which we call the law of nature." The stone marked the boundary or limit of the law, which (as Bianchini knew from Plato by way of Plutarch) the ancients associated with the periods and harmony of music.[42]

During the Bronze Age, Vulcan developed the art of the blacksmith, which presupposed a knowledge of fire. His sister (or perhaps his wife) Minerva invented more arts. Bianchini supplied a family tree for Cicero's five Minervas; the eldest of them is Uranus's daughter and Vulcan's wife. The image for this fifth century after Creation boasts a forge, an anvil, two blacksmiths, and a set of bellows, all taken from ancient monuments. Bianchini conceded that the dating of the arts of Vulcan and Minerva II, and of the inventions of the preceding metallic ages, gave only a rough indication of the true sequence. Some began later than Century V and all continued to develop down to the Flood. Having no information about the true sequence, Bianchini assigned to Centuries VI through XVI inventions in accordance with his estimate of their proportions of necessity and luxury. In Century VI humans discovered the uses of fruit (disfavored, no doubt, by their experience in Eden) and the milk of animals; in VII, they dressed in animal skins; in VIII, they improved astronomy, in IX, measures and arithmetic; in X, they built their first permanent shelters. Thus ended, in relative comfort, the first millennium of human existence.[43]

During the next six hundred years our ancestors learned how to use fire (Century XI) and invented the arts of the potter and the mason (XII), improved those of the smith (XIII) and the hunter (XIV), continued to seek discipline in the marketplace with weights and measures (XV), and discovered iron, war, and navigation (XVI).[44] The easier their lives, the more men inclined to luxury and violence. It became necessary to drown them. That happened halfway through Century XVII. Bianchini knew many Flood stories, from the Middle East, China, Japan, Greece, Europe, and America. He deduced that the Flood was universal and catastrophic. Even without written testimony, the scattering of animal remains outside their natural habitats, like marine fossils on mountain tops, sufficiently demonstrated the power and extent of the Deluge. To which Bianchini added the impeachable testimony of the liar Berossus, who reported that in his day remains of the ark could still be seen in Armenia. The available illustrative material was as abundant as Flood stories. Bianchini chose a Neptune sitting in a cave with his feet in the water; coins with aquatic scenes; and, stretched over it all, the upper body of a bearded, winged figure with outstretched arms resting on a rainbow (Fig. 14). He is not the Christian God, but Jupiter Pluvius as depicted on a column of Marcus Aurelius. Between the rainbow and the cave hangs a human skull.[45]

The second millennium concludes with the division of the earth into three main nations (Century XVIII), the origin of idolatry and writing (XIX), and the rise of the most ancient kingdoms in Egypt, Babylon, and China (XX). Bianchini's account of the triune inventions of sculpture, idolatry, and writing is worth recounting. Prometheus, a man of the nineteenth century, invented sculpture, perhaps in Egypt, where idolatry probably began. The connection between sculpture and idols is clear enough—Cicero had observed it. As for writing, we have it from Hyginus that one Mercury or another interpreted the words of man and thereby divided the nations and sowed discord. "Interpreting" no doubt signified writing; by which means this literate Mercury, known in Greek as Hermes, made possible the endless squabbles of "hermeneutics." Kings then came along to promote idolatry and commerce with the technique that

Immagine Decimaſettima.

Fig. 14. Century 17: The Deluge. FB, *IU* 186.

Mercury had invented to preserve the rites and words of humankind. *Quid clarius?* "If this narrative is not history, certainly it contains very little of the fabulous." It corresponded perfectly to the idolatrous first stage of theology identified by Varro as mumbo jumbo invented by priests and kings to bamboozle primitive people.[46]

In a second theological stage recognized by Varro, philosophers fell for the false gods of fire and atoms and disputed about their age and number for centuries. In the third and final stage, the civil, in which Varro lived, the philosophers dropped the gods but kept up schools and banquets, and official priests prescribed rites, plays, games, and dances that preserved the old fables and public order. These relics had (and have) inestimable value for historians able to unscramble what the priests obscured by "imitating the vices rather than the virtues of their ancestors and placing among gods and genii living persons who inherited their waywardness and their guilt."[47] As the wayward proceeded through Varro's stages, the writing they invented to preserve the rites of worship diversified to perpetuate philosophy and codify law. Cicero's sceptic, who regarded the tales of the gods as "a composite of sheer inanity and arrant stupidity," but tolerated their worship as the basis of civil society, might have studied with Varro.[48]

We have crossed a deluge of obscurity to reach a beachhead of history—that is, we have landed in the third millennium. We have found relics of the search for the necessities of life even before the invention of writing and have confirmed the order of development discoverable in Greek myths by the records of other civilizations. In all, we have found astonishing agreement about creation from chaos, ordering of the world, states of innocence and guilt, invention of the basic arts, ill will and dissensions among men, universal castigation by flood, political division of the earth, and the introduction of idolatry and writing. "Therefore, there shines forth from the first twenty centuries of the world so great a character of truth, even if framed in the dim light of natural tradition conserved by providence and mankind, as to suffice for the use of the histories most necessary to know, and of the erudition most worthwhile to seek." In continuing into the age of the heroes, we can call on astronomy to help unveil myth; for the third millennium saw the creation of the asterisms that remain on our celestial globes and that recount for those who can decipher them the political and dynastic history of the eastern Mediterranean.[49]

The poets and chronologists put the foundation of the Assyrian dynasty by Belus (a grandson of Noah) or Belus's son Ninus (Nimrod) in the twentieth century before Augustus (Century XXI). The chroniclers further almost agreed that Assyrian dominance lasted for thirteen centuries. Ninus made certain that the people understood the power and majesty of his lineage by placing Belus among the stars, as the constellation the Greeks preserved as Orion. Bianchini pointed out that Orion's weapons—a club and an animal-skin shield—show that the figure must be very old. He placed it, as drawn on the Farnese globe (an object made in Century XLII), in his image for Century XXI (see Fig. 15.1). Much of the rest of the image concerns Queen Semiramis (15.2), reputed to have built a great palace from whose roof Chaldean astronomers observed the stars (15.8). The pedestal (15.5) shows a young Jupiter standing on an eagle and

Fig. 15. Century 21: Assyrian Astronomy. FB, *IU* 253.

flanked by the sun and moon, symbols assigned to him by the Assyrians, "simultaneously worshippers and observers of these planets." The youths (15.6) present to Semiramis a plaque depicting the temple devoted to her relative Belus. The moon (15.7) was one of the two planets attached to the cult of Semiramis. The other, Venus, prompted the surmise that the Queen of Assyria may have been Venus, or perhaps Dione, who may also have been Venus. What is certain is that the fish goddess beneath her (15.3) was her mother.[50]

The use of astronomy appears to best advantage in the century of the Argonauts (XXVIII). Then Chiron and Hercules made the first celestial sphere, invented chronology, and fashioned the Olympiads. Hercules's role is indicated by his relieving Atlas of the weight of the heavens, which signifies, as Plutarch and others had perceived, that Hercules learned astronomy from Atlas and disseminated it further. "Thus the Greek Republic triumphed in arms in Asia and in learning in Europe." Much

Fig. 16. Century 28: The Argonauts. FB, *IU* 367.

of this history appears on the Farnese globe (see Fig. 16.6). Bianchini rightly decided that it dated from the time of Ptolemy the astronomer but supposed that it pictured the asterisms as they appeared at the time of the Argonauts. The supposition relied on the euhemerist arguments that the asterisms represented real people and events and that no one of importance after the time of Hercules has a place in the heavens.[51] The constellations, "a sort of hieroglyphs that conserve the meaning of heroic times," corroborate the conjecture: many Olympians are among them, Hercules of course, and Theseus, who slew the Minotaur and contributed to the design of the games, and also Orpheus. Hercules, Theseus, Orpheus, Chiron, Aesculapius, Achilles, and Castor being contemporaries, Bianchini united them and their attributes in his image for Century XXVIII. We find the constellation of Saetta (Hercules's arrow), Corona (perhaps Theseus's property), Lyra (certainly Orpheus's), Aesculapius or Serpentus, Chiron (Fig. 16.1–5), all taken from figures on the Farnese globe.[52]

Bianchini did not have the time to analyze the globe in detail while writing his *Istoria*. Later he made a thorough study published posthumously. It incorporates a coin from AD 157 from Ottoboni's collection that bears an Atlas carrying a globe and symbols of Belus–Jupiter, who transmitted knowledge of astronomy to Babylonia. Bianchini explained that the coin recalls the invention of the sphere, with its constellations, poles, tropics, and equinoxes by Atlas and the earliest systematic observations of the Babylonians.[53] Interpreting one of the circles on the globe as the limitation of circumpolar stars, Bianchini referred it to a latitude of 40 or 41 degrees, in agreement with Jason's embarkation. He further deduced, by an irrefragable argument to be given later, that the astronomical information on the globe, and consequently the voyage of the Argos, dated from around 1275 BC. That was a most happy result, agreeing perfectly with traditional dating from Greek literature and with the euhemerist inference from the personages identified with the asterisms as analyzed in the *Istoria*.[54]

Bianchini was not conned by the argonauts' reputation for heroism. Freeing their voyage from its poetical veil, he diagnosed it as a hunt for gold and trade. Jason probably brought hides produced in Thrace and returned with precious metals; whence the earliest coins bore the likeness of a sheep or an ox.[55] Our cynical mythographer inclined also to regard the Trojan War as an episode in commercial competition and colonial rivalry. The gods stood for different regions or kingdoms; thus, Venus for Ethiopia, Juno for Syria, Minerva for Egypt, Neptune for Caria, Mercury for Phoenicia.[56] The flight of Daedalus and the appetite of the Minotaur also bear witness to the acquisitiveness of the age. The image of the century (XXVIII) refers to these fables by representations of the Trojan Horse (Fig. 16.10), the ground plan of Minos's labyrinth (16.8), and a portrait of its celebrated architect, Daedalus, shown fashioning his wings (16.7).[57] Bianchini's interpretation of the Trojan myths as a record of commercial warfare was much approved.[58]

To the capital question for chronologists, when did the siege of Troy take place, Bianchini gave a common-sense answer in the style of Mabillon.

Two dates competed. The authoritative Scaliger and Petavius followed the Greek astronomer Eratosthenes, who gave a date 507 years before his time on the strength of doubtful reports of eclipses. Bianchini preferred a date twenty-six years earlier on the evidence of the Parian marble, a chronology of Greek history from 1582 BC to 299 BC. As a practiced astronomer, Bianchini dismissed Trojan observations as too imprecise to support chronological inferences and, in any case, corrupted by copying. Hence, faithful to his method, he put his trust in the Parian marble, "still extant in the original . . . rather than in books transcribed by other historians."[59]

The third millennium, whose commencement saw the rise of the longest-lasting monarchies, the Chinese and the Assyrians, closes with the ascendancy of the most extended, the Egyptian or, as Bianchini preferred to call it after the birthplace of its king Sesostris, Ethiopian or Arabian. The image of Century XXX features a pyramid, a map, and a celestial globe (see Fig. 17.1, 5, 6).[60] Mathematicians described the empire on maps and modeled it on globes; and Egyptian architects piled up a part of it ("I would say a rather noticeable part") in the great masses of the pyramids.[61] To date Cheops and his great pyramid as late as *annus mundi* 3000 took some argument against Egyptian claims to antiquity and also against more sober estimates (the modern date is the twenty-sixth century BC). Bianchini proceeded to adjust the ancient sources he trusted so as to agree with one another and to invent etymologies to eke out a history of Egypt from the descendants of Ham to the conqueror Sesostris and the builder Cheops. The symbols and proofs of the Arabian and Ethiopian origins of Egypt are its obelisks and mathematics (brought by the Chaldeans) and its pyramids (inspired by the monuments of Abu-Simbel).[62]

The map (17.6) shows Asia Minor, "the birthplace of geography and geophysics," a region of travel, warfare, and commerce, and extends far enough to include all the sites that claimed to be the birthplace of Homer, the father of geographers (because of his account of Odysseus's travels) as well as of poets. Figs 17.2 and 17.3 depict reincarnation as taught by Pythagoras and Plato. The earth mother leaning on the globe (17.3)

Fig. 17. Century 30: End of the Third Millennium. FB, *IU* 413.

receives souls ordered by lot (17.2). The scene comes from an ancient funeral painting that could not stand the air in Rome and when unearthed "perished miserably under the learned eyes of our century."[63] Its symbolic value for Century XXX is to indicate the final place of virtuous souls among the stars: "I think that the same figure can express the history and superstition of the Greeks, since, around a thousand years after the first invention of idolatry, they consecrated . . . this second type of their gods and celestial beings, honoring with the title of heroes the Argonauts and the captains of Troy."[64]

Fig. 17.4 also represents the completion of the age of heroes: it is the Milky Way, which, according to the Pythagoreans, bears evidence of the thousand years elapsed from the beginning of idolatry to the end

of the heroic age. Fig. 17.5 confirms this reading, since none of the asterisms on it refers to anyone who lived later than Century XXX, "although
there was no lack of adulation aimed to introduce the images of new
princes into the heavens up to the time of Augustus and Nero." Although
we now have no interest in heroes like Belus–Orion, the ancient asterisms have survived through the inertia of teachers and students.[65] Chinese
astronomers, supported by a powerful king, followed a similar and simultaneous path, devising and propagating asterisms, "the exemplars of
monarchies and the sciences." To indicate this parallel, Bianchini drew
out two constellations (17.7) from a Chinese planisphere given to him by
Ciampini. Under the stars stands the celebrated observatory of Ceucun
[Chou Kung], "which as a testimony of that age and that art, the Chinese
still preserve after so many years, as I heard from father [Daniele] Bartoli
[SJ]." Bartoli thought that Prince Ceucun was the Ptolemy of China and
lived 1,200 years before the real one, around *annus mundi* 3000. Not close!
The tower Bartoli knew was first raised in AD 1276 and has nothing to do
with the prince or with Bianchini's drawing; and the prince himself has
been consigned to legend by modern authority.[66]

This fact may destroy for us Bianchini's stirring epitaph to the third
millennium of the existence of mankind:

> The two extremes of Asia, Ionia and China, and the two most ancient books
> of men, the heavens and the earth, thus close the decade of heroic centuries
> with admirable completeness: and human genius recognizes that it goes as
> far astray in searching for the future among the stars as it comes close to
> the past from reading in the asterisms of the globes the memorials of for-
> gotten events. And so, instead of feigning a suppositious language of stellar
> influence, search diligently to learn the plain and simple interpretation of
> chronology, and of those primary signs and characters that convey history
> to us.[67]

A book illustrating its message by images should itself have an overrid-
ing symbol or hieroglyph. Bianchini reached high for the occasion. The
frontispiece to *Istoria universale* is a particularly valuable example of his
symbolic logic as even he may not have known how it related to the
precious monument that stimulated it. This prize, the inspiration for

Fig. 18. The triumph of Christianity in the struggles of history. FB, *IU*, frontispiece.

the stone box covered with Greek characters in Fig. 18 and the veiled lady sitting on it, surfaced in Rome in 1551. The box was then a chair whose side bore calendrical tables in Greek and whose back recorded the titles of Greek books. The seated party then lacked a head, which was duly supplied by its excavator, together with arms and other bits, on the theory that the statue represented a philosopher. The reconstructed monument now sits at the entrance to the Vatican Library (Fig. 19) and, in

Fig. 19. St Hippolytus of Rome, the ancestor of the veiled lady and her annotated seat in Fig. 17. FB, in Hippolytus, *Opera* (1716–18), i, tab. 2ab, after p. 36.

a copy, in the Cancelleria, in the church of San Lorenzo in Damaso, where Ottoboni installed it in 1737.[68] The philosopher sports a beard, naturally, although some experts now think that the lost head, like that in Bianchini's frontispiece, was female.[69] From the titles on the chair, however, most historians accept the philosopher as St Hippolytus of Rome, a bishop and martyr who flourished during the first quarter of the third century and devised a canon for calculating the date of Easter.

The veiled lady in Fig. 18 would therefore seem to represent not only the Church triumphant, crowned by an androgyne Christian Roman soldier (note the XP on his or her helmet) while baptizing figures representing the four continents, but also the church intellectual and spiritual, represented by the calculations of the learned martyr Hippolytus. Embellishing the triumph, continent Europe offers up the Pantheon (shown with the towers it once had to help make it a church); other tribute, a terrestrial globe, crowns, coins, weapons, and the apparatus of ancient cults and religions, lie about. The spiritual appears further in the figure of St John the Evangelist, whose companion eagle supplies the water of baptism. And, as will be clear when we consider Bianchini's analysis of Hippolytus's canon, the writing on the box presented an intellectual challenge of the first order. Here it is enough to say that the Alexander mentioned on the box (illegibly here) is neither Alexander the Great nor Alexander VIII Ottoboni, as has been plausibly conjectured, but Emperor Alexander Severus, whose reign began, as did Hippolytus's cycle, in AD 222.[70]

The Reception

The fourth millennium brought Bianchini to historical times, relative reality, and exhaustion of mind and purse. He published what he had and reckoned that he would require another 250 to 300 densely printed pages to fulfill his forty centuries and promised to supply them.[71] He expected to turn to the job in the summer of 1699 and did do a little chronological work a year later; but he did not get far before he took on some confidential work for Ottoboni as Innocent XII began to fail and pre-conclave intrigues multiplied.[72] The project was never completed, and, although the Istoria universale made Bianchini's reputation as a man of letters, its first edition probably did not circulate very widely, even in Italy. A patrician of the Republic of Genoa, Paris Maria Salvago, who corresponded regularly with Bianchini about astronomical observations and instruments, did not know of the book until 1715, when he chanced to see a copy in the unlikely hands of a Capuchin monk; he applied to Bianchini, whom by then he had met personally, for a copy, which took four months to procure. Since Salvago also corresponded with Cassini and other astronomers closely connected to Bianchini, his ignorance of the existence of the Istoria for almost twenty years after its publication implies theirs.[73]

In Rome, however, the book had an immediate positive reception, and its author, no longer regarded as an astronomer with a good classical culture, was now a literary figure whose powerful learning had squeezed a little reliable information from the mass of myth and mystery descended from remotest antiquity. It was only to be expected, therefore, that Bianchini would be elected to the national literary academy, the Arcadia, another spin-off from Queen Christina's *Accademia reale*. Its initial purpose was to "restore correct and good taste to literature" by replacing the formulaic pomposity of baroque Italian poetry with the more direct style of Galilean science. They called themselves "Arcadian pastors" to indicate their embrace of the simple and wholesome ways of ancient shepherds; and they took bizarre names, such as (in Bianchini's case) Selvaggio Afrodisio, "the wild Aphrodisian," to emphasize their occasional departures from ordinary life. Bianchini became an enthusiastic member and, if he had had the time, could have advanced to *custode*, or chief shepherd, of the Arcadian flock.[74]

No doubt Bianchini was pleased to know that Gisbert Cuper, a great expert on coins, medals, and inscriptions, liked the *Istoria*'s choice and use of ancient monuments: "[Its illustrations] fill me with great pleasure whenever I consult the book, which is often."[75] And that Lodovico Antonio Muratori, librarian to the Duke of Modena, and, like Bianchini, an admirer of Mabillon's historical methods and Galileo's experimentalism, judged the use of artifacts as prime evidence to be "a very great and noble idea."[76] If only other Italian savants would emulate Bianchini's historiography informed by physical objects and his natural science derived from instruments they could attain the European fame they deserved![77] Publishers eventually saw the merits of the *Istoria*. If Bianchini had had the reputation he derived from writing it when he wrote it, he would not have had to pay for printing it. Despite its size, complexity, and incompleteness, it has been reprinted three times: exactly in 1699, faithfully though with redrawn plates at its semi-centennial in 1747, and romantically reformatted, in five volumes, in 1825, as "one of the most stupendous works that the Italian genius has ever produced." Among its direct descendants are prints by Giovanni Battista Piranese dating from the 1740s depicting Roman monuments in the style and under the inspiration of Bianchini and a truncated version of the *Istoria*, the symbols without the history, published in 1871 on the discovery of the eighty images prepared for the unwritten third volume of the *Istoria*.[78]

The secretary of the Paris Academy of Sciences, Bernard le Bovier de Fontenelle, wondered in his éloge of Bianchini that a mathematician could be so good a historian. "Naturally the genius for mathematical truths and that for deep erudition are opposed; they exclude one another, they scorn one another." And in those rare cases that the same person possesses both, he usually does not do justice to either. And yet Bianchini did. Fontenelle praised particularly the parts of the *Istoria* where a shrewd reading of myth toppled received knowledge, as in the interpretation of the voyage of the argonauts and the Trojan War. In Bianchini's hands, the *Iliad* is "an allegorized history in oriental style," transmitting via antagonistic gods personifying regions around the Mediterranean information about an ancient trade war, "which is a surprise." As for Bianchini's quasi-mathematical uniformitarian arguments for dating the Flood and the progress of humankind, Fontenelle wryly observed that "chance seemed often to be in agreement with M. Bianchini's designs."[79]

The most enthusiastic of Bianchini's latter-day panegyrists was Ugo Foscolo, a writer and patriot who rests near Galileo in the pantheon of great Italians in the church of Santa Croce in Florence. For Foscolo, Bianchini's *Istoria* was not only "a very great book, unworthily forgotten by us, who enthusiastically embrace what comes from abroad and take no interest in our treasures." In addition, its rescue of slighted people from the depths of time resonated with Foscolo's efforts to gain recognition for his countrymen's contributions, "buried by the forgetfulness of many centuries."[80] Similarly, in his authoritative history of Italian literature, Francesco De Sanctis underlined the resurrection begun by Bianchini. "Monuments are no longer dead letters . . . They stabilize dates, epochs, customs, thoughts, symbols, the prehistoric world is remade . . . facts and people flicker, fade out, become fables and the fables become ideas."[81] In the authoritative judgment of a German *Handbuch* of 1880, "Bianchini was the first to conceive of a universal history guided by art . . . but even now historians have not truly recognized the idea of artistic artifacts as knowledge sources in addition to literary ones."[82] The situation has improved.

There were, of course, detractors. Uncharitable people rightly pointed to the arbitrariness of the chronological scheme, the capricious euhemerist reading of myth, the cheat in using late artifacts to "prove" assertions about more remote times, and the echoes of older usages of symbols as mnemonic devices. The inspiration for this catalogue of demerits was Benedetto Croce, who railed that modern critics who praised the *Istoria* could not have read it; if they had, they would have seen, as he did, that it is inferior to Giambattista Vico's *Scienza nova* (1725).[83] That may be; Vico built on Bianchini; but we must evaluate the *Istoria* on its own merits, on its systematic exploitation of the physical evidence and its attempt, however flawed, at a rational history of human progress.

Immediately after seeing his *Istoria* through the press, Bianchini published an account of several items excavated in Antium (Anzio), a coastal town 100 km south of Rome favored for villas by emperors and other great men. Interpreting the items with the help of ancient accounts of the sojourn there of the disciples of the magus Apollonius of Tyana, Bianchini associated the coins and statues in question with the Emperor Hadrian, who collected Apollonius's books and hankered after the occult. A significant part of Bianchini's account was its dedication to Francesco Aquaviva, a rising star in the Vatican firmament, then just made a bishop, soon to be sent as nuncio to Spain, and, as the usual consequence of that assignment, to be created cardinal.[84] Aquaviva could be an important patron. Let us not forget that Bianchini, who was but a librarian at the Cancelleria and a canon at Santa Maria ad Martires in the Pantheon, entered upon his "epoch-making historical work by reason of the consideration of monuments" to get himself a better job.[85]

3

JUBILEE LINE

Pilgrim and Pope

A jubilee offers the opportunity to acquire a plenary indulgence directly from the pope. It will not restore your innocence or guarantee your salvation; it relaxes the rigor of penances imposed for past sins. Do not ask how it works. It does if you unite yourself in spirit with "the sighs, groans, mortifications, deeds, and sufferings of all martyrs and saints, and especially with the agony, abandonment, the passion and sacrifice of Jesus Christ;" and also pray for everything good, for the pope, bishops, and priests, the extirpation of heresies, exaltation of the church, remission of sins, and peace among Christian princes. This was the authoritative advice of the famous and wordy French bishop, Jacques-Bénigne Bossuet, to the pilgrims of 1700.[1]

Maria Casimira Sobieska, dowager Queen of Poland, widow of the Christian paladin Jan Sobieski, was the highest ranking of the myriads of pilgrims who flocked to Rome in 1700. Pious and spendthrift, she understood and needed indulgence. She began her pilgrimage in the winter of 1698–9 with a retinue of three hundred, including her father and two youngest sons. They were an unprepossessing lot. The adolescent lads when released in Rome ran wild. Casimira's father did not behave much better. A military man raised to the purple by Innocent XII at the request of Sobieski, the nonagenerian Cardinal Deacon d'Arquien devoted his pastoral work to keeping a good table and entertaining women.[2] That accounts for the immediate Sobieski family apart from Casimira's eldest son, James (Jakub) Louis, who remained behind in his palace in Silesia. He had failed in his efforts, uncharitably opposed by his mother, to succeed his father as the elected king of Poland. His leisure gave him time to produce several attractive daughters, one of whom was Wogan's princess.

Maria Casimira, though born in France, had grown up in Poland as a lady-in-waiting to its French queen. There she met Sobieski, whose election, in 1672, she may have influenced. She acquired so notorious a reputation for political meddling that Louis XIV would not allow her to settle in France after her husband's death in 1696. Nor was she wanted in Poland or the Holy Roman Empire. She went to Italy, where memory of Sobieski's great victories and Innocent XI's part in them were still fresh, and she could arrive as a penitent while intending to remain as a queen. With her rank, reflected glory, political connections, fleeting wealth, and self-confidence, she fancied that she would conquer the Holy City like another Christina. She made a better pilgrim than scholar, however, whence the pasquinade:

> Nacque da un Gallo semplice gallina,
> Vissi tra gli Pollastri e fui reggina,
> Venni a Roma Crisitiana e non Christina.
> Born from a cock, a simple hen,
> She lived among the chickens and was their queen.
> She came to Rome a Christian but no Christina.[3]

The fun turns on the puns *gallo*/Gaul (cock, France) and *pollastri*/*Polacchi* (chickens, simpletons/Poles).

Casimira received the welcome she expected. As she proceeded from Venice, where she arrived in early January 1699, to Rome, she was entertained royally and admired universally for the conspicuous acts of piety she performed along the way. Innocent XII, who had not long since made her father a cardinal, welcomed her with an extravagant banquet, for which she prepared by washing pilgrims' feet.[4] The pope's extravagance was out of character. The jubilee entertainments he had planned were what readers of Bossuet's account of indulgences might have expected.

Instead of creating imposing statues, buildings, and fountains to celebrate himself and his jubilee, Innocent had devoted the entire income from his patrimony for 1700 to the poor.[5] Nothing comparable to the great spectacles of the previous jubilee, that of 1675, when Christina had played a prominent part as a convert, occurred during the reign of Innocent XII. The most dramatic event of the jubilee year was his death on 2 September.

Casimira found herself well placed for meddling in the quest for a successor. With a specialist's insight into the affairs of the Kingdom of Poland and her ties to the court of France, she could be useful in preparing for the upcoming conclave. Pietro Ottoboni studiously cultivated her acquaintance. There was, of course, a rumor of a connection more intimate than political between the susceptible cardinal and the impetuous queen.[6] Ottoboni headed a party whose leading candidate was Noris, well liked, "capable to govern a world," and, at sixty-nine, sufficiently advanced in years to justify the hope that he would not last long. Outside forces killed his candidacy. The emperor vetoed him, for being Venetian, while the Jesuits, working through cardinal zealots, opposed him for being an Augustinian. Their previous attacks on his character and theology had helped propel him to nuncio, bishop, and cardinal; but this time their animosity worked for them, and Ottoboni's party turned to a man who had not been considered papabile at the outset of the conclave.[7]

The college of cardinals of 1700 included many members who had risen from local administrative assignments to senior offices in the Roman bureaucracy. Not all were priests. Many were former nuncios who had run a course that began with an assignment to such places as Cologne, Florence, or Naples; continued to Portugal, Poland, or Venice; and culminated in Vienna, Paris, or Madrid, from which promotion to cardinal usually followed. Other cardinals had developed administrative skills in running the major bureaus or congregations, such as the Index, Inquisition, Vatican Library, and the Apostolic Camera (Treasury). Finding a competent potential pope among them was easy; electing him was hard. The cardinals had national ties, and, if Italian, were further divided by allegiance to the family of the pope that had elevated them. Moreover, the king of France and the German emperor had an effective veto power through their national cardinals and Italian allies, who served, for a good fee, as their "Protector" in the Vatican.[8] And there were even religious considerations.

An evaluation of plausible candidates for the election of 1700, drawn up by Count Orazio d'Elci, a Vatican insider and nephew of a cardinal, reveals the considerations in play. D'Elci reckoned Ottoboni's default candidate, Giovanni Costaguti, to be the front runner. Why? Costaguti had done well at the Treasury, liked neatness and order, was genteel, sociable, sober, judicious, charitable, intelligent enough, not overly learned, and, most importantly, had no important enemies at home or abroad. In short, dependable, pleasant, a plausible contender "owing to a scarcity of Persons eminently deserving of that high Post." Apart from mediocrity, d'Elci could perceive no faults in Costaguti as a candidate except his youth (age sixty-four) and the relative obscurity of his family.[9] The count did not think Albani papabile. Nor Ottoboni, though "he has the Soul of an Emperor." He overspent his vast income, on himself, to be sure, but also on academies, operas, public entertainments, churches, alms, dowries for poor girls, and medical care for parishioners. The emperor vetoed him as too inclined to France; zealots thought his life a little loose even for a cardinal; and in any case he was far too young.[10]

News of the death of the childless king of Spain, Carlos II, changed the thinking in the conclave. The cardinals could see that the Spanish succession would bring a war between the two major Catholic powers, Austria and France. Carlos had left his entire empire, which included possessions in America, the Low Countries, and Italy, to Louis XIV's grandson, Philip of Anjou, against whom the Austrian Habsburgs, who had dynastic ties everywhere, pitted Archduke Charles, son of the reigning Holy Roman Emperor Leopold I. In the hope that the Holy See might intervene successfully in the expected war, the conclave sought a man of energy, judgment, political savvy, and steadfastness acceptable to the major belligerents. The list was very short. The conclave chose Albani, the Holy Ghost concurred, and neither Emperor Leopold nor King Louis objected. Luckily, Albani had taken the trouble to become a priest just before entering the con-clave, from which he emerged, on 23 November 1700, as Pope Clement XI (see Plate 1).[11]

For three days he had refused to accept the poisoned chalice. He under-stood perfectly that the Vatican would be very awkwardly placed in a war over the spoils of Spain. Ottoboni and others eventually convinced him that he was particularly well qualified to guide the Church through such a crisis. *Habemus papam!* It was time for congratulations. The irre-pressible queen of Poland came in person, unannounced, and found the new Clement conferring in his cell with Ottoboni; she adroitly joined her dissolute father for dinner. Clement soon received her cordially. She responded by sending an over-the-top gift composed of the Albani arms (three mountains and a single star) made of fruit plus two large aviaries containing a dozen live birds each.[12]

D'Elci's portrait of the new pope rings true. Albani is a fine politi-cian, he wrote, "and makes it his particular Study to displease no Man." He combines great learning in law, church history, and ancient authors with a deep knowledge of public affairs, "not only of Princes, but of the most considerable Families in the Universe." Further in his favor, he is "of exemplary Virtue, of a great Spirit, of great Parts, and of a most lively Behaviour; he's judicious, cunning, prudent, civil, facetious, and high-ly obliging," and, withal, "bountiful to the poor." What is not to like? Being unable to deny anyone anything, he makes contradictory commit-ments, and "at last determines himself in Favour of the Person to whom Fortune seems to be most propitious."[13] As the powers scurried to take sides in the upcoming war, Clement's openness to all was useful; but, as the war progressed, the parties deemed him shifty and unreliable, and sidelined him.

Clement's first local business was not the upcoming War of the Span-ish Succession, however, but the jubilee depressed for want of a pope. He immediately facilitated its main purpose. When, three days after his election, Rome suffered a great flood, he gave all pilgrims still in the city extra indulgence to palliate their inconvenience. His first bull, issued 25 February 1701, extended the universal jubilee to "implore Divine help at the beginning of [his] pontificate."[14] At Bossuet's suggestion, he renewed the celebration for two months in 1702 for stay-at-homes, who had only to say five paters and five aves at each of four churches to remit their punishment. These were by no means the only works that Clement undertook in the jubilee spirit. He made much of visiting hospitals, succoring pilgrims, and distributing alms.[15]

Being young for a pope and disdaining nepotism, Clement did not rush (though he did not forget) to advance men of talent who happened to be his friends. Bianchini guessed that his turn would come in two or three months and that Clement would probably offer him the rank of prelate and a position to justify it. Prelates typically were bishops and abbots; but clerics around the pope's person and high Vatican officials might attain a similar status and the title of monsignor. Bianchini was by then a cleric, having taken the orders of subdeacon and deacon in 1699 on becoming a canon at Ottoboni's church of San Lorenzo; he was punctilious in his observance of the canon law related to deaconship and to assistance at the altar; but, not being a priest, he could not expect a lucrative benefice. On the contrary, becoming a prelate would be costly.[16]

He would need access to a carriage, appropriate garb, and three servants, also a valet and a coachman. That would require an initial outlay of at least 500 scudi and large ongoing expenses, perhaps 300 scudi annually. (For comparison, a liveried servant in a good household earned about 60 scudi a year plus clothes.[17]) Bianchini's income consisted of a portion of 150 scudi of rent from Brescia, which he shared with a sister, plus 200 scudi from his canonry. He could expect other offices and benefices that would enable him to meet his regular expenses, but he would have to borrow setup costs from family members (he had given his share of his father's estate to a spinster sister) and draw the entire rental income from Brescia for a few years. The question whether to accept the prelature that might be offered therefore concerned several of his closest relatives. Apparently all consented. Had not the investment in the *Istoria universale* paid off? Bianchini accepted the office he anticipated, *cameriere d'onore* in the pope's household, on 16 February 1701.[18] The appointment brought a special obligation to live morally, as Clement proved in a homily hard to follow, and rooms in a papal residence, the Quirinal palace, reputed for its healthy location and extensive gardens. The quarters must have been larger than Meijer's one-room flat, since Bianchini needed accommodation for his many books and two servants. He also had space for telescopes, which he fixed in a window or deployed in the gardens, and for the installation of a meridian line.[19]

The Quirinal gardens boasted a perpetual universal sundial with four working faces. Pope Urban commissioned it in 1624, the year after his election. Its dials faced the cardinal points, all but the northern one with the stinger of a bee as its gnomon, that to the north employing a solar ray; all to be understood from the appropriate motto from Virgil, "sunt quibus ad portas cecidit custodia sorti," "there are those by fate made guardians of the gates." Which, unriddled, signified that Urban, whose family crest featured bees and who identified with the sun, had been chosen the protector of the Holy Roman Church.[20] That was the sort of learned play in which the new monsignor delighted.

Bianchini's first assignments as a top papal aide made use of his technical skills in one of the most pressing of papal responsibilities, furnishing Rome with ordinary water. Clement united two of its three most important aqueducts under the same treasury official and put Bianchini on the board that policed unauthorized taps and siphons and kept the old installations in running order.[21] The two waters—Paul V's Acqua Paola, which had (and has) its main terminus, or *castello*, at the famous Fontanone on the Gianicolo, and the Acqua Vergine, in operation since Roman times, with its *castello* in the still more famous Trevi Fountain—presented a range of technical problems. The low-pressure virgin system, updated in the sixteenth century, served the public at the piazze di Spagna, Navona, Venezia, Mattei, and Rotonda. The high-pressure Paola served the Vatican, Trastevere, and, via a large conduit under the Ponte di Sant'Angelo, parts of the Campo Marzio. Bianchini had to traverse most of Rome to study hydraulic and distribution problems. He and his colleagues did their job well; their two aqueducts ran much better than the newer third main addition to the water system, the Aqua Felice.[22]

The pope made another heavy demand on his *fisicomatematico*: to preclude another scandal like that of the Easter of 1700, which Rome, following the Gregorian canon, had celebrated a week later than specified by its definition (the first Sunday after the first full moon falling on or after March 21). Something had to be done; "all Europe was busy with calendar reform."[23] Cassini discovered an error in the computations made using the calendrical rules imposed by Pope Gregory XIII in 1582, but not in the parameters and cycles. Nevertheless, the new pope decided to reopen calendrical questions in the hope that openness might convince Protestant states that Gregory's reform was not a papal plot but a practical device improvable with ecumenical cooperation. He established a committee on the calendar under Noris and invited input from abroad.

The new calendar committee comprised three cardinals and twelve experts, including Eschinardi and Guillaume Bonjour, an Augustinian with a taste for calculation, and, as its secretary, Bianchini.[24] The Roman diarist Francesco Valesio did not have a high opinion of most of the committee, which, owing to unequally distributed ignorance, resembled "a family so poor that he who had a cap had no jacket and he with a jacket had no shoes."[25] The work fell on the secretary. It became full time when Clement commissioned the jubilee meridian line in Santa Maria degli Angeli ostensibly to support the committee's conclusions. Late in 1701 Bianchini could report that the sun, when observed at the unfinished instrument, occupied the precise spot in the sky where the best ephemerides told him to look for it.[26]

Preparation

In creating his *meridiana* Bianchini had as a model the heliometer at San Petronio he no doubt had visited with Ferroni during his attendance at the

Jesuit college in Bologna and one he had made himself in Padua when studying the sun with Montanari. Cassini's line runs within a whisker of two gigantic piers and yet deviates from true north by only a minute of arc. The slightest error in calculating the direction would have meant disaster (see Plate 2). The line itself, originally of iron but now of brass, is encased in marbles depicting the zodiacal signs indicating the sun's course through the heavens. The beauty of the installation and the faithfulness of the sun's rendezvous with it made a great impression. At its inauguration in 1655, Cassini boasted that "nothing has been done before under the patronage of princes so worthy of the majesty of astronomy or so suitable for advancing its dominion."[27] In fact no prince was involved. Permission and payment for the work came from the board (the *fabbricieri*) responsible for maintaining the structure (fabric) of San Patronio.

Cassini's boast implied that he and the *fabbricieri* had more in mind than measuring the length of the year. Their *meridiana* had astronomical work to do. The most romantic of its projects was deciding between the claims of eccentrics and half-eccentrics. The whole nuts followed Ptolemy's solar theory in offsetting the center of the sun's orbit C from the center of the earth E by the eccentricity EC = *ae* (Fig. 20). The half-nuts adapted Ptolemy's planetary theory, separating the center of the sun's motion X (the "equant point") from the orbital center C, putting EX = *ae* and placing C

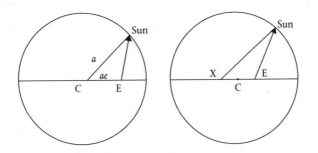

Fig. 20. Eccentrics versus half-eccentrics. Full eccentric to the left (CE = *ae*), half to the right (CE = XC = *ae*/2). Here C is the center of the sun's orbit, E the earth, and X, in the half-eccentric case, the center of the sun's motion (the "equant point"), about which it moves through equal angles in equal times.

midway between X and E: now EC = $ae/2$, half the eccentricity. Since on the second hypothesis the sun is farther away from the earth at perigee (their least separation) than in the first hypothesis (by an amount $ae/2$), and closer at apogee (their greatest separation), and since the size of the solar image depends on the solar distance, Cassini expected that measurements at San Petronio should settle the matter.

The half-nuts won. Why should anyone have cared? EC = $ae/2$ was the answer expected from Kepler's elliptical version of Copernican theory. A positive answer would not decide definitively whether God had chosen to fix the earth or move it around, but it would strengthen the Copernican side: for, if the earth were a planet, the half-eccentric orbit found for the sun by observations at San Petronio would in fact belong to it, and make it move like other planets. The *meridiana* could help also to determine mundane matters of astronomical interest like the effect on observations of atmospheric refraction; but enough has been said to assert that, in undertaking to build a *meridiana* in Santa Maria degli Angeli for the Pope of Rome, Bianchini had an opportunity to advance himself and astronomy. And also to fail at both. It is not easy to satisfy simultaneously the demands of science and magnificence.[28]

Even the great Cassini had not been perfect. His instrument had a movable part, which, contrary to intentions, moved. This was the horizontal plate containing the aperture through which the sun's rays passed into the basilica. It began to slip soon after installation. The mathematicians in Bologna, including Montanari, calculated corrections to take the slippage into account, but they did not trouble to reset the plate. Cassini could not repair it, as he was in Paris, leading the new observatory built by Louis XIV. The king had won him in a tussle with the pope and the senate of Bologna on condition that Cassini maintain his professorship in its university. In 1695 he returned briefly to San Petronio, reset the plate and the zodiacal marbles, and turned the iron rod into brass. The refurbished instrument, according to the repairers, offered "a secure and royal road to celestial observations."[29] Cassini's road then took him to Rome, where he visited the Ottoboni library. There he saw something he may not have anticipated: a meridian line or two in the Cancelleria installed by Bianchini.[30]

Immersion in universal history had not blunted Bianchini's enthusiasm for astronomy. Discovering another problematic euhemerism in the sky did not bring the excitement of being the first to spot a comet. He had not forgotten the thrill of discovering a comet in 1683 and the glory of publishing his report of it (the only printed report received from Italy) in an international journal, the *Acta eruditorum* in Leipzig. With the goal of enhancing his reputation as an astronomer, he sent the *Acta* a simplified description of a technique that Cassini had proposed for determining the parallax of a planet from observations made from a single point on earth.[31] Later he would use the technique, which will be described in its place, to measure the solar system to within thirty million miles. Had he known the magnitude of his error (over 30 percent), he probably would not have been upset. "I do not bother with the smallest fractions, in order not to be pointlessly precise in a business so uncertain."[32] Compare his declaration that "the conjecture of a historian is not the decree of a magistrate."[33] It is a counterproductive category error to insist on details not ascertainable in principle or in practice, and his astronomical instruments did not enable him to determine the distance to the sun by Cassini's method more securely than he could date the Flood by his own.

Montanari's telescope, which Bianchini used for his cometary observations, extended to 4.5 m. This was not a large instrument for the period. Since the longer the focal length of the objective lens, the less the effects of its geometrical imperfections, cutting-edge telescopes grew so long and heavy that they had to be rigged up like booms from a ship's mast (see Fig. 21). To avoid the consequent difficulties in operation, Bianchini sometimes did without the tube. By fixing the objective to a movable shelf attached to a high post and running a string between the housing of the lens and the eyepiece, an agile astronomer had a chance to align the lenses. Very few astronomers had the patience, strength, and dexterity to master the apparatus; Bianchini had them all and could manage telescopes tens of meters long. Around 1700 he invented a way to make

Fig. 21. A long telescope developed to minimize spherical and chromatic aberration. FB, *Hesperi et phosphori nova phaenomena* (1728), tab. VIII.

long telescopes portable. A collapsible stand supported on a tripod held the objective lens as high as 11 m. A smaller contrivance supported the housing of the eyepiece (see Fig. 22). Everything fit into two boxes that one man could carry.[34] Bianchini was no armchair theorist. He had considerable experience designing, making, and using astronomical instruments before the call to build the Clementine *meridiana*.

He also brought to the task the deep explorations of chronology he had undertaken to work up universal history. From constantly seeking significant astronomical configurations to fix the dates of historical events, Bianchini stumbled on a unique property of the jubilee year 1700. It was the only year in the entire history of the world when a new moon falling on an autumnal equinox would also fall on the vernal equinox immediately following; moreover, to advertise the coincidence, the playful Fates had arranged that the autumnal new moon would eclipse the sun. That made 1700 not only a jubilee year but a fit epoch for Easter calculations.

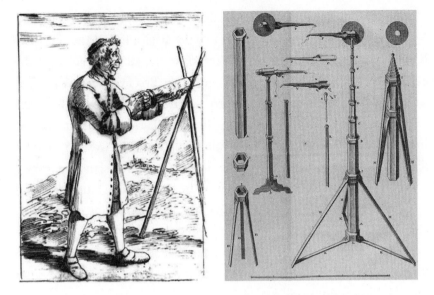

Fig. 22. Bianchini's long telescope. (a) The astronomer himself at the eyepiece of the instrument. (b): Bianchini's portable stand for the objective lens of a tubeless telescope. Pier Leone Ghezzi, caricature of 1720, and René de Réaumur, Académie des sciences, Paris, *Histoire et mémoires*, 1713, 299–306, resp.

Paschal mathematics then used liturgical tables based on the Victorine cycle, a period of 532 years after which full moons, and therefore Easters, repeat their sequence of days and dates. Why 532? Because 4×7 = 28, the leap-year cycle after which the sequence of days and dates repeats, and 532 = 28×19, nineteen being the Metonic cycle, a period of nineteen years after which the dates (but not the days) of full moons recur. Suppose that a Victorine cycle began in 1701. Go back ten cycles; you arrive at 3620 BC, 380 years short of Creation. But 380 = 20×19. In the year 1700, therefore, twenty Metonic and ten Victorine cycles had elapsed since the Creation of the world.[35]

There is more to this innocent erudite amusement. Saturn spent the jubilee of 1700 in Pisces. Exactly 532 years and one month earlier, it dwelt in Aquarius; for, in a Victorine cycle plus a month, Saturn goes eighteen complete circuits plus a zodiacal sign around the earth, or, perhaps, the sun. No matter. It follows that the earth, which goes through a sign a month, will be in opposition to Saturn at the beginning of the tenth Victorine cycle if it is in opposition at the beginning of the eleventh. (Opposition signifies diametrically opposite the sun as seen from the earth.) And so on, back to the beginning of the first cycle, 380 years after Creation, when the opposition occurred with Saturn in Taurus. The cycle can therefore be named after the sign in which Saturn stands at its commencement.[36] We can easily determine where Saturn was at Creation. Since in 380 years it goes back twelve circuits and eleven signs, Adam saw it in opposition in Gemini. The order of the signs, Aries, Taurus, Gemini ..., thus gives the loci of Saturn's significant oppositions if altered to Gemini, Taurus, Gemini ...[37] Why the double Gemini? Do not require perfection in everything. What counts is that the root of it all was the unique chronological significance of 1700, which Bianchini would build into the Clementine *meridiana*.

Although Clement had complete confidence in his *cameriere d'onore*, he wanted the most experienced builder of exact *meridiana* to undertake so important a work as a pontifical observatory. Cassini declined it. He agreed, however, to consult about any special problems posed by the project and sent his nephew, his *primo nipote*, Giacomo Filippo Maraldi, from Paris to Rome to help. Cassini advised, however, that the parameters of the Gregorian system were sufficiently accurate to underwrite an Easter canon off by less than a day in a thousand years.[38] Nonetheless, Clement and Bianchini desired to proceed for the advancement of science and the glory of Rome.[39]

To bring his *meridiana* to life, Bianchini needed workmen of high quality. The project being pontifical, he had his choice. Carlo Maratti, the most respected painter then active in Rome, Clement's artistic advisor and portraitist, was put in charge of most of the decorative scheme. Maratti adapted the exquisite pre-telescopic depictions of the zodiacal signs in Bayer's *Uranometria* (which Bianchini had used to record variations of stellar brightness) for the floor of the basilica.[40] Domenico Paradisi, who had labored to fix up the derelict Cancelleria for Ottoboni's occupancy, directed the artistic work.[41] After making full-scale colored drawings from Maratti's improvements of Bianchini's sketches, Paradisi oversaw their transformation into stone and brass, "at great inconvenience," as he claimed in bulking up his bills.[42] The zodiacal marbles, done in the manner of ancient mosaics, were the work of Francesco Tedeschi, "second to none [in Bianchini's opinion], in this way of sculpting, or should I say painting." (see Plate 3).[43]

Just after finishing at Santa Maria degli Angeli, Maratti and Paradisi pooled their talents for a project more demanding even than the pope's *meridiana*. They risked perpetual dishonor by agreeing to restore and complete frescos by Raphael in the Farnesina, a jewel of a Renaissance villa now owned by the Accademia dei Lincei, Italy's national academy of sciences. Success in the risky endeavor brought Maratti knighthood in the pontifical Ordine supremo di Cristo. He celebrated the honor on the tomb he designed for himself near Bianchini's *meridiana*, where he took up permanent residence in 1713.[44]

The Clementine *Meridiana*

Santa Maria degli Angeli occupies a portion of the ruins of the ancient Baths of Diocletian (erected around AD 300), which extended over thirteen hectares (thirty-two acres) and included the present spacious Piazza della Republica.[45] Pope Pius IV commissioned Michelangelo to work a church into the surviving structure, then, in 1561, the haunt of tourists by day and brigands by night. Michelangelo designed a spare church, which in the execution was made sparer, for three good reasons: there was little money for it, the Cistercians were to run it, and the Tridentine spirit informed it. While Michelangelo was drawing up his plans, the Council of Trent was winding down its affairs. During its twenty years of part-time deliberations it had clarified dogma and strengthened discipline to outfit the Catholic Church for spiritual battle against Protestants. Michelangelo's new basilica augured a Catholic victory in repurposing for Christian worship a space built for pagan indulgence by a great maker of Christian martyrs.[46] It also anchored the redevelopment of a deserted quarter of the old city. Pius IV celebrated his achievement in urban planning with a powerful exorcism of evil spirits that might have lingered in the Baths since Diocletian's day. It is still to be read there: "What was a pagan Idol is now a Temple of the Virgin—Pope Pius is its Founder. Demons Be Gone."[47]

The perfections of the basilica as the seat of a *meridiana* lay in its orientation and antiquity. Its age had completed its settlement; it offered a fixed coordinate frame against which ongoing repetitive observations could be reliably compared. The Lord tested its fixity the year after the *meridiana*'s completion with an earthquake that shook Rome. Nothing moved in the basilica.[48] Further to its perfections, the basilica's orientation, free from obstructions, allowed the sun's noon rays to pass through a hole in the south wall onto the meridian line throughout the year. A hole in the north wall made possible observation of the transits of stars. Since, in Michelangelo's Greek-cross plan, the main axis trends from southwest to northeast, Bianchini had to run the *meridiana* from a corner of the south wall diagonally through the crossing up to the main altar (see Plate 4). That gave it due prominence.

Bianchini published the layout (see Fig. 23) in 1703 in a little book, *De nummo et gnomone clementino*, gathering into one happy phrase the coin (*nummus*) he had minted to celebrate the completion of the installation, the *meridiana* itself (*gnomon*), and the pope's commission and expenditure. The coin bears on its upper face an image of Clement XI and the beginning of Hebrews 3.1, which the King James Version renders as "through faith we understand that the worlds were framed by the word of God." But the Latin Vulgate, which has *saecula*, ages, rather than worlds, agrees better with Bianchini's intent; for the *meridiana* has to do with time, particularly the time ordained by God for the celebration of Easter.[49] The order was challenging: to fix the time, the computer (a person so named) had to know the precise length of the interval between successive visits of the sun to the vernal equinox. The best way to fix the value of this "tropical year" was to locate the image of the sun as it crossed a *meridiana* at noon on the

Fig. 23. Layout of the *meridiana* at Santa Maria degli Angeli; FB, *De nummo et gnomone clementino*, after p. 20, bound with FB, *De kalendario* (1703).

day of an equinox and count the days until it returned to the same position the following spring. Repeating the measurement over many years gave a very good result.

Around the summer solstice, the sun's image is very close to circular; around the winter solstice, a large ellipse; during an eclipse, imperfect. The stretching-out of the spacing between the zodiacal marbles along the line mimics the lengthening of shadows as the sun's noon altitude declines between the summer and winter solstices, and its subsequent shortening as the altitude increases between the winter and the summer solstices. The process is perfectly symmetrical, whence the pairing of the intermediate zodiacal signs: the sun's noon altitude on entering Gemini is the same as that on entering Leo, the signs on either side of Cancer; similarly, the sun's noon heights when it enters Sagittarius and Aquarius, on either side of Capricorn, are equal. The marbles depicting the signs where the sun turns back, Cancer and Capricorn, whose entry points mark the summer and winter solstice, respectively, are unpaired at either end of the *meridiana*.

To make the scheme work, the marbles must be placed so that their *centers* indicate the position of the sun's image on the *meridiana* when the sun itself enters the corresponding signs in the zodiac.

There is much more. A ray from the pole star (Polaris) lands on the southern end of the *meridiana*, making an angle there equal to the latitude of Rome. Over twenty-four hours, Polaris's virtual image runs around an ellipse concentric with those that Bianchini picked out by twenty-four brass stars (see Plate 5). Their common center, marked by the lone star in the Albani arms, is the intersection of the earth's polar axis with the floor of the church—a nice if impertinent symbol of the pope as *polus mundi*. Each ellipse represents the diurnal image of the daily revolution of the pole star at the beginning of a jubilee year. The innermost ellipse corresponds to the separation of Polaris from the pole in 1700; the next, to the separation in 1725, and so on, through sixteen ellipses and four centuries, to AD 2100, when the separation, caused by the precession of the equinoxes, reaches its maximum.[50] Then, as Polaris comes back towards the pole, its diurnal image retraces the interior ellipses, in reverse order and twenty-five-year intervals, until it returns, in 2500, to its place in Bianchini's time. To follow up his universal history, Bianchini had intended to divide the period from Augustus to 1600 into intervals of twenty years; at the *meridiana* he took instead the interval between jubilee years, "as instituted by Roman pontiffs."[51]

The jubilee ellipses combine with the meridian line to constitute a grid on which history can be written. No larger frame was likely to be required, since the 1600 years specified by the ellipses exceeded the time that reliable eschatology calculated that human history had yet to run. Two drawings of the nest of ellipses, perhaps sketches by Maratti or Paradisi from which the masons worked, have survived. The mix of science, history, and liturgy that the artists thought Bianchini had in mind appears from the lengthy legends on their sketches. One reads:

> In order that the Catholic religion triumphing over its tyrants might celebrate the sacred rites of the church and the memory of martyrs . . . Clement XI, pontifex maximus, during the first year of his reign ordered placed in

the baths [of the tyrant Diocletian] ... a meridian line, a perennial mark
of the equinoxes, to preserve the rule of the Nicene and Gregorian year
fixed by law, and a polar *eortologio* [calendar of saints] for distributing 32
Jubilee years in 8 centuries, for the orderly praising of God in agreement
with heaven and earth.

The other sketch clarifies the reference to saints' days by including them
around the outermost ellipse. Its legend:

[Behold] the Clementine *eortologio* that divides the 8 Gregorian centuries
into 32 Jubilee years following the constitutions of Paul II and Sixtus IV, and
that counts single days from midnight preserving the Church's ritual for
celebration of the memory of saints according to the elliptical projection
of the diurnal circle that the pole star will describe for 800 years.[52]

Bianchini's explanation is clearer and briefer: "In this single instrument
not only astronomy but also sacred chronology and the Roman Catholic
calendar may be seen independently and united by the rays of the celestial
bodies."[53]

To the north of the Jubilee ellipses a line of stars NO (see Fig. 23) represents a portion of the diurnal path of the image of the star Arcturus
at quarter-hour intervals; and, further down, skipping for a moment the
two tracks between them, QR represents the path of the star Sirius. These
are the brightest stars visible through the hole in the north wall. Since,
unlike the sun, distant stars are fixed in the heavens, their diurnal track at
the *meridiana* does not change over periods short enough for the precession of the equinoxes to be ignored. That is long enough to give meaning
to the intermediate tracks, which record not astronomical but historical
events. GT marks the track of the solar image on 6 October 1701, the feast-
day of San Brunone, the founder of the Carthusian order, when Clement
visited the line in its raw state. Exactly a year earlier Clement had said his
first mass there before entering the conclave that elected him pope. His
presence at the building site answered critics who deplored placing an
astronomical instrument in a basilica.[54]

Clement returned the following year, again to commemorate San
Brunone, and saw the *meridiana* encased in the zodiacal marbles studded

with brass markers mimicking the positions and magnitudes of the principal stars in each constellation. It had taken about eighteen months to design and build.[55] For six of them Bianchini worked at the construction night and day, with and without artisans, powered by his robust strength and ability to survive on a few hours' sleep. "He had no complex except for having to be always at work."[56] The dimensions suggest the effort: the line is 46 m long and the hole in the southern wall about three-quarters as high as the roof aperture at San Petronio. There is nothing at the earlier instrument comparable to the studs distinguishing six magnitudes of stars or the polar ellipses at that "singular trophy of Italy," the *meridiana* of Saint Mary of the Angels.[57]

That leaves the puzzling track HI. It marks the path of the sun's image on 12 September 1702. On that day, carefully chosen, Maria Casimira came to observe Bianchini's celebration of jubilees. Her visit fell precisely nineteen years after Jan Sobieski's grand victory at Vienna. That made it a perfect historical–liturgical event, since, as we know, nineteen years complete the fundamental Easter cycle (the Metonic cycle) of 235 lunations. Bianchini recorded the queen's visit and the king's victory with two medallions placed on either side of the meridian line where the sun's image falls on 12 September (see Plate 6). The inscriptions, which make up a single message, read:

> Maria Casimira, Queen Consort of Poland, placed this inscription in Rome in 1702, in the second year of the pontificate of Clement XI, at the completion of the 19-year cycle after which the sun and the moon repeat the motions they had on that 12 September made joyful for Christendom by the victory of John III King of Poland over the Turks at Vienna.[58]

Symbolically, the Sobieski medallions linked the affirmation of Christian faith in the joyful celebration of Easter with the security of Christendom through the defeat of the Turks. To preserve that security was Clement's constant concern. "He would omit nothing, conceal nothing, that could contribute to [it]." Omitting nothing, Clement acceded to the request of the Queen of Poland to act as godfather to her granddaughter Clementina,

Wogan's princess, born two months before Casimira secured her place at Clement's *meridiana*.[59]

The pope too had his medallion at the spot associated with his visit in 1701 (see Plate 7). Further recognizing his patron, Bianchini decorated the holes he had made in the walls to admit solar and stellar light with the Albani arms (see Plate 8), which stood out more in Bianchini's design than they do now. The same is true of the star tracks apart from the jubilee ellipses. Rebuilding in the middle of the eighteenth century moved Michelangelo's original sober design into the baroque. The fading of the stars in the marble that replaced the brick floor was a minor loss compared with the injustices done by new wall paintings, reworkings of the windows and capitals, closing off the chancel, and sealing the entrance east of the line.[60] Recent preservation has further removed the *meridiana* from its appearance in Bianchini's day by placing guard ropes around it to prevent eager viewers from trampling on it.

Bianchini's earliest observations at the *meridiana* confirmed Cassini's prediction that the Clementine year would not differ from the Gregorian enough to make any change necessary. But they were a treasure in themselves, more exact, Banchini boasted, than any yet made of the latitude of the basilica, the inclination of the earth's axis to the plane of its (or the sun's!) annual motion ("the obliquity of the ecliptic"), and the effects of atmospheric refraction. More measurements supported his belief that he had "observed the equinoxes with the greatest certainty ever achieved by astronomical instruments."[61] Comparison with earlier measurements revealed the hard numbers necessary for computing holidays: the tropical year, 365 days, 5 hours, 48 minutes, 59.19 seconds; the average lunation, 29 days, 12 hours, 44 minutes, 3 seconds.[62] Clement informed the universities of Europe that his new solar observatory had confirmed the Gregorian values. It appears that Bianchini's instrument can fix midsummer noon to within a second of time.[63]

For a quarter of a century, Bianchini studied the solar image he projected on the basilica's floor. The information he recorded almost every other day about the state of the sun and the atmosphere contributed something to solar physics and meteorology. Perhaps his most important detection was a very small decline in the obliquity of the ecliptic, which revealed itself in a minute displacement of the solstitial images of the sun toward one another along the *meridiana*.[64] Since it seemed more likely that this small secular creep arose from a motion of the earth than from a complicated twist of the heavens, it added a bit of evidence to the growing pile favoring the Copernican theory. But no cosmological conviction was or is required to admire the artistry of the jubilee ellipses and zodiacal marbles, or the bright image of the sun fleeting across the floor of the church that Michelangelo built within the ruins of the Baths of Diocletian. Bianchini enjoyed showing it to the aristocratic, learned, and royal visitors who came to see it.[65] And onlookers enjoyed watching him, "in the complete costume of a prelate, kneeling on the ground, covered with dust, writing, calculating . . . fixing the position of the equinox."[66]

Bianchini's journal of observations was published after his death by his friend Eustachio Manfredi, professor of astronomy at the University of Bologna, who cooperated in synchronizing observations at San Petronio and Santa Maria degli Angeli. Typical journal entries around the time of the installation (1702–3) report observations of the sun, its limbs and its spots, often in the presence of notable people; Casimira, for example, who liked to "watch the passage of the sun's rays across a bronze ellipse inscribed to the memory of her late husband John III," and the Spanish ambassador with a train of twenty "noble and learned men."[67] Bianchini was by no means the only astronomer at work there. His journal mentions viewing sessions of transits of planets and stars with Maraldi, Manfredi, and Celestino Galiani, a priest with whom he studied Newton's work.[68] Entries in the journal continue until 19 February 1729, when, eleven days before his death, Bianchini tested the perpetual rendezvous he had arranged for the sun at Santa Maria degli Angeli for the last time.[69]

The Clementine heliometer received great praise. "The most exact of meridian lines . . . There is nothing of the kind more beautiful."[70] Protestants as well as Catholics took the time to notice it. Perhaps it is true that Leibniz never tired of extolling it to his friends.[71] Certainly it is true that the Committee on the Calendar commended its secretary for his achieve-ment, and its president, Noris, made him a bequest of 50 scudi a year. That started to flow in 1704, when Noris died from the press of business and the attention of doctors. Like a saint, he was divided into pieces, his body deposited in Rome and his heart in Verona, "for the consolation of his country." The busy queen of Poland was present at the parting, to rec-ommend Noris's shade to God "with the ratification of her piety . . . and esteem."[72]

Bianchini did his work well. Laser measurements show that the line runs off true north by 4.5 minutes of arc and deviates from straight by no more than 2 mm for most of its length except for a slight bow between Sagittarius and Aquarius reaching 5 mm. These errors do not affect the timing of noon. The change in the obliquity of the ecliptic and the conse-quences of the precession of the equinoxes since Bianchini's day are easily detected and agree with theoretical calculations. Bianchini obtained from it results as accurate as the instrument allowed; his measurement of the latitude of the basilica is good to 1 arcsecond. The installation still has value as a teaching tool.[73]

The calendar committee ceased its work around the time of Noris's death. It had not been free from bickering. Bianchini had to engage a squad of mathematicians to purge the mistakes of one touchy member while combating another who opposed every calculation likely to lead to an innovation.[74] On the productive side, the aptly named Father Bonjour followed Noris's lead to concoct a cycle of 1,932 years that returned the sequence of full moons to hour and minute, as well as to day and date, and one even more accurate, of 3,400 years, about half the expected life-time of the universe.[75] Bianchini found a cycle of only 1,184 Gregorian years, which, he claimed, achieved the same accuracy, and was good up to the year 4000, should anyone need it then. These extravagant propos-als were deemed too fastidious for implementation and contrary to the pope's intention to correct, rather than replace, the Gregorian reckon-ing.[76] Whatever the substantive value of this work, it paid off in part the initiative that created it. Several Protestant states came into chronological communion with Rome over the civil if not the ecclesiastical year. This communion in one kind began the year on 1 January and eliminated three leap days every four centuries to correct for the eleven minutes or so by which the Julian calendar exceeded the tropical year.[77]

The most useful and highly praised calendrical contributions were two dissertations by Bianchini prefaced to his description of his *meridiana*. One dealt with Caesar's calendar; the other made sense of the canon of Hippolytus, which Bianchini had compressed almost unintelligibly into the stone supporting the personification of the church on the frontispiece of the *Istoria* (see Fig. 18). Several antiquarians had tried to interpret the canon, among them J. J. Scaliger, who claimed descent from the family of fratricidal Veronese condottieri and attacked scholars who did not agree with him with a belligerence in keeping with the style of his supposititious ancestors. Without troubling to understand it, Scaliger dismissed Hippolytus's calculations as the puerile work of a mediocre intellect.[78] Correcting Scaliger no doubt added to Bianchini's pleasure in showing that Hippolytus had adapted a basic technique of the Julian calendar to develop a cycle of 112 years that had many subtle merits.

Over the basic span, 112 years, the calendar dates of full moons advance eight days toward 1 January; if the paschal moon (the full moon that falls on or next after 21 March) should come on, say, 13 April in the first year of the cycle, it would come on 5 April at the beginning of the next cycle. In seven such rounds, or 784 years, this same moon would arrive fifty-seven days before 13 April (a discrepancy of some three hours in the 112-year cycle accumulates to almost a full day in 784 years). This moon is about two months too early to be paschal; but we know, since the average lunation is very close to 29.5 days, that in fifty-nine days there will be a full moon on 15 April. That will be the paschal moon at the beginning of the eighth basic cycle. There is more delight to come. Since in 784 years the moons retrograde two days in the calendar and in 112 years they advance eight days, in four rounds of 784 years and one basic cycle they will be back where they started. So Bianchini (and perhaps Hippolytus) offered a grand cycle of 3,248 years after which the days of the week and the calendared paschal moons recommence their series.[79] The magic number 8 dominates the calculation. In eight years there are ninety-nine lunations to within an hour or two; in fourteen such periods, or 112 years, the agreement is even better; moreover, 112 is eight times twice the number of days in the week; and, as Bianchini pointed out, eight, being twice the interval between Olympiads, easily fit into a widely used ancient scheme for reckoning time.[80]

Let us play. Chronologists had come to agree that the Exodus occurred between 1,500 and 1,550 years before the time of Christ. Why not 1,548 years? Hippolytus had suggested such a figure. Then the number of years between the Passover Moses celebrated after the liberation of the Children of Israel and the Easter of 1700 would be, yes, 3,248, just the grand cycle Bianchini derived from Hippolytus. For all that, Bianchini did not recommend it, if for no other reason than it was built on the defective Julian calendar. As we know, his substitute cycle of 1,184 Gregorian years did not displace the Victorine; but it would have the same status as Hippolytus's if, as suggested, it had been inscribed on a stone and deposited somewhere in the Vatican library.[81] Bianchini used it to help the Arcadians fix their quadrennial prize-giving in the style of the Olympiads. They needed to know the date of the full moon that fell on or next after the day of the summer solstice. Bianchini's solution made an entertaining whole, a combination of historical research with the solution of current problems, of Christian and pagan computations, Easters and Olympiads, the serious and the jocose.[82]

The dissertations on the Roman calendar and the Easter canon of St Hippolytus won Bianchini as high a reputation among chronologists as his *meridiana* won him among astronomers. Domenico Ausilio of Naples, who was both, can speak for many, including Manfredi, in praising "the grandeur of the conception, exactness of observation, [and] subtlety of invention" of Bianchini's chronology and his "extraordinary ability to explain it."[83] Perhaps Ausilio's commitment to Pythagorean philosophy made clear to him number-juggling that ordinary people do not find easy. The business of *De kalendario*, alas, "can only be understood by reading all of it, which is not, however, accessible or useful to everyone."[84]

Post Jubilaeum

Termination of the calendar committee did not free up much time for its secretary. Responding to a suggestion by Bianchini that the Vatican assemble a collection of early Christian antiquities, Clement appointed him supervisor of all ancient inscribed marbles found in Rome. "Without a special written license in Our name given by our Monsignor Bianchini" no such stone could be moved or molested.[85] A big one soon came to the new supervisor's attention. It was a column sticking into the air 6 m above ground, which workmen preparing an area around the Circus Maximus for building soon discovered was less than half the whole: 9 m below the surface they uncovered its marble base, a huge lump of stone 3.4 m square and 2.5 m deep. An inscription disclosed that the column it supported celebrated Emperor Antoninus Pius. Clement decided to reset the base and the column on a little hill 2 km away, in front of the Palazzo Montecitorio, which then housed the papal law courts. A similar monstrous monolith had been lifted, rotated to the horizontal, moved, and re-erected a century earlier; it now stands in front of Saint Peter's. The engineer then in charge, Domenico Fontana, wrote a detailed well-illustrated book about his achievement, which required, for the lifting phase, the simultaneous pull of 900 men and seventy-two horses. Fontana had run ropes from capstans through pullies anchored on a scaffold built well above the obelisk and tied the other ends of the ropes to a wooden frame or carriage bound tightly to the monolith. In October 1704 lesser men tried the same method on the shaft of Antoninus's column. Some of the beams of the scaffold failed.[86] The pope and the cardinals in charge of the project called for Bianchini.[87]

For most of November and December the *fisicomatematico* in him was at work. Fontana had not been able to calculate much, not knowing how to scale the strength of beams from experimental results to practical situations. Bianchini could do it, for Galileo had written on the subject. He calculated the load that every beam in the structure would have to carry at various points along them; he took into account the changing slopes of the ropes as the column rotated into the horizontal; he observed that the ropes, pullies, and capstans should be of the same number and quality wherever placed; and he described, with a persuasive engraving, how the ancients had moved the colossus of Nero, a bronze statue 30 m tall, using twenty-four elephants and the system Bianchini proposed for Pius's column (see Fig. 24).[88] The system, when tried with manpower in September 1705, succeeded perfectly. The shaft and its heavy pedestal then sat for over a century in sheds on Montecitorio. The shaft eventually perished. The pedestal, after restoration, moved to the Vatican, into the Clementine museum, an institution created along the lines suggested by Bianchini.

The pedestal, not its removal, had attracted Bianchini's attention from the moment of its disinterment. As historian, he recognized the importance of its sculptured images; as draughtsman, he drew them for study and preservation; and, as superintendent of antiquities, he published his sketches as soon as he could.[89] They appear bound between his lengthy dissertations on the Julian calendar and the canon of Hippolytus. He explained to Manfredi that his other work did not leave him the time to make his usual thorough study; he was publishing what and where he could because the pope wanted a short quick account.[90] His analysis of

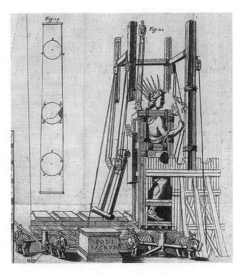

Fig. 24. Bianchini's method of raising monoliths adapted from the Romans; the pairs of elephants just visible at the bottom would have had a job of it. FB, *Considerazioni teoriche* (1704), second plate after p. 52.

the Antonine images was not altogether incongruous in its setting. The connection to astronomy appears in the celestial sphere in the left hand of the winged figure who is carrying the emperor and his wife to their apotheosis (see Fig. 25). The other images represent circus games (on the left) and the city of Rome (on the right).

The sphere shows a clip of the Zodiac centered on the spring sign Aries, above which stands a crescent waxing moon and a star. Another star shines above the sign of Pisces, and a serpent crawls around the north and south poles. Bianchini deduced from the position and phase of the moon that the sun must be in Pisces and that, with a little indulgence of the workmanship, their configuration commemorated the date AD 14 March 161. That was in order: Antoninus died on AD 7 March 161, and, as Bianchini demonstrated with lashings of learning, apotheosis customarily occurred on the eighth day after death. The Roman Senate established annual circus games in the emperor's honor, whence the reclining athletic youth. Rome gazes at him and his suggestive obelisk over some military equipment representing Antoninus's most important conquests. And the snake? A classic symbol of the rotation of the heavens.[91] The Farnese sphere, which Bianchini had studied with Cassini in 1695, also dates from

Fig. 25. Bianchini's sketch of a relief at the base of the column of Antoninus Pius unearthed in Rome in 1703, and an enlargement of the celestial sphere held by the central figure. FB, *De kalendario* (1702), second plate after p. 72.

the age of Antoninus and features a section of the zodiac around Aries.[92] Bianchini's posthumously published dissertation about these artifacts must have been completed substantially before 1712, when he was traveling in France and England, for he was able to bring printed drawings of the Farnese globe with him for the instruction of people interested in the astronomical achievements of the argonauts.

Bianchini understood his brief as superintendent of ancient inscriptions to go beyond policing the finds of others to digging and collecting for the pope and himself. He worked hard on his proposed museum for the early history of the Church and its relation to imperial Rome. The project perfectly fit Clement's policy of placing modern Rome at the center of Christian culture as a good in itself and as a substitute for the temporal power the papacy no longer possessed. To keep the pope keen, Bianchini would bring him the most interesting portable artifacts as they were found, bought, or copied.[93] But the pope, then in unequal contest with Emperor Leopold, could not afford both Mars and Minerva and terminated the project in 1710. Bianchini had already drawn up plans for installing the exhibits, which featured collages like those in the *Istoria universale* but now of physical items, not pictures of them. Busts of emperors and saints would be conspicuous in the display; St Peter, as the tie between Rome and the new Church, a focal point. The objects would be placed on walls, not in cabinets; note in the lower reaches of Fig. 26, which exemplifies the design, the image copied from the base of Antoninus Pius's column. The pieces Bianchini assembled for the museum rested in the Vatican until 1716, when most of them were sold.[94] His idea lived on, however, and became an enduring reality after his death.

Projects undertaken to furnish Bianchini's aborted museum included the first systematic excavations on the Aventine hill. Among the finds were oddments from Egypt—a representation of the heavens on a marble square and one or more statues—that Bianchini described with a rigor approaching modern standards, replete with measurements instinctual to an astronomer.[95] Clement bought the statue(s), which he eventually gave to the Capitoline museum. Bianchini may have acquired the heavenly

Fig. 26. A portion of Bianchini's plan for a Vatican museum, exhibiting items illustrative of the second century AD. Giuseppe Bianchini (ed.), *Demonstratio historiae ecclesiasticae* (1752–4), iv (plate volume), tab. II, saeculi II.

representation, as he often retained items bearing directly on his work.[96] He recognized that its damaged surface displayed the zodiacal signs as Ptolemy laid them out and in the manner of the Egyptians, and also in a third scheme, which he could not identify. He sent a nice drawing of the planisphere to the Académie des sciences in Paris. Lacking Bianchini's ability to extract science from rubbish, the academy did not appreciate the token. It smacked too much of astrology. "It is not that the history of the follies of mankind does not make up a large part of what we know . . . but the Academy has better things to do." Some antiquarian, perhaps Bianchini's friend, the French ambassador to Rome, Cardinal Melchior de Polignac, thought differently, and purchased the object. It can now be found reburied in the Louvre under the epitaph "the planisphere of Bianchini."[97] In the year that collecting for the museum halted, 1710, Bianchini added to it five large Egyptian granite pieces. He valued them highly for the sophisticated knowledge he imagined, but could not decipher, in their hieroglyphs.[98]

To compensate his chamberlain for the termination of the museum project, Clement enabled him to exchange the canonry in the minor basilica of San Lorenzo in Damaso for one in the major basilica of Santa Maria Maggiore. Bianchini was delighted to alter his allegiance from Lorenzo to Mary and to apply the increase in income to the purchase of scholarly apparatus. But in truth his income never permitted extravagant expenditures, perhaps no more than 40,000 scudi over forty-five years.[99] Cancellation of the project had a second important consequence: it freed Bianchini for diplomatic intrigue under the laissez-passer of an academician.

4

REPUBLICAN OF LETTERS

As Prelate

Bianchini structured his multifarious activities around noon observations at Santa Maria degli Angeli. After attending to his personal and religious necessities, he would wait upon the pope if needed or study or write. He did not have to bother with the common concerns of life: his servants kept his quarters, collections, and clothes in order, and, when he had to cut a figure or go a distance, arranged his transportation. Otherwise, he went about the city on foot, serving patrons, examining artifacts, looking up books, inspecting fountains, and composing homilies and histories as he walked, probably reading, between his station in the Quirinal (see Fig. 27) and Ottoboni's in the Cancelleria, or between his solar observatory and the Vatican. After lunch he might visit his cardinal patrons and spend some hours in their gardens and collections. No doubt he looked into the churches in which he held canonries during his perambulations and, on feast days, participated in their rituals.

Late afternoons and evenings offered a choice of high-level entertainment. Private musical performances had proliferated after Clement prohibited carnival shows and "profane spectacles" (public theaters) ostensibly as a safety measure against earthquakes like that of 1703. Bianchini attended Ottoboni's musical soirées, including operas, at the Cancelleria and at Maria Casimira's home on Trinità dei Monti; and, no doubt, he enjoyed the *serenata* Ottoboni gave for Casimira, memorable for lyrics the cardinal wrote himself. Since both the queen and the cardinal

Fig. 27. The Quirinal, the papal palace in which Bianchini had rooms overlooking the garden. Giovanni Battista Falda, 1676.

employed Händel, Scarlatti, and the Corellis, Bianchini's acquaintance included some of the best composers in Europe.[1] Naturally, he refused invitations to entertainments he deemed unsuited to an ecclesiastic; he was an informed stickler for clerical behavior and did not shy away from specifying the proper deportment of popes.[2]

At Clement's order, Bianchini combed Roman collections and catalogues for representations of ancient musical instruments with which to equip twelve angelic musicians the pope wanted painted on the walls of the Archbasilica of St John Lateran. The pope directed further that the instruments should be taken from the Bible and depictions of the muses as far as possible. Probably the project related to Clement's princely embellishment of the Lateran with gigantic sculptures of the twelve apostles designed by Maratti. Although Bianchini provided instruments enough for an orchestra, the angelic musicians did not materialize on the Lateran's walls. Nor did he ever find time to publish his descriptions and color illustrations of the fifty-nine types into which he divided the instruments. They were printed posthumously with no indication of their original purpose and no finer division than the manner of playing them, scraping, blowing, or striking. A pity that Bianchini had not seen his way to writing their history from their invention for religious ceremonies in the fourth century after Creation to their wider deployment by Nero and other classical performers![3] Nonetheless, with the help of a lengthy account of his results by Montfaucon, to whom he generously sent drawings and descriptions, Bianchini's inventory became an authority for eighteenth-century musicologists.[4]

At feasts and musicales our monsignor ranked high for conversation but not for position. In the many academies he frequented, however, though cardinals sometimes attended, no one outranked him. After distinguishing himself in Ciampini's academies, he became, around 1710, director of one he could claim as his own, an academy devoted, as was his *Istoria universale*, to ancient and ecclesiastical history and chronology "proved by authentic documents."[5] Its patron, Clement's nephew Alessandro Albani, was then a non-academic teenage officer in the pontifical cavalry. The first task of the Accademia Alessandrina was to expose its patron to the learning that Clement deemed requisite for an ecclesiastical leader. Under Bianchini's tutelage, Alessandro became a major collector, beginning with items from his uncle's aborted museum of antiquities; and a discerning patron, advancing such scholars as J. J. Winckelman, an attentive reader of the *Istoria* and a pioneer in art history.[6]

In Bianchini's academic calculus, wealthy and well-placed men, such as Alessandro Albani would be when made a cardinal by Clement's successor, had an obligation to support the sciences. This was a higher duty for them than pursuing their own erudition. "Who does not see that all the sciences that animate our spirit are kept alive by the patronage of the great?"[7] The greatest of patrons provided a secure post with minimal obligations in which a bold savant could advance the reputation of his Maecenas along with his own. To take a pertinent example, of whom more in a moment, Lodovico Antonio Muratori argued for, and received, freedom to spend time with other scholars and to write books unimpeded when negotiating his appointment as librarian to Duke Rinaldo d'Este of Modena. Withholding this freedom could only injure the patron, Muratori advised Rinaldo, "to whose honor the reputation and fame of his servants redound."[8]

Clement understood that supporting learned and literary academies was an obligation for him as Rome's greatest patron and an advantage to him as its ruling prince. It offered a way to keep track of thoughts and thinkers he might find troublesome. A tempting target was the *Conferenza dei concili*, the historical academy founded by Ciampini in 1671, which met fortnightly in the palace of the Propaganda Fide. It gave serious study to questions of papal authority, conciliar decrees, theological disputes in the early church, the lives of popes, ecclesiological discipline, and other potentially dangerous matters; Clement knew its methods and purposes well, as he had been a founding member.[9] In 1708, ten years after Ciampini's death, the pope took over the *Conferenza*, appointed its supervisor, required its members to have studied theology for at least two years, and cut down its membership to twenty-four full academicians and forty-eight associates. It continued and perhaps enhanced its role as a nursery of high churchmen. Of its 186 known members, twenty-eight were or became cardinals; another twenty-eight, bishops; nine, other prelates; fourteen, qualificators (senior consultants) to the Holy Office; and two, popes.[10] Bianchini could be counted twice, as prelate and qualificator. He might also be counted as a mole, as he no doubt kept Clement informed about the business of the *Conferenza* and other learned societies he attended.[11]

A larger and more representative body, the literary Arcadia, also fell under the supervision of pastor Clement through the strong hand of its founder and *custode*, Giovanni Crescimbeni.[12] The community of Arcadians had grown rapidly by addition of high churchmen, many of whom also belonged to the *Conferenza*, and, following their example and allure, young nobles, lesser aristocrats and clerics, lawyers looking for briefs, poets looking for audiences, abbés looking for places, and, at least at first, men of true erudition and science.[13] During the thirty-eight years of Crescimbeni's stewardship, 2,619 people joined the *Arcadi*. Although Rome remained the center, "colonies" spread all over the peninsula; in 1710 there were twenty-one, two in monasteries. By 1728, an interested public could read about Arcadian accomplishments in nine volumes of

Rime, three of *Prose*, and three of *Vite* of defunct pastors, all chosen and edited by Crescimbeni. The exemplary *Vite* taught the ingredients and rewards of a correct blend of literature, erudition, accommodation, and religion.[14] The series on poems and prose exhibited examples of the style the Arcadia prized but could not always meet: the plain style of Galileo and his disciples, who wrote "according to the laws of true eloquence in Florence, where the first light of the vulgar tongue penetrated the darkness that engulfed the rest of Italy."[15]

Bianchini assured his fellow pastors that they would prosper in the benign climate of Clement, from whom they should fear neither storms nor thunderbolts.[16] After all, Great Shepherd Albani had been one of them. The Arcadians promoted Clement's cultural and religious policies, and the pope rewarded their service, as a great prince should, with offices and benefices.[17] When the organization started to admit a large number of undistinguished people, Bianchini warned Clement that it might decline toward frivolousness, laziness, and apostasy, and spread "bad habits and false opinions."[18] Whether owing to his alert or to its inertia, an increasingly passive Arcadia fell under the guidance of church officials, so as to secure the "hegemony over Italian intellectuals" always desired by the Holy See.[19]

As Academician

Bianchini did not disdain the frivolous if it was learned, and among his contributions to the Arcadia are some clever riddles that earned him the reputation of a humorist.[20] Here is one concerning the Jubilee of 1700, which the Arcadians reckoned as the inauguration of the 620th Olympiad. Would it be a happy one? The correct answer with war coming was "no;" cheerful pastor Ottoboni responded "yes," in fourteen lines of veiled verse. Did the cardinal's verse refer to the new Olympiad? Bianchini: of course, because $14 = 2+4+8+0$, which, read as 2,480, equals the number of years in 620 Olympiads. "Now who could have imagined this mystery?" Having solved it, Bianchini interpreted Ottoboni's verse with the additional help of the relics of Antoninus Pius and arabesques drawn from the chronologies of Scaliger and Petavius. Monsignor had a sweet tooth for complicated numerological games.[21]

Muratori was among the pastors most insistent on improving Italian poetry not only as an intrinsic good, but also as an answer to French litterateurs who denigrated it. He observed, however, that, with its ridiculous names, silly verses, and inclination to frivolousness, the Arcadia was an unlikely instrument of betterment. By late 1703 he had in hand three remedies: lengthy instructions for reforming poetry, which he published after some delay as *Della perfetta poesia italiana* (1706); a lengthier treatise on tact (*buon gusto*) in the arts and sciences (published in 1708 and 1711); and a plan for a *Repubblica letterata d'Italia* that would cultivate the natural sciences and Maurist historiography and make it difficult for the rest of Europe to ignore or appropriate solid science from the peninsula.[22] He wanted Bianchini as its leader.

The dismissive attitude that irritated Muratori may be illustrated by remarks of a Dutch professor, Jacobus Perizonius, who taught eloquence and history at the University of Leyden, about an Italian book he did not like. Far from the truth, he said, and arrogant besides. "If learned men in Italy can be nourished and entertained with such fables and figments . . . we [in the north] believe only what we judge to be proved, not mathematically, but historically and probably. Certainly, no one here, except perhaps those given by frivolous vanity to paradoxes and unheard-of opinions, will fall for what Blanchinus has to say about the cause, reason, and history of the Trojan War . . ." Yes, Perizonius was writing about Bianchini's *Istoria universale* and its reception on the peninsula. His attack on the learned of Italy because of their approval of "the senseless and worthless writings of Bianchini" did not go down well.[23]

Muratori had begun cultivating acquaintance with Bianchini just before the jubilee of 1700.[24] He was then twenty-six or twenty-seven and an energetic disciple of the most important Maurist in Italy, Benedetto Bacchini, then librarian of the Este in Modena and a zealous defender of modern historical methods. Further evidence of his humanity appears from his publishing the sparse literary remains of Elena Lucrezia Corner and his championing the methods of Galileo, the crampons, as it were, that had enabled philosophy to mount "the glorious heights to which it has now climbed." Thinking it important to visit these heights himself, Bacchini had shown his students vacuum experiments of the sort performed by Galileo's disciples at the Accademia del Cimento. Bacchini's prime student Muratori consequently came to worry over that prime puzzle, why the barometer stands higher in fair than in foul weather. Since its height depends upon the weight of the atmosphere, should it not rise when the atmosphere is full of water? Muratori's dissertation left the puzzle a puzzle.[25] But it made a bond with the Galilean author of the *Istoria universale*. Another bond was forged when Muratori became Este librarian in succession to Bacchini, whose negligence had cost him his job.[26]

Armed with Galilean and Maurist concepts and worried that France and Austria would make Italy a battleground in the war over Spain, Muratori turned to refashioning Italian literature into a unified expression of the Italian genius. What he aimed at may be discerned from his book on good taste rather than from his rambling reformation of poetry, which retains much of the Arcadian style he professed to dislike.[27] *Buon gusto* opened with a friendly review of Bianchini's *Istoria*: erudite but not pettifogging, useful to politics, religion, and morals without numbing antiquarian detail. It showed *ingegno* and *giudizio*, inventiveness and judgment, the ingredients of good taste; want of the one brings stagnation, of the other, submission to authority; of both, abandonment of the search for truth, "the principal objective of humane studies."[28]

For matters testable by our senses, judgment based on reasoned conviction should prevail; although *buon gusto* cannot yet tell us whether the earth moves or phlebotomy works, it specifies how to find out. For the contingent facts of history we must rely on others' testimony duly vetted by experience and common sense, and attempt to find a reasoned position between credulity and scepticism.[29] Doubting keeps us from accepting the false; it should not prevent us from embracing the true.[30] In some matters there is no room for doubt. Catholics must follow the church whose authority is infallible when pope and councils pronounce on faith and morals; "we must accede without wishing to inquire into the rationale."[31] To all this and to Muratori's final directive to truth seekers, more easily preached than practiced, Bianchini would have agreed: "Humilitas, humilitas, humilitas."[32]

Bianchini's agreement with Muratori over the direction of the Arcadia and the elements of good taste in the sciences did not extend to implementation. Muratori called for a strict organization of the Italian province of the Republic of Letters that would ensure quality at home and equality abroad. It was the second aim that made the difficulty. Bianchini did not regard the Republic of Letters as an arena requiring defensive leagues; its citizens should compete freely, individually, and openly for its honors, not in blocks that reproduced the alignments in the rapidly worsening Spanish war. Muratori was already experiencing the war's effects. French troops were oppressing Modena as French savants had suppressed Italian scholarship. "What perils war imposes on innocent letters! How difficult it is to keep a stable mind among so many fears and troubles!"[33] Muratori had a protectionist plan that required Bianchini's cooperation as man of letters and czar of culture.

Writing under the pseudonym of Lamindo Pritanio, Muratori sent the initial plans (*Primi disegni*) for a new academy to potential members and requested that they direct their responses to Bianchini. He had not bothered to tell Bianchini anything about the plans. When responses arrived, Bianchini protested against "the shallowness of this unknown founder of a fantasy of an academy," "this unheard of way of setting up an academy without academic founders," "this imposture," and replied to Pritanio (whose identity he did not know) that he would have nothing to do with the enterprise.[34] That prompted a revealing exchange of correspondence between Bianchini and Muratori that ended (in March 1705) with an admission by Pritanio of his, that is, Muratori's, want of *buon gusto* in trying to form a defensive league of Italian *eruditi*. That effectively killed the project.[35]

Muratori's plan admitted as ordinary citizens of the *Repubblica letterata* everyone, irrespective of condition or degree, who had published at least one solid book and would work for the "reform and increase of the arts and sciences in furtherance of the Catholic religion, the glory of Italy, [and] public and private welfare." To inspire emulation, the republic would offer prizes; to exclude the incompetent, it should have tasteful gatekeepers able to diagnose the defects of scientific work in all fields.[36] The main subjects would be ancient and modern languages and natural philosophy; the goal, a middle ground between obstinate scholasticism and immoderate modernism. "Acute minds can then discover a thousand ways to benefit physics and the truth." Italians have an advantage here, according to Pritanio–Muratori: inheritors of the science of Galileo and the critical methods of the *eruditi*, they are positioned to correct the errors of the new ways and the old, and to lead the advance. "We must place our best hope for national glory in the philosophy we call experimental."[37]

The requirement to develop philosophy based on experiment and history based on material objects excluded many professionals who weighed down other academies. Only physicians who worked for the progress of their science would be welcome, but not lawyers, who merely carped and pled, and pure mathematicians with their worthless speculations. Applied mathematics, however, astronomy, geography, mechanics, navigation, fortification, hydraulics, architecture, music, in all of these "we desire our republic to advance by steady observation, experiment, and invention." The *Repubblica letterata* might help solve the problem of finding the longitude at sea and certainly could map Italy more exactly.[38] But it should not stop there. Although *fisicomatematica* and its applications might offer Italians their best chance to regain the van, they must not shirk more difficult topics. These were moral philosophy, then unhappily out of fashion; theology, still blemished with medieval barbarisms; and erudition both sacred and profane, with its demanding study of monuments, chronology, geography, government, laws, and military matters.[39]

114

Like all proper republics, Pritanio's would have police powers. It could instruct booksellers and publishers not to print the twaddle of literary academies. It could appoint scholars to review all worthwhile books, domestic and foreign. It should build up libraries (behold the librarian behind Pritanio!). Above all, it had to oversee the restoration of good taste in the schools of all religious orders dedicated to study, whence should issue a supply of manpower for the advancement of science. To assure quality in the supply, the *Repubblica letterata*'s disciplinary powers should extend to dilettantish students and incompetent professors. And, despite its censorship, instructions, and corrections, it should demand civility, criticism, competition, and harmony; "open warfare in opinions, but not in hearts." In this impossible way, "the empire of science and arts will increase with the reputation of the learned and with advantage to everyone."[40]

All citizens of Pritanio's republic were not equal. At the top were *arconiti*, acknowledged experts, authors of books, whom the recipients of Pritanio's *primi disegni* would nominate. The arconites would co-opt the rest of the membership, *studiosi* (upcoming scholars) and *candidati* (beginners). Too elitist and complex, protested Bacchini, who knew Pritanio's identity. Bacchini suggested a reduction to two classes, the first including important patrons of science, and the second mechanics and instrument makers as well as scholars. The admission to the higher class of patrons whose deep pockets compensated for their shallow scholarship was essential to cover "the immensity of the expense with which anyone who wants to advance in scholarship in our age is inevitably faced." *Eruditi* needed reference works, dictionaries, editions, and archaeological remains; experimentalists needed instruments and specimens. Ascending from organization to purpose, Bacchini believed that the *Repubblica letterata*'s best chance of success lay not in experimental science, as Muratori had supposed, but in history and literature, better suited to routine Italian genius than Galilean science.[41] Above all, Bacchini presciently warned Muratori, do not privilege the increase of Italian prestige over the quest for truth among the republic's objectives.[42] As for the next step, Bacchini agreed that it should be Bianchini's. Let him choose thirty arconites from the names proposed by Pritanio's correspondents and let the thirty appoint the rest of the republic.[43]

Bacchini had identified precisely the points that would put off Bianchini. And of these the main one lay at the heart of Muratori's initiative. Bianchini's reasons for not supporting Pritanio's *Repubblica letterata* is a significant piece of self-evaluation. He wrote to Muratori:

> I am as concerned as much as any other Italian with the true good and reputation of our nation. But personal and national ambitions should not be mixed with scientific matters. Where [Pritanio] tries always to advance himself and other Italians in the opinion of other people and in competition with the northern nations, I believe that without being jealous or desirous of the applause of others I ought to try to meet my obligation as a human being and Christian philosopher, for whom there is neither barbarian nor Scythian, free man nor slave. My obligation as a man requires me to perfect my understanding with truth, and my will with the moral virtues; and my obligation as a Christian requires me to direct both my understanding and my will to the supernatural end for which we are created and redeemed. Therefore, I cannot agree with the projector that we should form a literary league of nation against nation, or, to speak plainly, engage in a battle of wits with the *oltremontani* . . . Other nations have not succumbed publicly to our weakness of amour propre. To be sure, some of them have national academies, but for purposes very different [from those proposed], that is, they are aimed either at perfecting the national language or at assisting the country in some other way, but never to compete for praise above all others. Consequently, their academies admit foreigners, indeed, some explicitly reserve places for foreigners. Praise is earned by the quality of the work, not because the worker demands or receives it. . .

It is worth observing that a large fraction of Bianchini's correspondence, and only three percent of Muratori's, was directed abroad.[44]

The long letter just quoted answered one from Muratori disingenuously hinting that Pritanio's initiative came from Rome ("and not without the knowledge of the Holy Father"), and congratulating Bianchini on his election to secretary ("perhaps inspired by the magnanimous zeal of the reigning pope").[45] And more: he, Muratori, has discussed Pritanio's project with Bacchini (true enough!), and both are enthusiastic about the *Repubblica*, "particularly because the generous character of the Holy Father will give wings to the undertaking, as we have reason to believe considering his rare virtue and knowing that he has benevolently approved the proposals of Pritanio." The claim of support from high places, where Bianchini was at home and Muratori nowhere, was abnormally duplicitous. As for details of the plan, the grades and titles of the republicans, Bianchini dismissed them as an infantile way to stimulate competition. As for the attempt to involve him, it was just plain dishonest.[46]

For some months more Muratori maintained the charade before exposing Pritanio.[47] Then, in March 1705, he allowed the merit of Bianchini's criticism of the class names and the pursuit of national glory; but the choice of Bianchini as secretary, that was not open to criticism. Who else possessed similar distinction, modesty, humanity, and "many other virtues, intellectual and moral," and such strategic advantages as his friendship with the pope and access to the Vatican? The republic without this greatest of grandees was impossible.[48] Muratori apologized for what he now called his joke, his playful way to reform Italian letters. To be sure, he had not been truthful, but then neither had he lied; he had invented a harmless fiction to say publicly what all friends of science thought privately.[49]

Bianchini wrote to all those he guessed would have received the *primi disegni* to disengage himself. One was his calendrical collaborator Bonjour, who had written to him as Pritanio had instructed; Bonjour thought that the plan would be "glorious for Italy and very useful to the entire world," if it advanced religion and had the support of the pope. To which the reply: Pritanio's approach was impudent, uncivil, and puerile; "why pretend that the [proposed members] could unite in a body to which none of them had previously given a thought?"[50] And, more succinctly, to Manfredi, just after receiving the plan composed by "a writer with a fake name": "I am astonished that the inventor, claiming to want to promote the sciences, begins with a lie."[51] Bianchini was not sophisticated enough to accept fraud and deceit in the service of knowledge and truth.[52]

Bianchini soon received more than his virtue as his reward. He had been a corresponding member of the Paris Academy of Sciences since 1699 through the influence of Cassini. As a correspondent, he served the

Paris academy much as he and Montanari had the Royal Society, reporting astronomical observations sometimes printed in their journals.[53] His
contributions to chronology and calendrics, his terrifying erudition, and
his ability to transcend the inhibitions of the Italian intellectual milieu
earned him (and Bacchini and a few others) the praise of the *Journal des
sçavans*.[54] In 1706 he received from the Parisian academicians the highest
honor they could give a savant not resident in France: they nominated
him a foreign associate. Since no more than eight of these worthies could
exist at any time, the appointment was a very high distinction in the
Republic of Letters.[55] It required royal assent, and even choice, since the
king picked one among several candidates proposed. We might suppose
that Louis XIV, trying to preserve relations with Clement XI while occu-
pying Italy, reckoned that the pope would be pleased by the honor done
his chamberlain.

The appreciative chamberlain wrote a very polite letter to the Académie
in acknowledgement of the extraordinary recognition he had received.
And, ever conscious of protocol, he advised the Francophile Ottoboni to
tell a correspondent close to King Louis of "the great satisfaction he felt
on seeing [Bianchini], his librarian . . . being raised to the distinction of a
royal notice." Clement too had good reason to share in the satisfaction. He
had given Bianchini the opportunity to enter the high aristocracy of the
Republic of Letters and so to come to the attention of "His most Christian
Majesty, reserved by God for the advantage of our age and an example for
others."[56]

As Censor

Muratori could not easily let go of Pritanio's project. Its aims were merito-
rious, its supporters numerous, and Bianchini its obvious leader.[57] Why
then had he rejected it? Dislike of fraud and parochialism did not strike
Muratori as a sufficient reason. What then? The censorship! In November 1703, just as Pritanio was drafting his *primi disegni*, Bacchini asked
for permission to publish his latest exercise in Maurist historiography,

a scholarly edition of the eighth-century *Liber pontificalis* by one Andrea Agnello. Bacchini had found a manuscript of the work, which reports the lives of the early bishops of Ravenna, in the ducal library in Modena, and, as it presented an account, admittedly not very reliable, of the relations between church and state in Italy from the Roman Empire into the Middle Ages, he resolved to publish it, not only as an example of editorial method, but also as a contribution to the ongoing controversy over claims of princes to the temporal government of the church in their territories.[58] He would let the chips fall as they might. As he had written in his earlier masterpiece, *Istoria del Monastero di S. Benedetto di Polirone* (1690), "knowing that I have followed truths devoutly in treating my subject ... I believe that these pages deserve the name of history." But did his *Agnello*, with its assertions, however carefully edited, that Ravenna had been given the right by Emperor Valentine III to elect its own bishops without reference to Rome, qualify as history?

The censors asked Bianchini for his opinion. The super polymath advised that the manuscript should not be published until Bacchini had corrected, or properly glossed, significant errors in Agnello's work—especially the assertion of Ravenna's ecclesiastical independence from Rome. His fellow qualificator, the Vatican *custode* Zaccagni, agreed. So did the censorship. That created a double-blind confrontation. In his letter of December 1704, which hinted that the pope backed the proposals of the mysterious Pritanio ("it would take all the astrologers to identify [him]"), Muratori asked Bianchini for help to clear *Agnello*. He made it evident that he considered clearance essential to the defence of Italy's literary reputation. If the censorship did not approve Bacchini's *Agnello*, "I must say that I could not be very sanguine about the proposed republic."[59] Thus the rationale of the ruse of Pritanio stood revealed.[60] Wanting more protection for their work than the duke of Modena could give, Muratori and Bacchini designed their literary republic with Italy's most distinguished and Rome's most influential savant as its head.[61]

Bianchini's identification of the many weaknesses in *Agnello* resulted in its prohibition on 5 March 1705.[62] And so Muratori conceded the game. "Since Bacchini's *Agnello* has been prohibited ... I don't see how the project to make any literary republic can go on." Still, it was a good idea. Muratori reprinted his plans for it in the first instalment of his *Buon gusto* in the hope that an arconite bolder than Bianchini might come forward.[63] Bacchini too gave up and went to Rome to learn what was needed to clear his *Agnello*. He had the support of a circle of dissatisfied Arcadians who called themselves drummers, not just to make noise but also with an ear to the military use of the instrument to warn off opponents, an employment Bianchini spelled out in his book on ancient instruments. The chief drummers were two mismatched Albani protégés. The more flamboyant, Domenico Passionei, who would become a cardinal but was not yet a priest, went to Paris at the conclusion of the Bacchini affair to deliver a cardinal's hat to the nuncio, Filippo Antonio Gualterio, with whom Bianchini would work closely when abroad. Passionei would also be helpful, as a travel guide, since after completing his duty in France he went to dwell among the heretics of England and the Netherlands.[64]

Passionei's associate drummer and sometime teacher, Giusto Fontanini, was a priest educated by and disdainful of the Jesuits. He moved in a familiar orbit: he served as librarian to an influential cardinal, Giuseppe Renato Imperiali, who owed his hat to the Ottoboni and helped swing the conclave of 1700 to Albani. At the time of the Bacchini affair Fontanini was also professor of eloquence at the Sapienza. He had helped Muratori in his directions for improving Italian poetry; "reforming" it smelt too much of Luther for a paragon reactionary like Fontanini. He could not accept teachings in any sciences not set forth by a Catholic, or Catholic positions endorsed by a Protestant; he rejected Bianchini's calendrical reforms because Leibniz supported them. Nonetheless, Fontanini followed the historical methods of Mabillon and earned valuable perquisites from Clement for services rendered in a long fight with Muratori using archival munitions.[65] The *casus belli* will be disclosed after we free Bacchini from the censors.

Passionei had Bacchini's manuscript copied and circulated; everyone who might have an interest in the stale gossip of Agnello of Ravenna now had access to it. Moreover, Passionei advised Clement, prohibiting a book already available because it contained errors inimical to the material interests of the Holy See would give ammunition to Rome's enemies abroad.[66] Bacchini and the censors soon reached a compromise. He would add notes to undercut Agnello's authority.[67] That done, he got his imprimatur, in March 1706, a year after the book's prohibition.[68] Editor Bacchini now warns against his author, a barbarous writer, "incredibly ignorant of matters both sacred and profane." Agnello mixes up dates, things, and persons, scatters inanities among improbabilities, and gives ludicrous interpretations of scripture. "Of all these things I warn you, dear reader, so that you will know what faith such a writer merits when he rails at the most holy pontiffs of the Roman See." Bacchini then proves that the bishops of Ravenna never had the independence they claimed.[69] Modern editors, with no censors to please, take a similar view. Agnello is a poor writer, often unintelligible, confusing different people with the same name and carrying on, "with God's help," whenever he runs out of information.[70] "Truly Agnello bubbles over with so many anachronisms that a prudent man would be desperate if he had to correct it for use."[71] Bacchini's Agnello was a poor case for academic freedom.

Bianchini read many books for the censorship, always responsibly. "I read [Fontanini's *Life of Suor Camilla Orsini Borghese*] attentively from page 1 to page 1013, where it ends." A wry comment on its unreadability; a modern account of the venerable sister does the job in a tenth the space.[72] Most of Bianchini's qualificator's reports resemble good peer reviews in correcting errors and offering suggestions, and bad ones in condemning everything opposed to the theory of his school—that is, Catholic doctrine on faith and morals: a perfect mix of *verità* and *carità*, of sympathetic criticism and defense of the truth as he saw it.[73] Therefore he found nothing amiss in a work by Mabillon but several serious errors, which Mabillon corrected, with many thanks to his censor: "if the book is less unpolished than before, it is owing entirely to you." And he, Bianchini, found nothing redeemable in Pierre Bayle's erudite sceptical *Dictionnaire historique et critique*: "it must be prohibited altogether."[74] On borderline books like Francesco Orsini's account of basilicas dedicated to St Mary, which touched on the unintelligible doctrine of indulgences, Bianchini tried not to pronounce.[75]

When a good Galilean treatise came across his desk Bianchini was all enthusiasm. His report on one of them shows that he could urge his views about Galileo while advising the censorship system. He wrote of a manuscript that its author had adopted the correct approach to natural philosophy, Galileo's, by trying to "reduce hypotheses about real experiences to geometrical demonstrations of their causes." There may be no limit to the height to which an investigator can rise from such foundations. "We may hope [Bianchini's report concludes] that one day we shall have from the elevated and fertile mind of this most distinguished author the new system he introduced in this book."[76]

Bianchini disliked Bacchini's book because it lacked the critical editing required to protect the inexpert reader from an unreliable source. To be sure, the book was not likely to please the Vatican in any form. Agnello made claims inimical to papal authority; Bacchini's repudiation of the claims, though consonant with Vatican policy, raised touchy questions better left alone.[77] Whatever advice Bianchini gave the censors would have a political coloring. That does not mean that he did not base it on objective scholarship. The question, rather, was whether caution and restraint, or full academic freedom, should be privileged in airing sensitive historical material. Although a champion of critical historical scholarship, Bianchini thought that the publication of its results might not always be useful, and so took upon himself the judgment, which Bacon assigned to a college of Solomons, whether a discovery could be released safely to society. Bolder scholars, following Galileo and Mabillon, plumped for full disclosure of what they took to be true irrespective of reasons of state.

Muratori skirted the problem. In the second part of *Buon gusto* he wrote that he had kept back "many matters in my pen that might not be useless" because he would need to apply for a license to publish them, which would put his useful matters at the mercy of ignorant and zealous censors. But what if there were no impediments? Should every imbecile who wants to appear in print be allowed to? Certainly not: we must protect the public from madness, impiety, and the stupidities of slavish philosophers who give our enemies so many sticks to beat us. The remedy is *santa moderazione*, charitable forbearance, practiced by learned, prudent, openminded censors. Balance, moderation, and good judgment are the best ways to prevent the bombastic few from prevailing over "truth and the legitimate freedom of intelligent minds." What is moderate to one person, however, may be reckless to another. Which returns us to asking how to keep the press from poisoning the public while allowing savants to publish more than the paternoster.[78] Muratori and his censors managed to find a solution. None of his books was prohibited, although, as Pope Benedict XIV observed, they contained many things inimical to Roman practice and policy. "How many times have we ourselves found here or there in Muratori and other respectable writers opinions that are certainly condemnable; but, in the interest of peace and scholarship, we have done nothing about it!"[79]

It is hard to stay pure. Muratori and Bacchini were soon bending their critical historical investigations, the first to favor, the second to oppose, their duke in local issues in church and state. In 1708, the year that Bacchini's *Agnello* saw daylight, Muratori became embroiled in the dark business of proving that Comacchio, a medieval town on the Adriatic north of Ravenna, belonged to what was left of the Duchy of Modena and Ferrara after Ferrara had been ceded to the Holy See in 1598. In practice in 1708 Comacchio belonged to the emperor, whose troops had occupied it in response to a weak attempt by Clement to arrest their progress in Italy. "The Guelf dream of the *Primi disegni* gave way to the Ghibelline defence of the Este rights in Comacchio"—that is, imperial considerations now trumped Italian ones, in the hope that the peace negotiations to end the Spanish war would restore the territory to the Este.[80]

Against Muratori Clement pitted *custos* Zaccagni, who, however, proved too good a historian to provide a knockdown punch. A more effective paladin, the reactionary Fontanini, then entered the arena against Muratori; their fight, with no documents barred, lasted seventeen years while the Empire retained the real estate. Clement made Fontanini a *cameriere d'onore* for his performance. Although Muratori argued his case in strict faithfulness to the facts, according to the historiographer Nicolas Lenglet du Fresnoy, "a judgment all the more valuable as coming from a French savant," he lost it in 1723 when the Empire returned Comacchio to the Holy See.[81] Meanwhile the circus had acquired a third ring. Using his favorite weapons, documents found in the Modena library, Bacchini insisted that his monastery had certain rights that his old employer Duke Raimondo preferred to retain. Raimondo had him transferred to a remote convent where he could reflect harmlessly on the nature of historical truth.[82]

The *Repubblica letterata italiana* faced obstacles harder to budge than the political divergences and allegiances of its projectors. Many of its utilities were already available to erudite ecclesiastics through the Roman Catholic Church in the libraries and academies of its cardinals and prelates, its patronage systems, its teaching orders, even its censorship. Muratori's dreams of combining papal resources with those of lay princes to better the material circumstances and establish the scientific authority of the proposed republic, and of replacing the Church's censorship of books with the republic's, did not appeal to the many who benefitted from the existing system. Nor was the propelling power of Muratori's plan, the insistence on Italy's return to the van of European culture, realizable in 1700. Leadership had passed beyond the Alps. To achieve international distinction, Italians had to adhere to the norms and expectations of the larger Republic of Letters.[83] And they had to do so while avoiding the obstacles their church and culture put in their way. Bianchini was a master at overcoming these handicaps. With mild self-censorship, he could do astronomy as a Copernican, discourse on natural philosophy as an atomist, write history as a Maurist, admire the scholarship of heretics and Frenchmen, and occupy a position of honor in the pope's household.

5

IN PARTIBUS

Europe at War

"The arts of peace have found an asylum in this temple of religion in these difficult times, which disturb all of Europe." With these words Bianchini began his account of the Clementine *meridiana*. He ended it with a prayer: may the pope be able to protect "all sacred things and the arts of peace."[1] Prayer did not suffice. Pope Clement had not been able to persuade Savoy to block the French, or Venice the Austrians, from crossing their territories to fight in and over the Duchy of Milan. The only major success of papal diplomacy in the first year of the war over Spain was securing the neutrality of the king of Poland. This achievement involved some people we know and need to know.

Sobieski's successor as king of Poland, Augustus II the Strong, was also Friedrich Augustus I, elector of Saxony. Among those who had favored him over the French candidate was the papal nuncio to Warsaw, Bianchini's friend Davia. At the outbreak of the Spanish war Davia was reassigned to Vienna. There he continued to prefer the Habsburg to the Bourbon empire while leaving to his successor in Warsaw, Cardinal Fabrizio Paolucci, the impossible task of reconciling the French faction in Poland to the apostate King Augustus, who had had to convert to Catholicism to secure his throne. The French faction held strong under the leadership of the country's cardinal primate and his mistress.[2]

Rome's greatest expert in Polish affairs, Maria Casimira, had backed her son-in-law, the elector of Bavaria, rather than her son, as nature and custom expected, as successor to her husband. She despised strong Augustus but could not support the French faction against him; for she held an irrational grudge against her native land for neglecting to make her undeserving father a peer of the realm.[3] In sum, Davia had good reasons to push Poland toward Austria, Paolucci to placate Polish grandees sympathetic to France, and Casimira to hover between perfectly balanced animosities. Clement did well to keep Augustus trapped in this political–religious maze. At the same time he entered one himself.

On 16 September 1701, James II, the exiled king of England, Scotland, and Ireland, died in France, where he had enjoyed the hospitality of Louis XIV for a dozen years. Louis quickly recognized James's teenage son, another James, as the rightful king of Britain. The Roman establishment also propped up the House of Stuart. Clement created a fund for supporting young James, thus identifying more closely with French policy than was compatible with his pose of neutrality.[4] The curia sang James's praises, the architect of Rome made a little temple to him, a coffee-table book advertised the temple, and so on. Under pressure of such attention and the teachings of his pious mother, young James declared to Clement his unswerving loyalty to the Roman Catholic faith, thereby enrolling the Pope of Rome in his cause and committing himself to a religious position that guaranteed he would lose it.[5]

Meanwhile the French and the Austrians were fighting in Lombardy and the Spanish were trying to secure Naples against internal factions favorable to the Habsburgs. The pope and the papal states languished between the devil and the deep-blue sea. The French and Spanish interests insisted that Clement recognize Louis's grandson as Philip V of Spain. The Austrians clamored for recognition of the younger son of Emperor Leopold as King Charles III of Spain. Philip further perplexed the pope by showing up in Naples on Easter Sunday, 1702, on an inspection tour of contested Italian holdings. Rather than invite him and trouble to Rome, Clement dispatched a high-level legation to Naples led by a senior cardinal, Carlo Barberini, who had perfected his diplomatic skills by acting as "Protector" (paid agent) of Sobieski's Poland in the curia. To give the expedition even greater gravitas, the pope added to it, as its "historiographer," Monsignor Bianchini.[6]

Bianchini's report of the expedition is a masterpiece of diplomatic dissimulation. Although it refers to Philip as king of Spain, it ignores the political implications of the title. It dwells rather on ceremony: the cavalcades, caparisons, vestments, uniforms, and banquets; the carriages and artillery salutes, trumpet blasts and drumbeats; the squads of nobles and hordes of people; the condescension and piety of the prince; the generosity and solemnity of the legate. Gifts there were aplenty, notably a sculpture by Bernini of Hercules strangling a snake, which belonged to the Barberini and is now in the Getty Institute, and indulgences too many to count, drawn on the treasury of the saints. The papal legate had entered the city in great pomp, preceded by thirty-six mules; wherefore the legate's historiographer, who followed on foot, no doubt kept his eyes discreetly on the ground. The later meeting of the king and the legate in the open air, with their trains of followers parading as in a *passeggiata*, ranked among the greatest spectacles of the age.[7] For such prose and other service to the legation, Clement made Bianchini a senator of Rome and nobilized his family.[8]

After King Philip had left to visit his war-torn domain in Milan, Barberini and his historiographer did some sightseeing. Bianchini sketched Neapolitan collections and reported without a wink the famous miracles of the occasional liquefaction of the blood of San Gennaro and the more dependable performance of John the Baptist's. He visited Anzio, while Barberini returned to Rome against contrary winds to a hero's welcome at the Quirinal palace. There he told Pope Clement that he had been as evasive as the situation allowed. Clement replied that he hoped that all the belligerent princes would be as sensible and "transfer the arms with which they are now fighting among themselves to the lands of the Infidels, and do battle for God."[9]

Despite Philip's visit to Milan, the Austrians soon beat the French in Lombardy, prompting Britain to ally itself with the emperor and the duke of Savoy to change sides. The Papal States thus faced an unstoppable incursion of imperial forces marching from Milan to Naples. Nonetheless Clement, "hiding under a profound and impenetrable Dissimulation, his own real Sentiments,"[10] would not recognize Charles III as king of Spain. That displeased the Holy Roman Emperor, now no longer Charles's father Leopold, but his elder brother Joseph. Although the late Leopold had been Rome's enemy number one, protocol required shedding a pontifical tear over his passing. Clement assigned the delicate task to Bianchini. The monsignor asked the imperial ambassador what imperial virtues he should mention. The reply was not helpful: "patience with the court of Rome."[11] Four months later, on 18 October 1705, the diarist who supplied this tidbit recorded that "Monsignor Bianchini, prelato domestico di Sua Santità [a new honorific title]," was on his way to Vienna, probably not out of pure curiosity, "given the many differences between that court and ours." On his return on 22 December, the Vatican explained implausibly that Bianchini had undertaken the journey "to set in motion the Clementine emendation of the calendar."[12]

Although his mission did not change minds in Vienna, it seems to have brightened ones in Rome, for a year later Bianchini was again instructing the dull Austrians in papal "calendrics," this time in Milan.[13] They still declined to respect the neutrality of the Holy See, and in 1707 an imperial contingent heading for Naples camped unwanted in Rome's Piazza di Spagna. Still the pope would not recognize Charles III. In 1708 Emperor Joseph increased the pressure by seizing troublesome Comacchio; Clement replied by fielding a show army of ragtag recruits he soon disbanded.[14] A peace forced on Clement in 1708–9 made him disarm, accept the loss of Comacchio, and recognize Charles. Although France and Spain complained, Clement had the luxury of ignoring them with impunity. For by then Louis XIV had recognized that he could not win the war and had begun to discuss terms of peace with Britain and the Empire.[15]

The armies of the Empire, Britain, and Holland controlled much of the Spanish Netherlands and were threatening to march on Paris; the emperor occupied Spanish possessions in Italy; and Philip V was just holding his own in Spain. Discussion of peace terms, unlikely to be favorable to France, were well underway when, on 17 April 1711, the tough young Habsburg emperor Joseph died suddenly of smallpox. That transformed the situation. Charles III would now succeed his brother. The prospect of the re-creation of the Habsburg holdings of Charles V, which had stretched from Transylvania to Ecuador and from the Netherlands to Sicily, frightened belligerents on both sides. On 8 October 1711, the precarious king of Spain, Charles III, was inflated to Charles VI, the unchallenged emperor of the Germanies. Four days earlier, France and Britain had signed preliminary articles of peace. Philip V would have Spain and its American possessions, Charles VI the Spanish Netherlands and territories in Italy. Both winners would renounce claims to more: Charles would give up Spain and Philip the possibility of the throne of France. These would be the main territorial settlements concluded at Utrecht and Rastatt in 1713 and 1714, apart from the award of Sicily and promotion to king of the dexterous duke of Savoy.

Clement did not have much leverage in this diplomatic game—no castles, as it were, but lots of bishops and a would-be king. For a time, the bishops, or, to speak plainly, papal emissaries including Bianchini, judged that the settlement might make the would-be king James one in fact. But, although James was the natural successor to his half-sister Queen Anne, who had no living children, the big powers treated him not as a credible king but as a strategic pawn. An "Act of Settlement," passed in 1701, had preempted his succession by barring Catholics from ruling England. That made the next scheduled sovereign of England a granddaughter of James I, the dowager Electress Sophie of Hanover. There was little enthusiasm, however, except among English Whigs, for a Hanoverian succession certain to bring in foreign forms of government and etiquette. Some Tories favored continuation of the Stuarts regardless of religion, and many more, and non-Tories too, would have welcomed a Protestant James.

The maneuvering of 1712, when Bianchini acted as an irregular agent in France, was complicated by the usual divergent interests of former allies. While plenipotentiaries for the main belligerents spent the year in Utrecht, France and England were perfecting their bilateral settlement. The Tory government was quite willing to throw over its allies the Dutch and the Austrians, who had not fully resigned themselves to an end of hostilities, and to discipline belligerent Whigs, who preferred to fight on until there remained no foreseeable circumstance in which a Bourbon would rule both France and Spain. The elector of Hanover, who looked forward to joining his electorate to the kingdoms of England, Scotland, and Ireland, and was beholden to the emperor and the Whigs, shared their intransigence. The English peace mongers held firm. Led by the earl of Oxford (the Lord Treasurer), Viscount Bolingbroke (Foreign Secretary), and Queen Anne, they showed their steel by cashiering the formidable ever-victorious duke of Marlborough, who wanted to continue campaigning. In his place they put the pliable duke of Ormonde. On Oxford's orders, Ormonde informed the French minister, the Marquis de Torcy, that he was withdrawing from combat. With Ormonde's disengagement in May 1712, France and England reached an armistice and French armies turned to retaking territory captured by imperial forces.[16]

Disease continued to have a seat at the table. Measles took away the dauphin (Louis XIV's heir apparent) and the dauphin's elder son in February and March 1712. That left only the younger son, a two-year-old infant, between Philip V and the throne of France whenever nature forced Louis XIV to vacate it. The union the English feared seemed not only possible but likely. Further negotiations with Torcy forced Philip to choose: keep Spain and renounce all claims to France or bet on France and enjoy a few principalities while awaiting the death of the little dauphin. In June Philip chose Spain, wisely, as the infant declined to die and reigned as Louis XV for fifty-nine years. Having disposed of the French succession, the negotiators turned to the English. The French agreed to acknowledge the Hanoverians and expel the Pretender from France. Torcy began the eviction by prying James out of Paris in September 1712.[17] In exchange, Oxford and Bolingbroke intimated that, after signing the peace, they would do something for James.[18] Completing the loop, Torcy kept James informed, and James so advised his supporters. Consequently, Jacobites in England had no reason to make trouble for Oxford's government. Clement protested loudly and uselessly against this cynical use of his king-pawn.[19]

Monsignor in France

One of Louis XIV's favorite projects was the domestication of Strasbourg, which he had acquired in 1681 and managed to hold during subsequent wars. To help secure it he had promoted a well-spoken sprig of the well-connected house of Rohan–Soubise, Armand-Gaston-Maximilien, as the underage coadjutor of the city's elderly bishop. Clement permitted the arrangement in one of his first acts as pope. In 1704, the young coadjutor, then just thirty and living mainly in Paris, became Strasbourg's bishop at a distance. The obvious next step to strengthen his position in Strasbourg was to make him an absentee cardinal. On 29 May 1712 a courier arrived from Rome with the good news, which Louis relayed: "The pope has made us wait rather long, but finally it is over, and a cardinal's robe will suit you well."[20] With the robe went a hat and a ceremony. Clement deputed Bianchini to do the honors. His orders included minutiae about the ceremony and the dress he should wear on official occasions and a reminder that, since he would be an Apostolic Nuncio, he should consult knowledgeable cardinals and ministers about French affairs "and follow their directions in everything."[21]

Gualterio's six years as nuncio in Paris and service as Cardinal Protector of Scotland made him the most useful of these informants for both French and Jacobite affairs. Bianchini had known him for years. They had much in common: age, bibliophilia, history in the style of *Istoria universale*, and enjoyment of Roman academic life. Gualterio introduced Bianchini to Rohan by letter as "a Prelate whom his Holiness regards very highly for his singular virtue and knowledge"—an evaluation in remarkable agreement with an introduction written by another cardinal administrator the same day to Torcy ("a prelate held in extraordinary regard by everyone for his many and rare virtues, especially mathematics").[22] The most important of Bianchini's cardinal directors was Paolucci, who had returned from his assignment in Poland to become the Vatican secretary of state. They had known one another as members of Clement's inner circle and Roman learned societies.[23] Paolucci would be Bianchini's primary link to the Vatican during his time in France and England.

The conspicuous honor of hat-bearer obliged Bianchini again to borrow money for fancy dress and incidental expenses, and for the books he intended to acquire from the shops in Paris.[24] (The total expense ran to 3,000 scudi, of which the Vatican defrayed only 360, the standard fee for hat delivery.) Bianchini's baggage included medals and prints and a crate of his own books, intended as gifts or trades, and optical instruments intended for use.[25] It also included at least two ciphers, for he was not expected to limit his reports to sights and ceremonies. He would exploit the scholar's cover, as recommended by Passionei: "I make every effort to have it appear that the only reason for my being here is curiosity about the business of the Republic of Letters." As a foreign associate of the Paris Académie des sciences, Bianchini had immediate access to the learned circles in Paris that complemented the ecclesiastical and ministerial elite with whom his mission and inclination obliged him to associate. Making astronomical observations with the ailing Cassini, who died during his visit, added to his camouflage.[26] The disguise still works. According to a modern account of espionage during the period, Bianchini's trip was "simply honorific," a reverse of the usual dissimulation, "a scientific trip under a diplomatic pretext."[27]

Bianchini left Rome for Marseilles in July 1712 by papal galley, unpacked his instruments, and taught his fellow travelers how to find their latitude. During a stop at Livorno, he paced out the main piazza (250 by 300 Roman feet); in Aix, which he reached by litter from Marseille, he did the main street (600 by 60 paces) before setting off for Lyon. Everywhere he was taken to see and draw ancient monuments. In Lyon he witnessed pneumatic and hydraulic experiments at the Jesuit College that taught him more about fluids in half a day than he had learned in Italy in fifty years.[28] Bianchini's surprise at these revelations anticipated the shock experienced half a century later by another authoritative *fisicomatematico*, Roger Boscovich, when he saw how far Rome had fallen behind Paris in excellence and variety of experimental apparatus.[29] On 19 July Bianchini was greeted at the Seine outside Paris by two coaches, one bearing old Cassini, blind and moribund, the other young Rohan, robust and rising. The new cardinal installed him in the Hotel Soubise, a paradise with quantities of servants and 30,000 books. Bianchini spent many hours in Rohan's library, where he ran across Montfaucon and other scholarly friends.[30]

Ceremonies connected with Rohan fill the earliest of Bianchini's reports to Paolucci.

> Last Saturday after lunch I attended theses at the Sorbonne dedicated to Cardinal Rohan, and witnessed the dexterity of its professors and bachelors, and the manner of the exercise ... It was not a mere pretense as is usual with theses in our universities, where in just a quarter of an hour some difficulty is proposed for solution under some restriction, and the crowd prevents any serious discussion. Here the defendant has books with him and reads the conciliar claims and patristic and other texts and is able to produce in writing the authorities on which he bases his argument; and he is allowed to develop it civilly and decorously without interruption.

The candidate had to demonstrate his competence in theology, dogmatics, ecclesiastical history, Latin, and Greek, and the application of them all to heresies ancient and modern. The examination lasted four hours.[31] The clergy were as diligent as the professors. Bianchini delighted to see that, unlike Italian prelates like Ottoboni, they could sit through services and sermons without a word, wink, or smile.[32]

Visits to the Académie des sciences and individual savants were a pleasurable necessity. Bianchini presented the Académie with a version of his portable telescope, which he had brought all the way from Rome; his colleagues gave it to René-Antoine Ferchault de Réaumur (Réaumur of the thermometer) for an evaluation, which he published, with an illustration of the instrument, in their *Histoire* for 1713.[33] Bianchini spent as much time as he could at the Paris Observatory with his friend Maraldi and their mentor Cassini. In his *éloge* at Cassini's funeral, Bianchini characteristically praised the departed's piety as much as his astronomy.[34]

Life is not all play, however, and our prelate's instructions demanded arduous socializing. Gualterio's contacts opened all doors, even to the royal palace, and overfilled Bianchini's days with visits paid "punctually and frequently." He was a hit at Versailles, where he made an address, and gave a painting, to the king; the painting, by Giulio Romano, was loosely interpretable as an allusion to Louis as another Constantine. For this performance Bianchini received the right to assist the infant dauphin at bedtime prayers and to see everything unusual in the royal palaces relating to waterworks. He spent the last week of July looking at treasures accumulated by the kings of France. The only reported hitch in his first pleasure-filled days in Paris was a squabble over protocol. Bianchini insisted on his right to sit next to Rohan when the two rode in a carriage together, to honor not himself but the representative of the pope.[35]

Occasionally Bianchini ran into the Pretender, James III, whom he judged able to perform successfully "when it shall please God to put him on his throne."[36] Overrating James's merits as a potential ruler, Bianchini advised that the Holy See give him whatever support it could. The English demand that he leave Paris could not be evaded; but let him leave without appearing to be driven away. Bianchini suggested giving him a conspicuous part in an important ceremony, like godfather at the baptism of the infant dauphin; a good idea, but too late, as the new little Louis had already entered the Christian life privately.[37]

The pressing question was not the manner of James's going, however, but where he should go. Bolingbroke and Torcy preferred Lorraine, then a duchy within the empire; but that required permission that the emperor hesitated to give. Still a pawn in the big game, James went to Châlons in September to await permission, which he had not received by the year's end. The English pressed; the French declared that they could do no more; rumors spread that James had rallied his supporters and Queen

Anne favored his cause. Plots, conspiracies, and spies multiplied around him. It was said that in Lorraine, where he took up residence at Bar-le-Duc in February 1713, everything he did would be reported to Vienna and thence to "all the Whigs in England."[38]

Meanwhile Bianchini had drawn close to the Pretender. He dined with James and his mother, ex-Queen Maria, whose hopes for her son he encouraged.[39] He sent four portraits of James to Rome, one each for the pope, Paolucci, Gualterio, and Alessandro Albani, in which, so the sender said, James's virtues should be obvious to all viewers. The value was in the viewing and in James's knowing it. "Please ask His Holiness to place this image where it can be seen. The thought of it will be a very great encouragement and consolation to [the king] now and perhaps for the next few years while we wait to see his constancy crowned on his own throne."[40] As we know, Bianchini ascribed great power to images. There were more substantial straws to clutch at too. Bianchini reported that James was now known as "the Papist Pretender"—an upgrade, according to the optimistic reporter, as it conceded that a Catholic could be a "legitimate pretender" to the British thrones. Bianchini advised Rome to receive James's recommendations of ecclesiastical appointments as those of a proper Catholic king, "lest people think that His Holiness does not bear the Pretender the affection we all know he does."[41]

Bianchini urged the same message on young Albani, "to give him a way of obtaining merit with God and His Vicar on earth." The way: suggest to the pope how to express "his paternal affection toward the King of England who so much deserves it." The how: assign James to a monastery run by the monks of Saint-Maur, "where piety and learning are exemplary, as His Holiness knows." Nothing better for young James than the society of erudite celibate monks! "I cannot imagine a more secure asylum for a secluded king, and a more salubrious school." The monks would see to it that His Majesty would continue to live virtuously free from the temptations to which a healthy young man of twenty-four could so easily yield. Bianchini followed this autobiographical hint with another: "Manifestations of affection by His Holiness toward the Pretender during his . . . time of need are worth a thousand times what he could do for him in the time of his prosperity."[42]

Summing up his impressions of France for the pope, Bianchini wrote that Louis was an exemplary monarch, putting down Protestants and promoting qualified bishops friendly to the Holy See. Here Bianchini walked a narrow path, for, on the one side, Louis liked and the Vatican opposed Gallican bishops, who regarded the king as the head of the temporal church; while, on the other side, Louis disliked and the Vatican equivocated over Jansenism, an Augustinian version of Catholicism developed by Cornelius Jansen in the seventeenth century and decried by Jesuits as a near-Protestant heresy. Bianchini had observed Jansenists at work in France and elsewhere and recommended appointment of sound priests to combat their "plots and deceit."[43] Continuing his judgments of Frenchmen in power, Bianchini ranked Rohan as a potential ally distinguished by his connections if not his piety. "A few months in Rome spent with Your Holiness and the sacred college would incline him to be more useful." Bianchini thought that Torcy was well disposed toward Rome and suggested that he and a few cardinals take over the running of France after the death of Louis XIV; they would be "entirely devoted to the Holy See . . . and enthusiastically accepted by the nation." But help for the stateless Pretender was Bianchini's main concern. He urged that the pope dispatch a special envoy to France, perhaps Cardinal Annibale Albani, who could declare Rome's support for James openly. Much can be spoken that cannot be written.[44]

Further to viva voce diplomacy, Bianchini proposed the unusual step of sending a Venetian bishop as nuncio to Vienna to try to recreate some unity among Catholic powers. An intermediary would be necessary. For this job he proposed himself. As a native of Verona, he would not need a passport to enter Venetian territory; as a savant, he could pass without anyone guessing his mission; and, as a man, he enjoyed playing politics. He omitted the last of these arguments in his approach to Clement, to whom he wrote that any service he could perform for the Church, his dearest spouse and beloved mother, and "the sacred person of Your Holiness," was for him a duty. Despite his secular knowledge and worldly wisdom, Bianchini always and seriously had religion in mind; as we know, he accepted the dogma, observed the rules, and loved the liturgy of the Roman Catholic Church.[45]

Much of Bianchini's non-academic time in Paris concerned his patron Ottoboni's business. Always in need of money to run his palace, theater, opera, and charities, to keep up his library, collect paintings, and patronize artists, the cardinal obtained, in 1709, with the help of the gift of an ancient statue, the lucrative office of Protector of France. (The previous incumbent, a Medici cardinal, had resigned the office to marry a Gonzaga princess styled the duchess of Guastalla.) Accepting the office was an offence, not in Rome, where even non-Catholic, non-existent kingdoms like James's had protectors, but in Venice. Ottoboni had either overlooked, or having looked thought himself above, a Venetian law prohibiting citizens of the Republic from acting as agents of foreign governments without express permission. For his oversight the Serenissima deprived him of his patrician status and sequestered some of his belongings. He did not recover them until 1720.[46]

The bereft cardinal begged his friend in Paris to move the French authorities, and, if Bianchini had the opportunity, Louis XIV himself, to agitate for reinstatement of his position and possessions. Yet at the same time he asked Bianchini to promote his candidacy for the open position of Protector of Spain—that is, to help him repeat his crime while seeking pardon for it. Ottoboni had great faith in Bianchini's diplomacy, a combination (as he saw it) of "the prudence of the serpent and the simplicity of the dove," but Quetzalcoatl himself could not have carried the day against Philip V's loyalist Francisco Aquaviva.[47] Ottoboni lost and lamented. "Where once I was an object of envy, I am now one of compassion."[48] The overindulged princeling could now hope only that the upcoming peace would bring Venice to its senses, free Italy from its Austrian fellow travelers, and fully restore him to the good life of "religion and liberty" he had enjoyed before dynastic ambitions threw Europe into war.[49]

The time has come to introduce the elector of Bavaria, Maximilian II Emanuel, who threw in his lot with France and lost his bet at the battle of Blenheim. He had fought with Sobieski against the Turks and won Sobieski's eldest daughter, Theresa Cunigonde, as his wife. With his meddlesome mother-in-law as his advocate, he had competed unsuccessfully against the Elector of Saxony for the throne of Poland. The Spanish war opened a more dazzling possibility. Cunigonde was Max Emanuel's second wife. Through his first, a daughter of Emperor Leopold and a Spanish princess, he had a claim to the Spanish throne.[50] It proved costly. In the defeat at Blenheim, he lost both his claim and his electorate. He retreated to Paris. Among the few to whom he granted interviews was Bianchini, who understood that, like James III, Max Emanuel was a shuttlecock among the great powers. Not until Austria and France agreed to stop fighting did he regain his electorate.[51] He and his electoral family stayed in touch with Bianchini.

Max Emanuel may have been the Highness whose help Bianchini enlisted in late October 1712 to obtain passports for himself and his two servants for travel within the empire to Frankfurt and Düsseldorf. He took the opportunity to stop in Strasbourg to complain to Rohan about his treatment in Paris. Since complaining was not in his nature, especially to a cardinal patron, he must have felt sorely tried. The reason: Rohan's servants in Paris had neglected orders to allow Bianchini the use of four horses and a carriage and so had forced the pope's man to go to Versailles drawn by two nags so bedraggled that the guards would not let him pass. Bianchini was mightily embarrassed by the refusal and irritated at having to hire proper equipage at his own expense. It made an additional reason for going to Holland and England: "to give Sig. Card. de Rohan the opportunity of my absence to calculate the dishonor done his house by this stingy way of treating the *cameriere d'onore* of the Pope who has given him so much."[52]

There is more to the story. Cardinals were required to pay a fee to the Holy See after promotion. When Rohan heard that a payment given to Bianchini for transmission to Rome had not been received there, the new cardinal, showing "little respect for the order of His Holiness, and not much for the person of the messenger," declined to entrust Bianchini with anything. Now the *cameriere d'onore* was very correct in money matters. He repudiated the insults to his rectitude and mission so effectively that Rohan apologized, promised the gift of a valuable ring, and requested that Bianchini carry to Rome whatever was owing for the cardinalate. This he declined, not wanting to carry a large sum into the unknown dangers of Protestant countries.[53] Having resolved his grievances, Bianchini could admire the steps Rohan had taken to catholicize Strasbourg. Following another hobby, he inspected the city's fortifications, which he thought would do.[54]

The way forward passed through Düsseldorf, where the elector of Cologne kept his choice collection of medals. Bianchini duly showed his appreciation of them and of the elector's art collection, which he extolled as possessing more great pictures than any other gallery he knew, with rooms full of Rubens, Titians, Veroneses, Correggios, Renis ... to say nothing of a Raphael or two.[55] En route to this artistic feast he spent a night in a country jail. The locals around Rastatt, seeing him wield an astronomical instrument, concluded that he was a wizard, and the imperial army locked him up. After he had talked himself free, he was treated with every honor. The commander apologized profusely; the famous monsignor good naturedly replied that he had not minded at all; he was an astronomer and used to sleeping on straw. A trip down the Rhine, whose breadth and depth he measured, took him at last to Catholic Cologne and its cathedral full of relics, whose value even he could not calculate.[56]

The next destination was "Egypt," the Calvinist Netherlands, for which Bianchini abandoned his clerical dress. He journeyed to Utrecht, where he met Passionei at the peace conference and also Polignac, through whom he obtained a passport for England. Returning towards "the promised land"—that is, Catholic Antwerp—he stopped in Amsterdam, where he heard a shockingly worldly sermon, and Leyden, where he took care to avoid Jansenists. In Antwerp he luxuriated in the Jesuits' library of 48,000 books before going on to Brussels and accommodation with the nuncio.[57] That was cheery. The then recent military successes of France against Austria in Flanders and elsewhere owing to the ceasefire with England made for optimism and, for Bianchini, another reason to cross the Channel.[58] He wanted to see for himself "the full accomplishment of this most glorious change [the desertion of allies!] in England." He planned a trip of eight or ten days and a return to Paris just after Christmas. He soon decided, however, to delay and extend his trip. The reason for the delay does him great credit: he did not want to sail with a shipload of Protestants.[59]

It had been a close thing. The breeze was fresh, the water pacific, the departure imminent. "But I did not think it right to leave from Ostend with heretics, and make the long journey from there to London, when I hoped to find at Dunkirk or Calais some Catholics in the train of the [French] ambassador the duc d'Aumont." A college of Jesuits, whom he consulted, agreed that it would not be wise to risk drowning with people who did not recognize the pope. At Dunkirk he found another quick boat but no seafaring Catholics; again, he decided with Jesuitical help not to chance it. He returned to Paris, having celebrated Christmas on the road, picked up the valuable ring Rohan had promised him, and set off for Calais. He found a ship he regarded as religiously safe and sailed for England on 8 January 1713.[60]

Savant in England

Arriving at Dover on 14 January 1713, or 3 January 1712 as the heretics counted, Bianchini proceeded up the Thames dumbfounded by the activity around him. As was his usual practice, he controlled and confirmed his wonder by measurement. He counted a hundred ships on one side of his boat in an eighth of an hour in a stretch above Greenwich; his journey lasted an hour and a half; whence, he calculated, he must have passed $2 \times (3/2) \times 8 \times 100 = 2,400$ ships in all.[61] "Certainly a prodigious number, which ... shows sufficiently the strength of this kingdom." Later in the week he took his latitude with his portable quadrant (51°26′), counted the steps in the Monument (346), gave its height in Roman (180) and English (170) feet, paced out London Bridge (900 Roman feet), reported the difference between high and low tides, made a thousand measurements of St Paul's cathedral, and, to put an end to it, worked out the area of London.[62]

Uncertain of his welcome, he stayed with Catholics and associated with the ambassadors of Catholic countries—France, Spain, Venice, and Florence—to whom, as a papal representative, he had easy access. He took an eager interest in British life and institutions, acted the tourist in Whitehall, visited Parliament, Westminster, and the Tower. The cultured Protestant aristocrats, *cavalieri inglesi eretici*, treated him with a "kindness infinitely beyond [his] expectations."[63] Soon after his arrival he attended a large reception given by the queen. Was it then that he had the courage to give her the copy of *De kalendario* inscribed "to the family of Her Majesty and all the Stuarts" that is now in the British Library? At the reception he met several leading Tories, perhaps including their master propagandist Jonathan Swift, and certainly their major general, the duke of Ormonde, back from sitting on his bayonets in France. The duke happened to be the Chancellor of the University of Oxford. He gave Bianchini letters of introduction to the Vice Chancellor, Bernard Gardiner, and other important members of the university.[64]

Though a fierce Whig, Gardiner entertained Bianchini almost royally. He gave a lavish dinner, an evening of musical entertainment in the Italian style, an armful of valuable books, and the offer of an honorary degree without compromising oaths of allegiance, which Bianchini declined, alleging "unworthiness." He was nonplussed to find on leaving Oxford after a stay of five days that Gardiner had paid for his lodgings; and, probably, equally astonished when, in an after-dinner toast, Gardiner proposed Queen Anne and Bianchini's prince, "whoever he is." John Keill, Oxford's professor of astronomy and a prominent Newtonian, came to the rescue. Keill was a high-church Tory inclined toward Jacobitism and, as Royal Decipherer, may have known something about Bianchini's business. At least he could inform the Vice Chancellor that Bianchini's prince was Pope

Clement, "a very learned man, and, like many Italian prelates and nobles, more generous toward men of learning than English aristocrats."[65]

The doors of the Bodleian flew open at the recommendation of Sir Andrew Fontaine, a great scholar (he owed his knighthood to a Latin oration he had given welcoming William III to Oxford) and traveler, whom Bianchini may have met in Italy. He is "very learned in medalls," Sir Andrew wrote to Bodley's keeper, John Hudson, and wants particularly to study the Parian marble he had used in his *Istoria*. Hudson warmly welcomed the former Ottobonian librarian with an eloquent harangue in Latin and a bundle of manuscripts.[66] Despite their difference in accent, Bianchini found that he could communicate with the learned in Latin and with the fashionable in French, and seldom needed an interpreter. His erudition astonished the professors, who fell over one another trying to talk with him. He spent the rest of his time admiring coins, books, manuscripts, and marbles, and enjoying town walks and college gardens. He recognized Oxford as the abode of the muses. There seemed to be even a muse of maintenance, "since in other places where there is so much traffic of young people the upkeep of public places they frequent is usually neglected."[67]

There are eyewitness accounts of Bianchini's Oxford stay. When on 14 January he visited the Bodleian for the second time, the captious antiquary Thomas Hearne took him to see the anatomy theater. There Bianchini spied a crucifix recently excavated from the gardens of Christ's Church. "He took it up several times. 'Tis the oldest,' says he, 'I ever saw.'" Hearne agreed with Bianchini's estimate that it came from the time of Charlemagne. "'I have,' says he, 'a very old one myself; but nothing near so old as this.'" Then an antiquarian's slice: "This priest's name is Bianchini . . . He is an ingenious Man too, as it is said, and hath the Character of being learned." And another:

> This Gentleman was very lately in Oxford. He is an old Man [he was 50], and from the little conversation I had with him seems to have good skill in Greek and Roman Antiquities . . . but whether any of his Books be done with that Depth of Judgment, and with that Accuracy, as is requisite in this

Kind of Study I cannot, in the least pretend to determine . . . being perfectly ignorant in the Italian tongue.[68]

All in all, our monsignor liked Englishmen and very much regretted that they were heretics. In thanking Hudson, he observed that he had some acquaintance among the noble English who were drawn to Italy "by the desire to study ancient sites or [Bianchini could joke] to purify our religion." He worked out that good nature had made the English easy victims of heresiarchs. Unfortunately, they were about to acquiesce again in their undoing, so Bianchini wrote in his diary, no doubt thinking of the Hanoverian succession. "Would that by Divine grace the light of truth will soon again illuminate the minds of these good people in name so close to Angels."[69] Later he reciprocated the kindness of his Oxford hosts by supplying information about Italian books and Roman artifacts.[70]

Back in London, Bianchini could inform his English friends that Westminster Abbey is half as long as London Bridge. He took the latitudes of prominent places in the city, kept his watch correct by checking sundials, and inspected rarities of all kinds.[71] He examined the collections of Lord Pembroke and Hans Sloane, whose hoard of recent European coins could in itself document the history of the previous two hundred years; visited John Flamsteed at the Royal Observatory in Greenwich and used its instruments; ate and drank well, attended Catholic services where he could, and, of an evening, conversed happily "about philosophical systems and physical experiments."[72] Of course he went to see Isaac Newton, a goal so eminently worthy that some Newtonians conceived that seeing their leader was his only object in coming to England.[73]

A week after Bianchini arrived in England, Keill had taken him to the great man, "very illustrious [Bianchini reminded his diary] in the more serious subjects." It was not likely the two would get on. To the one holy Yahweh (Iaoue sanctus unus, an anagram Isaacus Neuutonus devised for himself), Bianchini represented everything most hateful in politics and religion. He not only lived among Catholic—that is, "Idolators . . . Blasphemers & Spiritual Fornicators"—but he lived intimately with the

Whore of Babylon, the Pope of Rome. Newton was a closet Unitarian. Bianchini gravitated toward Tories and Jacobites. Newton was a staunch Whig and enthusiastic supporter of the Hanoverian succession, with the reservation, however, that he detested the favorite philosopher of the Electress Sophie, Leibniz, with whom he was then contesting the honor of the invention of the calculus.[74] Bianchini was a friend and admirer of Leibniz. Despite all this, Newton immediately warmed to Anti-Christ's lieutenant and came to rank him with Galileo and Kepler among a very few "candid promoters of truth."[75] Being a Newtonian swamped religion and politics.[76]

Bianchini had studied Newton's work for years together with friends who gathered periodically in the cell of Celestino Galiani. They began in 1707 after lengthy engagement with Galileo, Descartes, Gassendi, and Locke. Owing to a fortunate error of a bookseller, or because no copy of Newton's scarce *Principia* (1687) could be found, the group first tackled the Latin edition of his work on light and colors, *Optice* (1705). The book was controversial. No one in Europe had yet claimed success in repeating the experiments on which Newton had based his startling deduction ("the most considerable that has ever been made heretofore in nature") that white light is a mixture of colored rays. Within a year or two Bianchini and Galiani had become dexterous enough to perform the contested experiments before Alessandro Albani's academy. Bianchini described their performance at his first meeting with Newton. The great man rewarded the pope's man with his portrait, a glass of canary wine, and a promise of copies of his books when Bianchini returned from his excursion to Oxford.[77] Books bond. Probably the copy of *De kalendario* in Newton's library came from Bianchini.[78]

On 2 February 1713 Bianchini attended the Royal Society, where he met Newton again and saw Francis Hauksbee perform the sort of experiments that provided ammunition for Swift's lampoons of natural philosophy. These experiments, which Bianchini rightly judged to be important, showed how threads hanging from hoops oriented themselves within and around electrified, evacuated globes, exhibiting the centrifugal and centripetal forces of the Newtonian system. Five days later, Bianchini paid Newton a visit just as Sir Isaac was on his way to deliver his books, with his own hands, to Bianchini's residence. This attention, which apparently was not the first, alarmed its recipient. "I beg you again and again, for the sake of your health, whose value to letters you and your friends surely are aware of, to receive me nowhere but at your own house."[79]

On this second visit Bianchini brought with him one of his very best friends and most generous creditors, Count Giovanni Antonio Baldini, a diplomat in the service of the duke of Parma. Baldini was a lesser savant in Bianchini's style, interested in astronomical apparatus, practical devices, and historical artifacts; in the last category he assembled a collection of oriental *objects d'art* of considerable cultural importance.[80] Newton proposed Baldini for election to the Royal Society along with Bianchini. The three had been parted for scarcely twenty-four hours when they reassembled on 9 February at the Royal Society's weekly meeting, where Hauksbee performed experiments on capillarity more elaborate than those Montanari and Bacchini had shown their students. Baldini and Bianchini were duly elected, signed the register book, obliged themselves to cultivate the natural sciences, and repaired to an instrument maker to ask the price of an air pump.[81] And so Bianchni became a worthy member of the scientific society to whose work he had already contributed more than most of the other fellows ever would.[82]

Among the examples of his work that Bianchini brought to England was an engraving, after a drawing made by himself, of the Farnese Atlas and globe.[83] This monument, which he correctly dated to the time of Ptolemy, is one of the oldest surviving presentations of the constellations on a sphere; and it is further distinguished by bearing the chief reference circles of ancient astronomy, the equator, the equinoctial colure (the great circle running through the celestial poles and the equinoxes), the ecliptic, and the zodiacal band (see Fig. 28). Exploiting these unique characteristics, Bianchini had pinpointed the voyage of the argonauts to 1,425 years before the time of Ptolemy. Since the representations of the constellations agreed generally with their poetical descriptions in Aratus's *Phaenomena*

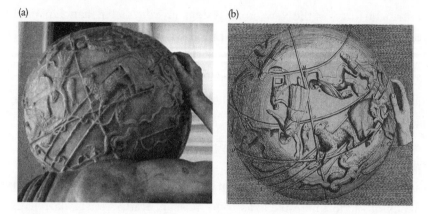

Fig. 28. The Farnese globe, which played a major part in Bianchini's periodization of history, showing the spring signs, as it now exists (a), and as depicted by him (b). Museo Archeologico Nazionale, Naples, and Giuseppe Bianchini (ed.), *Demonstratio historiae ecclesiasticae* (1752–4), i:1, resp.

(say 275 BC), he inferred that the globe transmitted information at least as old as the poem. We know that he argued for a much older date on the ground that every individual and symbol on the globe had some connection with the argonauts and concluded that the globe preserved the celestial knowledge of Jason and his comrades.[84]

Bianchini's later thorough study of the globe begins by placing the maker of its original in Thessaly, whence the *Argos* sailed. The globe has a circle parallel to the equator at an angular distance that Bianchini took to be 50°. If it indicated the limit of circumpolar stars at the place of observation, the latitude there must have been 90°–50° = 40°. The fortieth parallel runs through Thessaly. (Modern measurements put the limiting circle higher than 50° and the latitude of the observer as low as Athens or even Alexandria.) The most important feature of the globe, however, is the placement of the leading horn of Aries at about 6°40' to the east of the vernal equinox, the geometrical point where the two great-circle paths of the sun intersect: the annual path, the "ecliptic," which runs through the zodiac, and the diurnal path, the celestial equator, which the sun describes on the spring and autumn days when daylight and nighttime are equal.

The precession of the equinoxes can be represented on a celestial globe by a uniform eastward progression ("precession") of the stars parallel to the equator. Bianchini estimated the rate of precession at one degree in seventy-one years. If the 6°40′ between the ram's horn and the vernal equinox was the amount of precession over the time he sought, he would have to embark the argonauts at (6 2/3) × 71 years before Ptolemy, or 323 BC, about fifty years before Aratus's poem. Aratus probably took his star lore from Eudoxus, a contemporary of Plato. And where would Eudoxus have placed the equinoxes? According to an ancient source, Eudoxus arranged the constellations so that the equinoxes fell in their centers, rather than (as was later practice) at their western extremities. Jason would have set the vernal point in the middle of the ram! Where was that? From inspection of the globe, Bianchini estimated that Jason's spring equinox lay 13°24′ east of the further horn. "Thus" the vernal point as determined by Ptolemy was some 13°24′ + 6°40′ = 20°4′ west of its place when the argonauts learned to navigate from the original of the Farnese globe. The *Argo* began its voyage in 1275 BC, or perhaps earlier or later. It is not easy to locate the middle of a ram.[85]

Bianchini could adduce a nice coincidence in confirmation of his calculation. According to legend, the Trojan War took place a generation after Jason's adventure, therefore around 1250 BE. According to Eratosthenes, like Bianchini an astronomer and a librarian, Troy fell 407 years before the first Olympiad, therefore in 1207 BE. Bianchini suspected that Eratosthenes had used the report of an eclipse to anchor his estimate. As we know, Bianchini put little faith in observations reported at second or third hand from remote antiquity. He would stick with the genuine article, the Parian marble, even though it too came from a time well after the historical events it described.[86]

Bianchini gave his traveling drawing of the Farnese globe and a copy of *De kalendario* to Flamsteed, who approved the drawing ("'tis remarkable that the places of the constellations answer to Ptolemy's age . . . the designers were excellent draughtsmen") but not the observations at Santa Maria degli Angeli ("'tis a piece I should look upon as trifling if it had not in it some Roman observations of the sun's Meridian heights ... that may be of use to me").[87] These gifts were given just before the week of frequent communion between Bianchini and Newton and hence before, if the following reconstruction is correct, Bianchini learned that Newton would have wanted the drawing of the globe.[88] For Newton had been engaged for years in trying to date the voyage of the argonauts using Aratus and Eudoxus, whose texts were conveniently available in a book by the chronologist Petavius printed in 1630 and again, with other works, in 1703. Newton had a copy of this second edition.[89]

Newton had kept his research on argonauts secret, as he did much of his intellectual work, although for him it belonged to a task more important than the making of world systems. He strove to puncture the claims of ancient gentile peoples to histories far longer than the biblical reckoning allowed and to secure temporal priority of the Jews in civilized pursuits. Jason and his friends evidently knew something about navigation and shipbuilding, and so of astronomy and architecture. Since by hypothesis they derived these arts from the Jews, Newton had to show that they sailed sometime after the flowering of Hebrew civilization under Solomon, understood to have reigned about 1000 BC.[90]

Newton had been engaged in knocking down the claims of ancient historians for forty years or more before he met Bianchini. He shared his visitor's reservations about texts not corroborated by material remains and his enthusiasm for euhemerist interpretations of the ancient gods and asterisms. Myths could reveal real events to an analyst able to unscramble them. The pivotal real events for both our analysts were the argonauts' travels and the battles over Troy. In addition to astronomy, each of them had an axiom to help with dating. Newton stipulated that memory of events could not survive more than three generations without being written down; consequently, any happening reliably remembered could not be much older than the invention of letters, which he dated to just before the argonauts' expedition. Bianchini assumed that material objects, even copies, could accurately reflect the circumstances of their creation, since, being continuously on display, they could not disagree altogether from common knowledge. We already know his principle of equal progress in equal time, which Newton also employed when he found it useful.[91]

When, then, did the *Argos* set sail? Following the Hebrew scriptures, Newton put Solomon's death in 980 BC. To give the Hebrews priority in navigation, Newton could not let Jason sail before the middle of the tenth century BC.[92] He secured this result by adopting an assimilation accepted by some erudite chronologists: the quasi-fabulous Egyptian king Sesostris, who according to Herodotus conquered lands as far afield as Asia, was also the biblical pharaoh Sesac, who conquered Judah and rifled Jerusalem during the reign of Solomon's son Rehoboam. The spurious identification of Sesostris with Sesac fixed the date of the first gentile empire to the time of Solomon and so the date of the argonauts. For Sesostris-Sesac had appointed Hercules to guard the passes of Mt Caucasus and Hercules became an argonaut. The preservation of this information required the prior invention of writing. Newton noted that Thoth, who served Sesostris as secretary, was related to, or perhaps identical with, that mischievous Mercury who perfected letters. For full understanding, we must know how the Egyptians became indebted to others for the arts of civilization. Newton explained that the news came to them from the Edomites, who occupied territory from the Dead to the Red Sea. Military arts were not among the Hebrew accomplishments they mastered. After decisive defeat by David's son Solomon, surviving Edomites wandered about disseminating the civilized ways of the people who had driven them from their homeland.[93]

Newton confirmed this dating by playing with the sphere described by Aratus. Lacking the Farnese globe, he had to find the vernal equinox of Jason's time by tracing the equinoctial colure described poetically as running through the left hand of Boötes, the right hand and a knee of the Centaur, the head of Cetus, and the head and right hand of Perseus. There were no ancient globes from Aratus's time to help place these body parts, and the modern charts Newton consulted differed among themselves in the layout of stars and figures enough to enable him to anchor his colure in the ram's middle where it best suited his chronological calculations. The result: *Argos* sailed between forty and forty-three years after the death of Solomon. Eventually Newton settled on forty-one years and 939 BC.[94]

Bianchini rejected the identification of Sesostris and Sesac as a confusion introduced by Josephus and promulgated by later chronologists.[95] To agree with Herodotus and the Parian marble, Sesostris had to live in Century XXVI, half a millennium earlier than Newton put him, a time when navigation and martial arts were well developed and Jerusalem so insignificant a place that Sesostris would not have stopped to rob it. The breakup of Sesostris's empire brought forth the gangs of gods and heroes that struggled at Troy. Since a grandson of Sesostris lived during the struggle, we must date the fall of Troy two generations after Sesostris, say in 1350 BC, and the argonauts a generation earlier.[96] For subsequent Greek history, Bianchini followed the Parian marble, of which he gave a useful Italian translation, without the compression of intervals Newton required.[97] To find the date of the argonauts astronomically, Bianchini had the advantage of the Farnese sphere. As we know, his application of precession dated the original of the globe to 1,525 years before Ptolemy, or around 1275 BC.

Newton had thrown himself into astronomical dating after the Royal Society, energized by a question from Leibniz, asked him for advice about calendar reform. It might be a good idea, he replied, to reckon the year from 1 January, but certainly a bad one to adopt the entire Gregorian apparatus. The question apparently awakened him to the possibility of an astronomical confirmation of the amputation of the Greek calendar he had then already achieved by arbitrarily truncating regnal spans. Whereas the chroniclers and Parian marble made the reigns of Greek kings three to a century, Newton calculated a generation at nineteen years on average. That authorized him to cut down Herodotus and Greek time by a factor of 19:33, over half. Then, in 1704, having freed himself from physics by the completion of his *Opticks*, he went to work on Aratus.

Why did the Jews fall behind the Greeks after Solomon? The last paragraph of *Opticks* offers a clue. It advises that its methods applied to morals allow us to define our duties to God and one another by "the Light of

Nature." How did the light go out? Why, after so many centuries, are we still not fully acquainted with the true religion that obtained "under the government of Noah and his Sons"? Wicked Egypt! There the generations after Noah "corrupted themselves" with idolatry. After the Exodus, the Jews tried hard to purge their religion of false gods, but did not succeed.[98]

Bianchini referred to the discrepancy between his and Newton's dates for the argonauts in his posthumously published treatise on the Farnese globe. Although his dates agreed better than Newton's with the usual calculations of the extent of Greek history, Bianchini allowed that a defensible choice for the boundary of Aries and the course of the colure could move its center three degrees closer to the vernal equinox than he had made it. That would displace his date of the voyage of the argonauts (1275 BC) two centuries later, but would still leave a gap of 126 years, equivalent to almost two degrees of precession, between his date and Newton's.[99] Bianchini probably did not see Newton's chronological work in print. He might have heard about a manuscript version after his return to Rome, since, as we know from an enthusiastic Newtonian resident in Italy, many people there knew enough about the great man's historical research to think it a pity his "inflexible Modestie" kept him from publishing them.[100] Still, it seems more likely that Bianchini first learned of Newton's closet historical studies from the man himself.[101]

This timing and certain coincidences of detail in Newton's accounts of the origin of astronomy among the argonauts and of idolatry among the Egyptians suggest that Newton might have gained something from his conversations with Bianchini. Did they discuss the size and shape of the asterisms on the primitive sphere and the course of the colure of the argonauts? Did Bianchini concede the leeway of three degrees he offered in his posthumous publication? That made the latest possible date of the voyage of the argonauts around (to pick a memorable number) 1066 BC. Newton might have toyed with this date as just sufficient to delay the arrival of astronomy among the gentiles until after Jewish civilization had begun to flourish. In the end he stayed faithful to the calculation he had based on Aratus informed by Eudoxus and his discounting of reigns recorded on the Parian marble. After some fiddling, he retained 939 BC for the argonauts and a comfortable seniority for the Jews.[102]

Newton's removal of a few centuries from Greek history outraged historians and others not mesmerized by mathematics. "Newton as a chronologist had to be as eccentric as Newton the geometer and *physicien*." Mathematicians were delighted, however, and antiquarians as well, to see that their studies, "often regarded as useless, also attracted Newton, and his interest was a sufficient response to [such naysayers]." But, as uncharitable French chronologists charged, Newton's estimates were nonsense. One need only read the treatise of Nicolas Fréret, of the Académie des inscriptions, to calculate how far the English genius had gone astray. According to his editor, Fréret's objections were lethal and unrelenting, balanced, reasonable, scientific, unanswerable, devastating.[103]

Newtonians naturally took an interest in an accurate description of the Farnese sphere, which Bianchini had promised in his *Istoria universale*. Martin Folkes, a mathematician and antiquarian, hurried matters along by commissioning a half-size plaster cast of the globe, which he exhibited at both the academies of which he was president, the Royal Society and the Society of Antiquaries, and had published in planar projection in 1739.[104] That did not improve the fit between Newton's chronology and everyone else's.

The dating of the Farnese globe is again a subject of passionate interest. The astronomer Bradley Schaefer stirred debate by dating the globe's Hellenistic predecessor to three centuries before Ptolemy. Immediate and fatal refutation.[105] Another of Schaefer's initiatives, based on the writings of Eudoxus and Aratus, awarded the globe's star lore to observers living in 1130 ± 80 BC.[106] Without reference to Bianchini and by an entirely different method, he found the same era for the creation of the Greek asterisms and the celestial globe. Mere coincidence? No doubt. The era is far earlier than any date for which there is direct evidence.

Bianchini returned home via Genoa, where he stayed with his old correspondent Salvago and played with *meridiane*: one that Salvago had placed in the baroque Basilica della Santissima Annunziata del Vastato when the

Clementine gnomon was being built and one that Bianchini helped plan for Salvago's villa.[107] After this recreation he proceeded to Rome and distribution of the treasure Newton had given him. It included two copies of *Optice* and several of the *Commercium epistolicum*, in which the Royal Society claimed on Newton's behalf that Leibniz had plagiarized the calculus. Bianchini kept one copy of the *Optice* and gave the other to the Vatican Library, whose keeper had not known that Newton "excel[led] in this branch of learning." Galiani and Manfredi each received a copy of the *Commercium*, and, as Bianchini coyly phrased their response, "they much approve of the Royal Society in asserting the rights of the author of this outstanding discovery by means of documents." When pressed for his own opinion, however, Bianchini refused as usual to join a clique in the Republic of Letters. Leibniz was an older friend of his than Newton.[108]

More useful to science in Italy, Bianchini rejoined Galiani in the study and promotion of Newtonian philosophy. They had hesitated to embrace the ideas behind the *Optice* because they conflicted with the Cartesian physics Galiani had absorbed from study and Bianchini had learned from Montanari. Descartes had remained for them "the prince of mathematicians [and] the creator of the new philosophy."[109] Nor, at first, did their reading of the *Principia* reveal the fatal blow that Newton had dealt Descartes. They had trouble conceiving how gravitational and centrifugal "force" could be independent of the magnitude of the body affected.[110] Their difficulty, which does them honor, arose partially from confusion (Newton often used force, *vis*, for acceleration) and partly from an intuition of an undisclosed assumption. This was Newton's silent identification of inertial mass (the measure of a body's indifference to motion) and gravitational mass (the measure of its gravity). The equality made acceleration under gravity independent of mass and underlies Einstein's general theory of relativity.

Soon after Bianchini's return, Galiani's group accepted Newton's destruction of Descartes's cosmology and prepared to do battle against its adherents. Galiani composed an "Epistola de gravitate et cartesianis vorticibus," a letter on gravity and Cartesian vortices, which made the rounds in manuscript from Naples to Venice. It explained the basis and success of Newton's theory of universal gravity and its dire consequences for the whirlpools of subtle matter centered on the sun that Descartes invoked to sweep the planets around. No reasonable philosopher could stick to a system so obviously undercut by the facts.[111] But unreasonable philosophers might take comfort in Galiani's "Epistola" by misreading it as an attack not only on the prohibited physics of Descartes but on modernism in general. Many scholastic birds of prey, "who do not even know the definitions of Euclid," flocked to Galiani's cell to pick up his critique; "but then, seeing it dirtied with figures, [they realized . . .] that it was nothing they could digest." Colleagues urged Galiani to moderate his words to prevent such misunderstandings and avoid undercutting other champions of modernism.[112] But Galiani regarded his "Epistola" as the opening shot in an extensive campaign, not as an amendable tactic. In May 1714, two months after he had completed it, a perfect ground for continuing the offensive opened in Rome.

Cardinal Gualterio, the friend of up-to-date French savants and mathematicians, joined with Bianchini and Galiani to set up an academy to spread the latest knowledge from the North. It had its first official meeting on 5 May 1714. Bianchini was there, to explain Gassendi's atomistic concept of light, and Galiani, to demolish Descartes's and champion Newton's. The cardinal himself, or rather God, chose light as the subject of the inaugural session; for, when asked the order in which the academy should investigate the natural world, Gualterio replied, "in the order in which it was made."[113] At subsequent meetings, Galiani discussed his "Epistola," and Bianchini performed Hauksbee's experiments with the glass globe he had brought back from England. News of these experiments spread to other academies, in Rimini, Bologna, and Pisa, preparing the way for the assimilation of Hauksbee's experimental illustrations of Newton's philosophy and the Italian translation of his *Physico-Mechanical Experiments*.[114] Among Bianchini's papers there is a partial translation of about a third of the book apparently made at his direction. It differs verbally from the full translation published in Florence in 1716.[115] This was

PLATES

Plate 1. Pope Clement XI as rendered late in the War of the Spanish Succession by his friend Per Leone Ghezzi.

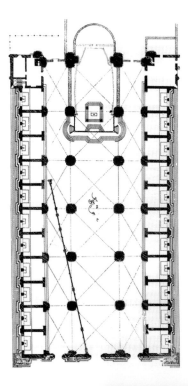

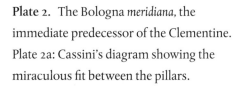

Plate 2. The Bologna *meridiana*, the immediate predecessor of the Clementine. Plate 2a: Cassini's diagram showing the miraculous fit between the pillars.

Plate 2b. Noon at Cassini's *meridiana* not long ago.

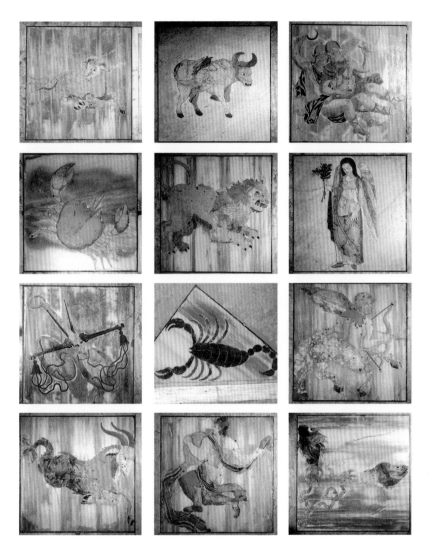

Plate 3. The zodiacal marbles around the Clementine *meridiana* designed after the figures in Johann Bayer's *Uranometria* (1600) by Carlo Maratti and Domenico Paradiso: Aries, Taurus, Gemini, Cancer, Leo, Virgo, Libra, Scorpio, Sagittarius, Capricornus, Aquarius, Pisces.

Plate 4. The Clementine *meridiana* looking south from markers for the earliest and latest possible dates for Easter ("terminus paschalis") toward the jubilee ellipses; the section is about a third of the entire line.

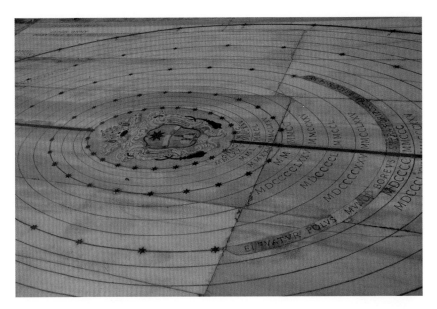

Plate 5. The jubilee ellipses beyond the south end of the Clementine *meridiana;* the brass-studded ellipses placed around the Albani arms as pole indicate the diurnal orbit of the pole star at twenty-five-year intervals beginning with the innermost for the year 1700.

Plate 6. Section of the Clementine *meridiana* north of the equinoxes at Aries/Libra showing medallions honoring the Sobieski inserted on the nineteenth anniversary of the victory at Vienna.

Plate 7. Section of the Clementine *meridiana* around Taurus/Virgo showing medallions placed to commemorate the pope's visits on the feast day of San Brunone, founder of the order of monks (the Cistercian or Carthusian), which had charge of Santa Maria degli Angeli.

Plate 8. The arms of Pope Clement XI decorate the hole in the south wall through which the sun's rays enter the basilica.

Plate 9. The Palazzo Ducale, Urbino, where James III held court and did astronomy with Bianchini; the *meridiana* ran along the ledge in the shadow to the right of the right-hand tower.

Plate 10. Charles Wogan as a merry young man.

Plate 11. Maria Clementina as she looked at her wedding. Copy by Francesco Bertosi of a painting by Francesco Trevisani, 1719.

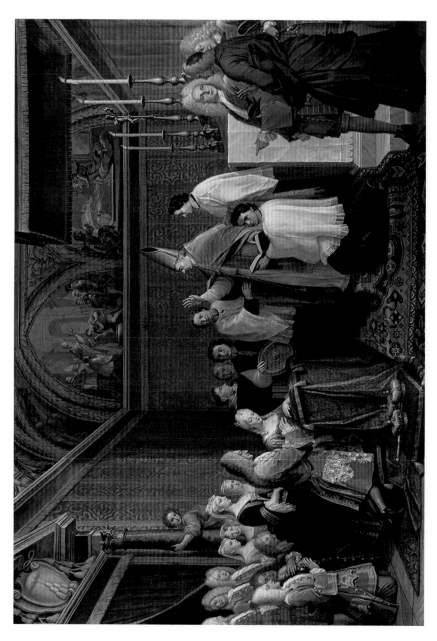

Plate 12. The joining of Stuart and Sobieski, 1719, by Agostino Masucci, 1735. The figures include the Hays, Murray, Wogan, O'Brien, and Eleanor Misset, all of whom are known from other sources to have attended the wedding. An educated guess by Edward Corp identifies Marjorie Hay as the woman standing behind the royal pair; Hay, Wogan, and O'Brien as the three men kneeling behind her; and James Murray as the man standing and pointing on the right. The register was signed by the king's men Hay and Murray, the princess's protectors Wogan and O'Brien, and a priest.

Plate 13. Location of the medallions celebrating the union of James III Stuart and Maria Clementina Sobieski. Plate 13a: Section of the meridian around the summer signs Gemini / Cancer / Leo looking north, with the medallions in the lower left. Plate 13b: The medallions and the jubilee ellipses looking east, with the crab to the right of the medallions.

Plate 14. James III immortalized in a marble in the pavement west of the Clementine *meridiana*; the sun's image falls on it on every recurrence of the date and hour of his birth.

Plate 15. Frieze on the tomb of Alexander VIII Ottoboni, Bianchini's first major patron; the protégé is the only person looking the pope in the face (the second figure standing to the pope's right). St Peter's, Rome.

Plate 16. Bianchini's monument in the Cathedral of Verona.

the labor of Thomas Dereham, an English Catholic resident in Florence, who with it initiated a project to make English natural philosophy easily available to Italians. Among Dereham's products were five volumes of papers translated from the Royal Society's *Transactions*.[116]

Having placed Newton in competition with Descartes, Galiani worked to assure all interested parties that Newton, though Copernican and Protestant, was a friend of Christianity. Newton did not err, nor did his followers who ascribed the power of spontaneous action, of innate gravity, to matter. What gravity might be, Newton did not say. "I feign no hypotheses . . . It is enough that gravity does really exist, and act according to the laws I have explained, and abundantly serves to account for all the motions of the heavenly bodies, and of our seas." This passage comes from the "General Scholium" or summary commentary that Newton added to the second edition of the *Principia*. The Scholium goes on to declare the existence and attributes of God, "to discourse of whom from the appearance of things does certainly belong to Natural Philosophy."[117] To all of which Galiani and his circle must have said "Amen." For them the idea of innate gravity (gravity as a primary quality of matter) was falser than a Cartesian vortex; the cause of gravity, unknown; and the *Principia* and *Opticks*, religious works.[118] What role if any Galiani, Bianchini, and others in their circle may have played in diverting the Inquisition from Newton is not known. They did intervene, successfully, in other cases involving books on the Newtonian or corpuscular philosophy.[119]

Although the heretic Newton's teachings were regarded more favorably in Rome than those of the reprobate Descartes, Galiani worried during the first meetings of Gualterio's pro-Newtonian academy that the authorities would try to close it.[120] But very soon he could judge that the battle against the inquisitors over the new world systems had turned. That by no means signified victory, however. A more serious enemy than inquisition was indifference. "What is to be feared is not so much the Inquisitor as the beliefs of men who hold [experiments like Hauksbee's] to be little more than the tricks of acrobats." The task remained of showing educated people that cultivating natural science does not diminish a person in divinity or erudition.[121] Bianchini was an exemplar for the cause. In 1718 Galiani became abbot of his seminary and began the rise that elevated him, in 1731, to bishop of Taranto and *capellano maggiore* (commissioner of education) of the Kingdom of Naples.[122] That gave him the opportunity to inculcate the great truth that the study of natural science does not necessarily stultify intellectual growth.

Or dull the taste for serious politics. Again Bianchini was exemplary. Early in 1713 Clement had given in to pressure from Louis XIV and issued a bull, *Unigenitus,* condemning 101 propositions thought to be Jansenist. Many French bishops refused to accept the bull, although Rohan had explained its merits in a lecture that lasted six and a half days. Bianchini suggested a shorter way to bring the dissidents around. He suggested that a direct conversation between them and a high-profile papal envoy (he proposed his protégé Alessandro Albani) might be able to win them to the pope's side. To avoid sabotage to his mission, Alessandro might pass through Avignon on the pretext of seeing James and continue easily into France. The young man would need advice and additional camouflage. Bianchini proposed himself for the role.[123] The trip did not take place. Bianchini had to content himself with pushing the cause of the Pretender in Rome.[124]

6

JACOBITE ADVENTURES

With peace secured, the British government pressured the European powers to push the Pretender further from England than Lorraine. At the same time, its chief ministers, Oxford and Bolingbroke, insinuated that, if he changed religion, Anne would recognize him as her successor. They maladroitly conveyed this message in February 1714 through a double agent, the abbé Gaultier. James was outraged by the approach. Was Gaultier not a Catholic? A Catholic attempting to lure him into heresy? In religion James was not a Pretender. He alerted the Holy See (in the person of Gualterio) to the frightful scandal. The See smiled as James scornfully rejected apostasy and remained in Lorraine, protected by its duke and his own poverty, which, he said, prevented him from moving to Switzerland, the only country in western Europe that the English would accept.[1]

News of the flirtation of Oxford and Bolingbroke with the Pretender leaked out in April 1714. Then events speeded up. In June, Sophie, the Hanoverian heiress, died, and on 1 August Queen Anne followed suit. Had she survived a few more years, the throne would have been James's, or so said the Duke of Ormonde: "This was for some time her Majesty's opinion, and also that of all true Englishmen, who did not want to punish the son for the sins of the father."[2] Ormonde turned his opinion into treason by urging "all true lovers of the Church of England, and their country" to refuse allegiance to George I and support the return of the Stuarts.[3] This he wrote in a letter he had carried to the mayor of Oxford. The shocked mayor ran with it to the vice chancellor, the good Whig

Gardiner; the vice chancellor summoned the heads of the colleges; and the heads voted a reward of 100 pounds for the apprehension of the letter carrier. They also deputed Gardiner to assure the new king that the university would have regarded the death of Anne as "unsupportable" had God not supplied the Hanoverians to relieve its distress.[4] Ormonde soon quit the country, forfeiting his estates and relinquishing his debts, to help James prepare for an attempt to regain his throne by force. Bolingbroke also joined the Pretender when the dominantly Whig Parliament elected in 1715 began proceedings against him and Oxford.

The new George did not yet sit safely on his throne. Was Hannibal at the gates? Common sense with a dash of Stuart sympathy, said no; Hannibal and his elephants, or the pope and his bishops, were "Chymerical wild Fancies, Embryo-Giants," bogies to scare the masses with one more papal plot. How could anyone fall for it? The clergy, army, and most of the country were irrevocably Protestant, and the Pretender had no resources. "*Popery* cannot gain Ground in these Kingdoms let who will succeed Her *Majesty* in the Throne."[5] Nay, nay, Hannibal is nigh, replied the Whigs' polemicist Daniel Defoe. Many of the clergy favored James and divine right, prayed for a French-backed invasion, and contentedly contemplated the "Restoration of *Popery* and *Slavery*." Jacobite clubs abounded, the enemy was already among us, beware! If the "Notorious Bastard" (James's legitimacy had been contested) returned, he will be armed with "a truly *Italian* Revenge, bred up in all the destructive Principles of a Bigotted Mother, and the Tyrannical Notions of a *French* Master."[6] Thus Defoe served up the usual papal plot seasoned with xenophobia to the English taste. He was right in this, however: Hannibal was nigh.

The King's Road to Rome

On 28 October 1715 James slipped out of Lorraine disguised as an abbé to become a knight-errant, wandering, quixotic, and risk prone. His goal was a channel port from which to sail to Scotland. His traverse of northern France, which took twelve days, proved the character that had attracted Bianchini. The route went via back country, to foil pursuit; through rain, hail, and mud, to test resolve; through vile inns, worse food, straw beds, vermin human and animal, to fulfil God's purpose. The journal from which these details come, kept by the king's traveling companion, a resourceful Irishman named O'Flannagan, ends with an assessment of James.

> I never knew any have better temper, be more familiar and good . . . never the least disquieted, but with the greatest courage and fermness resolved to goe through what he had designed on . . . satisfyed with the worst Diet and lodgings . . . and enfine possessing eminently all the qualityes of a great prince, with those of a most Honest private Gentleman.[7]

In the real world, the prince suffered under the anxiety of discovery as he scurried along the coast waiting for news from Britain and a change in the weather that prevented his embarkation. His untrustworthy new acquisitions, Bolingbroke and Ormonde, had planned the attack on Britain the preceding August in Paris together with Torcy, who hinted at possible French support. That possibility vanished along with the sun king, who had set permanently on 1 September; nonetheless, the plotters had gone forward, pushed by the preemptive raising of the Jacobite standard in the Highlands by a late convert to the cause, an incompetent indecisive shuttlecock, the earl of Mar. Part of Mar's army advanced far enough into England to be defeated decisively in November at Preston in Lancashire. Other defeats that might have been victories occurred elsewhere in Scotland.[8] Meanwhile, English countermeasures had prevented Ormonde's invasion. When James arrived in Scotland on 22 December and joined Mar in Aberdeen, his situation was untenable. Nevertheless, he made plans for his coronation.

In the clarity of the new year James recognized defeat and in February fled to France with Mar and other loyalists. He hoped to return to Lorraine, but the duke would not have him. In April 1716 he took up residence in the papal enclave in Avignon. By then the Vatican had decided that James had the makings of a martyr, and Clement invited him to Italy. James rejected the offer because it would take him further from Britain and compromise him with papal entanglements. Moreover, he had opened negotiations with Swedish representatives for support for another attempt at his father's thrones.[9] In preparation for his imagined reign, he constructed a court full of Protestant aristocrats that did not include the undependable Bolingbroke, whom he replaced with the undependable Mar. James's judgment was at its worst when evaluating character. "I never met with a more able nor more reasonable man [he had written of Mar to Bolingbroke], nor more truly disinterested and affectionate to me." He spent nine months thus shifting, recruiting, and plotting before Britain, France, the Empire, and the Dutch Republic decided that his presence in Avignon violated provisions of the Utrecht treaty. They demanded that he leave. On 6 February 1717 he set off for Italy.[10]

James arrived in Rome on 26 May and stayed for forty days and forty nights.[11] On 29 June, on the Feast of the Holy Apostles Peter and Paul, Clement preached a homily praising the Pretender's resolve. "[He] has shown with a sublime and truly Royal Soul an entire Contempt and Disdain for all the August Glories and splendid Honours that this World can bestow, that he might keep that pure, that strong, that invincible Faith which Christ so gloriously extoll'd in St Peter, and Rewarded with so many Benefits." In James's case these benefits might still be theoretical; but by his example he conferred a gift upon us all. "Let us . . . take care . . . least we, having so bright and illustrious [an] Example before our Eyes, be found sluggish and lifeless in the Duties of our most Holy Religion, and behold this shining Image of Heroick and Consumate Vertue (now before us) with looks of Indolence, Indifference, and Inaction."[12] The shining image asked the pope to appoint Gualterio, with whom he was residing, Protector of England. To this Clement assented; but he did not think that the shine on the image, or the heroic and consummate virtue, reason enough to increase James's papal pension above 10,000 scudi a year.[13]

During this first visit to Rome, James toured the city with a guide assigned by the pope. The cicerone, Bianchini of course, had perfected this loop of the grand tour the previous year, when, also on papal orders, he had shepherded the son of his acquaintance from Paris, Max Emanuel II, the restored elector of Bavaria, around the capital. The boy, Charles Albert, had to see "the most considerable things in Rome, either for their artistic excellence or for their memorials of history," or both, as presented at the *meridiana* at Santa Maria degli Angeli, where he could admire the medallions of his Sobieski ancestors. The tour culminated in a kaleidoscope of art and history, feast and fun, ceremony and sincerity at the Villa Borghese on 27 May 1716. Bianchini had arranged a party of sixteen members of papal families augmented by twenty-two relatives and retainers of the electoral princeling to inspect the Greek and Roman statues at the villa, feast on all the delicacies earth, air, and sea could provide, drink the wines of the Canaries, Spain, Italy, the Rhine, and Hungary, tour the gardens, listen to music, and dance for hours. Bianchini had insured the success of the dancing by inviting the wives and daughters of the papal families. He knew his way around Rome. His report of the event dwells lovingly on the status of the invitees, the details of the gold and silver utensils, the variety of food and drink, and the principal sculptures in the villa. He had good etchings of a dozen of these printed together with his report as a keepsake for the guests.[14] The elector acknowledged the honor shown to his son with a large emerald ring Bianchini wore for the rest of his life.[15]

A year later, Charles Albert returned to Rome with his younger brother Clemens August and their tutor, one abbé Filibert. Elector Max asked Bianchini to improve them all, the boys by example, the abbé by a benefice, which, Monsignor, you could easily procure, "since you enjoy so high a status in the affection of His Holiness."[16] Indeed, the entire Albani clan stood ready to serve Bianchini in reciprocation of the many errands he had run for them. No doubt Filibert got his benefice, as did others who presented their suits through Bianchini.[17] As for the Bavarian boys, or rather young men (Charles was twenty and Clemens seventeen in 1717),

Bianchini did what he could to instill "maxims of religion, prudence, and decorum" in them.[18] They would need it. Charles succeeded his father as elector of Bavaria and Clemens their uncle as archbishop of Cologne. Since the archbishop was also an elector, Clemens could vote for and crown his brother Holy Roman Emperor in succession to Charles VI. (This lapse in the Habsburg hold on the empire did not last long.) As elector and emperor, Charles Albert improved his capital, Munich, by completing his palaces in ways that perhaps owed something to Bianchini's passion for symmetry in architecture.

But we have left the Pretender waiting. Bianchini worked for him as for a dozen Bavarians, "for thus," he wrote in his courtliest style, "I [could] admire from close up the most glorious monarch that Divine Providence has given us for defender of his church, not only in our time, but for many centuries."[19] One of the first stops on their tour was Saint Mary of the Angels. It suited Bianchini's scheme of history to memorialize the visit of the king with a medallion close to where the jubilee ellipses measured the centuries. And so, on the morning of 21 June, three weeks after his first visit, James returned to the basilica to admire a copper plate put into the pavement to catch the sun's image where it fell on the day and hour of his birth.[20] That blessed unexpected event had occurred at mid morning on midsummer day in London in 1688. An easy reckoning discloses that James reached the age of 29 in Rome in 1717 as the sun, rising to the highest point in the ecliptic, cast its rays a little to the northwest of the jubilee ellipses. Quite a coincidence, since, as a puzzled Jacobite courtier reported it, "according to Monsignor Bianchini . . . only after an interval of 29 complete cycles can the sun return exactly to this point at the precise hour of His Majesty's birth."[21]

Although the pope admired James's character and piety (or so Bianchini reported), he did not want the king for the ages to make his eternal headquarters in Rome. As spiritual leader of the Catholic world, Clement worried about Protestants worshipping openly in his capital and otherwise disturbing the peace; and, as a secular ruler, he had no wish to annoy the great powers by appearing to condone Jacobite plotting under his nose. Still James had his uses, and his royal pretentions needed burnishing. Clement granted him several long audiences; Gualterio, Albani, Ottoboni, and other leading cardinals made much of him; fireworks, music by Scarlatti, and other regal entertainments completed the charade.[22] On 4 July the pope sent the Pretender off to Urbino. Not knowing the drawbacks of the place, James had chosen it over others as a proper abode for the court of an independent prince. Bianchini would join him there in August, bringing as a housewarming gift a quantity of a fresh cheese characteristic of Piacenza, which he obtained from his friend Count Baldini. From Gualterio, James received the even more useful gift of a chest of antidotes to the usual poisons deployed in Rome.[23] The separation of the pope's Rome from the king's Urbino was also a gift for natural philosophy, for during the many trips Bianchini subsequently made between them he stopped to take the barometric heights and determine the geographical coordinates of the places he passed through.[24]

The Jacobite court at Urbino numbered eighty, fifty-five of whom had salaries while the balance had pensions. The upper echelon consisted almost entirely of Protestants, the duke of Mar being dominant, although, for his behavior during the uprising of 1715, discounted and disliked. As secretary of state, he tried unsuccessfully to secure the support of Sweden or Russia for which James was angling. His easy way with loyalty—Whig, Tory, or Jacobite as circumstances dictated—caught up with him in 1719. Thereafter he lived in Paris and dealt mainly doubly, with the British and French courts, until discharged by both after half a dozen years of distinguished duplicitous service. Mar's brother-in-law, John Hay, raised to the Jacobite Scottish peerage as earl of Inverness in 1718, became the Pretender's main advocate and confidant. Other trusted royal advisors were Hays's wife Marjorie, one of two women at the Urbino court other than servants, and Marjorie's brother James Murray, who, as earl of Dunbar, would take over Mar's duties as secretary of state.[25]

The new Protestant recruits such as Mar, Hay, and Murray drove away many courtiers and servants who had been with James in France and Lorraine. A prominent exception, the Catholic David Nairne, who had charge of James's correspondence with Catholic princes and prelates, was on terms of mutual trust with Gualterio and Bianchini. Filling out the ranks at Urbino were several Irish Catholic officers. Their chief, Arthur Dillon, commanded a regiment of Irishmen in the French army; he became James's main agent in Paris. Another was Charles Wogan, whose adventures are soon finally to be told.[26] But first a word about how, with some help from Bianchini, the Jacobites marooned in the ducal palace in Urbino dealt with the boredom of confinement in an isolated Italian hill town.

They discovered walking as an exercise, walking in the hills in good weather, around the castle in bad, and practiced the noble sport of hunting small animals as long as there were any. The place consequently was very good for the king's health, Bianchini wrote to the queen mother, and might continue to be, if properly prepared for the winter.[27] Evening entertainment featured music, even opera, and, for some, astronomy under the stars. Bianchini almost made an astronomer of James. Together they observed lunar eclipses and measured the coordinates of Urbino and otherwise saw much of one another during the late summer and autumn of 1717 and 1718. Bianchini's friends wondered at his attachment to the king, and some resented it. There will be no vacations for you, Baldini complained, "for the King of England will not be able to deprive himself of your learned company."[28]

His Studious Majesty asked Bianchini to install a *meridiana* in the ducal palace, at which courtiers who rose in time could watch the sun's image notify noon.[29] The instrument was not as exact as its builder would have liked, for he had to place it on a railing mounted on a balustrade that did not run perfectly from south to north. A hole in a plank mounted at the top of one of the narrow towers flanking the main entrance to the palace cast the solar image (see Plate 9). Bianchini deduced the latitude of the place from the sun's midday altitude, diligently corrected for refraction and parallax, and, with his royal assistant, found its longitude by eclipses of Jupiter's moons. In both cases they were out by several minutes, which put Urbino even further from Rome than it is.[30] Deducing longitude from eclipses of Jupiter's companions is not easy. But it pales in difficulty with the problem that James and his courtiers encountered in telling time the Italian way. When the pope asked James whether he lacked anything at Urbino, he replied that he had everything he might wish for but a clock he could understand.[31]

Bianchini had a second project at Urbino to divert intelligent courtiers. The celebrated duke of Urbino, Federico da Montefeltro, he of the famous hook-nosed portrait by Piero della Francesca, had embellished the basement façade of his palace with seventy-two marble bas-reliefs. These displayed ancient and Renaissance technologies, mainly war machines and building apparatus designed by fifteenth-century engineers, to publicize Federico's blend of condottiere and humanist. The duke's concept of warfare as a science fit very well with the monsignor's fondness of fortification as applied mathematics. Clement assigned him the task of interpreting the bas-reliefs. Though welcome, the job was not easy, since, in the 250 years between their creation and Bianchini's work, they had deteriorated and, perhaps, had moved, since he found them disquietingly disordered ("I cannot excuse this lack of method"). Using his own method, he repaired the ravages of time in a Latin tract still regarded as judicious, accurate, and indispensable, for it gives details, recorded with the help of an engraver, Gaetano Piccini, that survive only owing to their work.[32] Bianchini did more than describe the marbles. He explained that Montefeltro's architects decorated the basement rather than making a frieze higher up because, as witness the pedestal of the column of Antoninus Pius, the Romans had done it that way. For technical details about the working of the machines, Bianchini referred readers elsewhere, to Montanari's treatise on waterspouts, for example, for the Archimedean screw.[33]

The duke of Mar (so elevated by James into the Jacobite peerage) took an interest in the Urbino marbles and their expositor. He had developed some architectural skill designing improvements for his and his friends' estates in Britain and had visited Bianchini in Rome to see his collections of inscriptions and witness his experiments. He liked his host. "He is exceeding civil and obliging," Mar wrote to his king, who agreed, with a reservation. "Bianchini is a good body and a friend of mine, but I would not say to him what I should not care *the Pope* should know."[34] Nairne shared both views. He observed to Gualterio that Bianchini displayed a "modesty and zeal to match his affection" but also a tendency to meddle and inform; "he is known to like writing letters and involving himself in politics." All true. Bianchini kept abreast of the doings of the great powers and had considerable confidence in his political acumen.[35] He continued to show his Jacobite zeal by attending James whenever asked; his affection, by drawing the king's portrait in pastel; his higher allegiance, by reporting the doings in Urbino regularly to the pope; and his politicking by constantly pushing James's cause.[36] He wrote of their friendship in accounts of the astronomical observations made at the palace and the seventy-two marbles decorating its basement. Both pieces appeared in a volume put together by the Albani to commemorate James's stay in Urbino. It is a pity, "a great testimony . . . to the unhappiness of our times," Annibale Albani wrote in the dedicatory preface to the book, that His Majesty had had to keep court in Urbino instead of London. He might also have said instead of Rome.

In answer to James's insistent requests to leave Urbino, Clement had proposed resettlement at Castel Gondolfo. Bianchini drew up a plan of the building for James's consideration entirely from memory, as he was often there attending the pope and enjoying the hunt.[37] Mar, acting the architect, thought that the place might do if the pope installed thirty-seven new fireplaces and made some other trifling improvements. This was in August 1718. Although Bianchini and Alessandro Albani recommended meeting the requirements halfway, in the end the task and expense of adapting the popes' summer palace to the year-round needs of the coddled exiled court were too great. Clement yielded. He allowed the Jacobites to inhabit a building, which became known as the Palazzo del Re, on the north side of the Piazza dei XII Apostoli, a short walk from Bianchini's quarters in the Quirinal. Bianchini had helped bring about the settlement. "The good man . . . is truly zealous and a sincere well-wisher of the king's and willing to give himself any trouble to serve him."[38] The move, begun in October 1718, was completed in November.[39]

Clement also granted James the right to allow his many Protestant retainers freedom to exercise their religion. "As the Pope himself has said to me, I am not an apostle. I am not obliged to convert my subjects other than by my example, nor to show partiality to the Catholics, which would only serve effectively to hurt them in time following." James could not see the relevance of this insight to his attempts to force himself on King George's subjects, which could only hurt the Catholics among them. Clement saw it and consulted prudence as well as his purse in keeping James's pension at 10,000 scudi. The total of other income reaching the Palazzo del Re from gifts, loans, and "honoraria" arising from James's nominating rights to ecclesiastical positions, including cardinalates, probably did not exceed another 10,000.[40] Still, his poverty did not tether him. Perhaps a wife would. It was time for King James to seek a rich Catholic consort with whom to live royally in Rome and create a family to annoy King George. Clement urged him forward, probably with a girl in mind.[41]

The Queen's Road to Rome

The bride was hidden in plain sight in the remote principality of Ohlau in the eastern reaches of the Empire. There she dwelt with her father, James Louis Sobieski, son of the king of Poland; her mother, Hedwig Elisabeth of Neuberg, aunt of Emperor Charles VI; and two older sisters. James had begun looking around in 1714, but not in Ohlau. He started at the top, with a daughter or niece of Charles VI, before reducing his market value by the failure of the uprising of 1715. He then fished in vain for a daughter of the Palatinate, his cousin of Modena, and a daughter or niece of Peter the Great. Meanwhile the duke of Lorraine and other realistic marriage brokers had suggested a trip to Ohlau, but James thought himself above a family whose claim to royalty was a lost election rather than a forfeited inheritance. James Louis seemed rather to perceive a kinship between their frustrated kingships and offered a large dowry with one of his daughters. By the end of 1718, pushed by Jacobites wanting heirs and alarmed by refusals of more eligible maidens, James dispatched Wogan to inspect a princess of Baden and, if she should prove unsatisfactory, to drop in at Ohlau.[42]

Wogan combined the talents of Dumas's D'Artagnan and Rostand's Cyrano de Bergerac (see Plate 10). Like d'Artagnan, he was an adventurer, a man of many wiles, cool and clever in tight spots, and a leader of three musketeers; like Cyrano, he was a poet as well as a swordsman. The Cyrano type ran in the family. Wogan's great uncle, the earl of Roscommon, was a famous duelist and man of letters, an early champion of Milton's blank verse against traditional rhymers. Wogan could rhyme if he pleased but came to despise "snarling in Doggerell;" he wrote his lengthy commentaries on the seven penitential psalms (which his picaresque way of life gave him many reasons to recite) in blank verse. He was painfully aware of the ways and miseries of the world. Had he not been brought up Catholic, he would have chosen to be a Jew because, he wrote, the "Obscenity, Free Thinking, Prophaneness, and Party Rage" of literature scarcely supported a rational belief that the Messiah had come.[43]

Wogan's poetic talent had a chance to develop together with the greater genius of his boyhood friend Alexander Pope. The connection with Pope permits an informed guess at the year Wogan entered this world, the first of his father's thirty-two children. He wrote that he had lived "in perfect harmony and intimacy [with Pope] for two or three summers" and that he had "the honour to bring [Pope] up to London, from our retreat in the Forest of Windsor [Pope's family home] to dress à la mode." An authoritative biographer dates this refashioning to 1705, when Pope was 17. We may conjecture that his more sophisticated friend was a year or two older.[44] They probably came together through connections among leading Catholic families; the robust literary nephew of Roscommon would have been a good companion and foil for the summer exercise regime that doctors prescribed for the physically weak but powerful poet Pope.

Wogan knew London well. His extended family had a presence there in his uncle, the infamous earl (and Jacobite duke) of Tyrconnell, who, after plotting unsuccessfully to murder Cromwell, found a place with the future James II. James made the earl his agent in Ireland for raising a Catholic army that might prove useful. It did not. Tyrconnell went to France briefly with his exiled king and worked hard for his restoration. Wogan was proud of Uncle Tyrconnell and of a more distant kinsman Colonel Edward Wogan, who had helped save the Stuarts by abetting the escape of Charles II after the battle of Worcester. By inheritance and pluck, Charles Wogan despised people and parliaments too pusillanimous to try to bring back their rightful king. He would do what he could for the cause. Until called, however, he frequented Will's literary coffee house in London.[45]

At this period of his life Wogan was a pleasant companion and, though Catholic and Jacobite, no bigot.

> I love religion, with all my soul, where it is sincere; but abhor, above all things, the pretence or abuse of it, to advance any purpose but those that regard the other world . . . I have studied all the present religions with care, and if my creed did not determine me to be a Catholic, I freely own I should be troubled by none of them.

Rome, he acknowledged, suffered from "extravagances of popery" and most religions from violence, cruelty, and superstition. He recognized that his religious tolerance based on intolerance of religion, and his profession of arms despite his dislike of violence, involved some inconsistencies. "I own I am a little mad."[46] He showed it by helping to organize a rising in the north of England to support the Pretender's attempt on Britain in 1715. By then he was no neophyte in war. He had gained some military experience fighting in Dillon's regiment during the last phase of the Spanish war.[47]

The regiment was an anomaly, Wogan's enrollment in it predictable. Lieutenant-General Count Arthur Dillon's peculiar unit was one of three formed around "wild geese," remnants of Irish armies defeated in the wars to re-establish James II on his throne. Dillon's father had formed the regiment, and, until it died during the French Revolution, it was a family fief officered by Dillon relatives loyal to the House of Stuart. Dillon's men helped support the Jacobite court in France by deductions from their salaries, and Dillon's senior officers, though serving in the French army, owed their appointments to the ousted Stuart kings. Wogan was welcome to the regiment not only as a Jacobite but also as a Dillon, for his poetical uncle Roscommon, otherwise Wentworth Dillon, was one.[48] Armed with Dillon training, Wogan played a prominent part in the critical battle, and Jacobite defeat, at Preston in Lancashire in 1715.

Wogan was captured, imprisoned in London in Newgate, and menaced with a charge of high treason and probable execution. Perhaps anticipating this result, his jailers indulged him with pen, paper, and publication, and he whiled away the time writing dreadful doggerel to generic ladies ("What Dread can Gaols or Gibbets shew | To Men who've died so oft for you?"). But he did not wait to discover whether, as in the case of one ringleader, James's boyhood friend the earl of Derwentwater, God would show His anger at his decapitation by a terrifying display of fiery meteors.[49] Instead he and other leaders rushed their guards on the eve of their trial. Half of the group, including Wogan, succeeded in their escape. His knowledge of the city helped him to elude the crowd eager to seize him and the £500 on his head until he found safe passage to France. He joined the exiled court in Avignon and accompanied his king to Italy.[50] Early in 1718 he went with Ormonde to Russia to try to gain Peter the Great as an ally and to persuade him to give one of his daughters to James. It was, therefore, with some experience as a marriage broker that Wogan undertook his survey of brides for the Pretender. Among his preparations before leaving Rome was to say some prayers at Santa Maria Maggiore, where he ran into Bianchini, who knew him and very likely also his assignment.[51]

Proceeding according to instructions, Wogan examined the candidate from Baden, found her unsatisfactory, and continued to Ohlau. Through an Irishman (and consequently a relative) who was the military governor of the place, Wogan had immediate entrée to the court without revealing his tender purpose. He judged the eldest of James Louis's three daughters,

Casimira, already engaged, to be "ill-humored;" the middle one, Charlotte, frivolous; the youngest, Maria Clementina, at sweet sixteen, just right, apart from being a little short. The broker: "[she has] sense, discretion, evenness of temper and a very becoming modesty," also "light brown hair, very pretty black eyes and genteel little features, with a good shape and behaviour . . . very devout and no manner of airs or variety of humour . . . a good mixture of haughtiness in her composition, but cunning enough to disguise it upon occasion" (see Plate 11). And her father promised a dowry of around a million scudi![52]

Wogan returned to Urbino praising Clementina above her weight and urging James to make her his queen. His Frustrated Majesty still pursued the figment of a Russian princess. When he came to his senses in the summer of 1718, however, he saw there was no reasonable alternative to Clementina. He then made one of his many damaging misjudgments of people. Instead of sending Wogan back for final negotiations, he deputed Murray with the instruction to keep everything secret until the little princess had reached papal territory, so that "it can be in nobody's power to prevent or interrupt" her journey. But boastful Murray could not keep a secret, and those who would interrupt her journey soon knew about it. Murray was also to try to negotiate a loan on top of the dowry, to be repaid after the conquest of Britain; to ascertain that Clementina would have her share of the famous Sobieski jewels, one of which, a ruby, was valued at more than half the promised dowry; and attempt the elder sister, should her engagement fall through, "[if] you think her the more agreeable and more desirable character." After both sides had signed the marriage contract with a stipulation that the goods (Clementina and her dowry) must be delivered in Italy within three months, James sent John Hay to escort the princess to Italy.[53]

There were delays. What clothes should the princess and her mother bring? What jewels? So many visits to make, goodbyes, farewells! By the time the train set off from Ohlau in October, "incognito" with four carriages, thirty-six horses, ten footmen, and other necessary attendants, the new King George had decided to prevent Clementina's marriage to the Pretender.[54] The proposed union alarmed him because of the bride's dynastic ties to the Habsburg emperor, the king of Spain, the electors of Bavaria and the Palatinate, and who knows how many other slight friends of Britannia and Hanover; and her fine dowry would encourage her monomaniacal spouse in his plots against the Hanoverian succession. George reminded Emperor Charles of the presence of English warships in the Mediterranean, where they might easily harass Austria's hard-won possessions in Italy, and asked for the slight favor of detaining Clementina and her mother in imperial territory. Charles protested, as they were his relatives; George insisted, as he did not want them to become James's; the emperor conceded and had his aunt and his cousin held up in a stronghold, the Schloss Ambras, in Innsbruck (see Fig. 29). James was awaiting the arrival of his bride in Bologna when Hay returned ingloriously empty-handed.[55]

Clement protested the insult to his reign and his goddaughter; several electors urged the emperor to relent; Bianchini railed eruditely that not even Alexander the Great or Scipio Africanus would have committed such a crime against the sacred laws of Europe; James urged his fiancée to

Fig. 29. Schloss Ambras in Innsbruck, the castle in which Maria Clementina was imprisoned. Matthäus Merian, c. 1650.

keep her troth and inquired when and how he might seek another if she wavered.[56] Mar, Murray, and Hay had nothing useful to add. Wogan did: he offered to free the princess. Impossible! She was under constant guard, and imperial spies abounded. That did not deter Wogan. He had escaped from prison the hard way, from within, and planned to recruit some of his relatives to help him from without. d'Artagnan therefore betook himself to the winter quarters of Dillon's regiment to find musketeers eager for dangerous demented derring-do in the Jacobite cause. He soon enlisted a major, Richard Graydon, and two captains, Luke O'Toole and John Misset. He also needed a woman to attend the liberated princess and another to take her place. Easily done: Misset's wife Eleanor liked adventure and had a female servant, Jeannot, who could serve as substitute. Eleanor was four months pregnant and Jeannot twice as tall as Clementina; but what is an adventure without difficulties? Two servants would round out the party, which would disguise itself as a count and countess and their retinue. Wogan made his clandestine way to Ohlau to charm James Louis into abetting the kidnapping of his daughter by a band of crazy Irishmen.[57]

Naturally the Prince of Ohlau refused to place his daughter in such a plight. He offered Wogan a ring for his trouble, a remarkable turquoise ring, picked up by Jan Sobieski in the Grand Vizier's tent after some disagreement with the Turks. Wogan refused it: he was no mercenary. It had been a test. The prince thawed to the gentleman–adventurer and together they worked out how to communicate with the captives through their dependable majordomo, a monsieur Chateaudoux. James Louis gave Wogan a letter to Clementina directing her to do as her rescuers told her. "Adieu, my beloved Clementina . . ." Wogan later recorded his amazement that a father could write such a letter to his daughter. "It will be difficult to find in any story, even in the most extravagant novels, that any prince has ever given a stranger so absolute a power over those dearest in the world to him." That was to put it moderately. "[In fact] all this, in the case of a Private Gentleman, appears above the strain of Romance, and even the highest flights of DON QUIXOTE."[58] The prince would pay for permitting it.

On 16 April 1719, in the sixth month of her captivity, the band of rescuers left Strasbourg, where they had assembled by different routes, for
Innsbruck. They stayed in a poor inn in a village a day's journey from the
town, giving out that the "countess" (Mrs Misset) was ill, while they made
connection with Majordomo Chateaudoux. Clementina was to feign sick-
ness, take to her bed, and be ready to hazard all on the night of 27–8 April.
The rescuers moved to The Lamb, a hostel across the river Inn from the
Schloss. By unhappy coincidence, on that very day, the 27th, the Prince
of Baden left Innsbruck for Italy by the route Wogan planned to follow.
The prince had come to woo Clementina, now, except for her incarcera-
tion, a most desirable marriage partner, owing to George I's offer to top
up her dowry (by some £100,000) if she jilted James. The disappointed
prince's extensive retinue would require the engagement of all the post
horses between Innsbruck and Verona.[59]

The weather on escape night was perfect for melodrama, the worst
in spring for years, icy cold and sleeting. Chateaudoux chatted up the
guards: why soak and freeze? There would be no visitors on such a night.
The guards went off to warm themselves, Jeannot slipped into the castle
and Clementina slipped out, wearing two shifts, a furred waistcoat, a pet-
ticoat, and an apron filled with jewels, all under Jeannot's overlong wet
cloak. It was 1:00 a.m. Wogan waited in a dark alley. The princess found
him, and the pair reached The Lamb at 1:30, the princess wet through and
thin and tired from Lenten fasting. The fake countess dried her off. By
2:00 a.m. the party had reached the well-provisioned carriage in which
the women would travel and the men take turns sleeping. As they were
about to embark, Clementina noticed that a packet of jewels had been
left at the inn. O'Toole rushed back to retrieve it. Not wishing to awaken
anyone who might give the alarm and being a muscular musketeer, he
lifted the front door off its hinges, hunted in the dark for the precious
bundle, found it, and replaced the door. His return to the carriage raised
all spirits, even the horses', for by daybreak they were picnicking on the
Brenner Pass, 40 km from Innsbruck.[60] A curious variant of the escape
has Clementina and Wogan gliding by the abattoirs of Innsbruck to avoid

the watch, which took no interest in them after determining that they were not students, "its enemies from the university."[61]

From the Brenner, the party descended into good weather and the valley of the Adige, threatened in their precipitate course only by sheer drops into the river far below. They traveled all day and night on the 28th, sleeping as they could, and at dawn found themselves within three stages of Trento. They had come 150 km. They then suffered the consequences of steaming in the wake of the prince of Baden. Not a single fresh post horse could be had. At Trento they begged and haggled for two carthorses and the two best horses from the last stage, and made do with the four, although the carriage ought to have had six. Perforce continuing more slowly, the musketeers entertained their princess with battle stories and English lessons. And so they crept along the precipices on the road to Revereto, the last garrison town of the empire before Venetian territory. But at Revereto, where they arrived on the afternoon of the 29th, not even a fresh carthorse was available. Crawling on then toward the border with the beasts they had, they were suddenly arrested by gravity.

An axle had broken. Jerry-rigged with a spar nailed to the broken axle, the carriage was on the road again at 10 p.m. It hit a stone about 15 km on the wrong side of the border, fell apart altogether, and threw the princess into a stream. The musketeers took out the horses and staggered to the border town of Ala, where Wogan found a decrepit cart that accommodated two. It was 2:00 a.m. on the 30th, exactly forty-eight hours after the recovery of the jewels and departure from Innsbruck. The horses, now attached to a vehicle they understood, pulled the ladies to the border, while Wogan and Gaydon made their way on foot. Misset and O'Toole remained a stage behind. Around sunrise the weary four were relatively safe in Venetian territory. Though desperately needing rest, they drove to church, it being Sunday, and encountered the bunch from Baden on the same pious errand. Their prince seemed to recognize something in one of the ladies in the ridiculous cart guarded by two dirty unshaven lackies but could not imagine that she was the object of his late affections. After the service Wogan allowed his band a little rest, in one of the poorest inns in town or (in another account) in the modest house of the postmaster, as the Germans had taken all the good hostelries. After a little sleep they braved the last 35 km to Verona, where they spent the night.[62]

And where were the pursuers? The authorities discovered soon enough that the party in the princess's bed was not the princess. Jeannot spent some unpleasant time hiding in a privy before managing to pose as a servant in the entourage of Clementina's aunt, the duchess of Parma, and disappear into Italy. Others were not so lucky. Chateaudoux was imprisoned and released, and soon died. Clementina's mother was harassed as an accomplice; she returned to Ohlau, uselessly, as the emperor, to show his loyalty to Britain, had ejected James Louis from his little principality. In fact, Emperor Charles had done little to catch Clementina. He may have sent only one messenger to alert officials in Trento and Revereto to the escape, for O'Toole and Misset, who had remained behind for the purpose, intercepted only one. The man did not know enough not to drink with Irishmen. They laced his wine and put him to bed so drunk that he could not mount his horse before Clementina was in Italy.[63]

The great journey, some 450 km, ended in Bologna, in papal territory, on the morning of 2 May 1719. At last the princess could break her incognito. A week later Murray and his sister, Lady Inverness, arrived to greet their queen-elect; the next day, the 10th, she was married to Murray standing in for James, who lingered in Spain on his usual business. Philip V had not reconciled himself to the loss of Spanish territory mandated by the treaties of Utrecht and Rastatt and to amend things declared war on Britain. James offered Jacobite might under Ormonde and Mar, and himself and Hay on their own. Mar contrived to get himself arrested, twice, perhaps intentionally, in trying to reach Genoa.[64] The Spanish invasion fleet was wrecked by a storm, as usual; the unsupported Jacobite force in Scotland, again as usual, had to surrender.[65] James would not be a happy man when he wed his wife on his own behalf in September 1719. As a dashing hero he did not compare with Wogan, for whom Clementina had developed some regard. Murray saw competition and advised James to order Wogan to join the mobilization in Spain. Clementina insisted that he stay; he was the only strong person near her whom she knew and trusted. He remained in Rome and so fell into disfavor with the pseudo-king.[66]

On the day of her proxy marriage, Clementina left Bologna for Rome. She arrived on the 15th to an enthusiastic welcome from a crowd of cardinals, ambassadors, and Roman nobility. Carriages lined the way to the Ursuline convent in via Vittoria, where she was to lodge. The following day the pope appeared in the form of a hundred baskets of sweets, flowers, and ribbons, conveyed by his unofficial ambassador to the Jacobites, Monsignor Bianchini. An English paper reported that fifty-two men were needed to carry the stuff.[67] Clementina became famous overnight. Her virtues and shortcomings, if she had any, were magnified. To the Jacobites, she was the cynosure of "female fortitude," "happy in all the Charms, both of Mind and Body, that her Sex can boast of;" to the Hanoverians, a nobody, a mock-queen for a mock-king, and maybe not so virtuous as claimed. Could there be anything in the rumor of her attachment to Wogan, who referred to her in a letter to James as "our admirable, I could almost say adorable, Queen"? Or in the insinuation of a romance with the pope, who had had a medal coined showing the princess as a goddess? The newsmonger who circulated this last scurrilous snippet literally lost his head over it. The subject of these romances had just turned seventeen. A birthday feast at the Ursuline convent celebrated the event.[68]

Bianchini was there. He had written a cantata for the occasion that adapted the extravagant speech at Christina's academy in which the future Pope Clement XI had lauded Clementina's future deceased father-in-law, James II. The cantata revealed that Clementina's innocence and purity had attracted the attention of the virgin goddess of justice Astraea, who was then watching the earth from or as the constellation Virgo. Astraea had deigned to live with mortals in the golden age, but by the iron age she had become so disgusted with humankind that she fled to the heavens to await a fit occasion for bringing justice to earth. Clementina's purity and the merits of James's cause prompted her to see to his rightful recovery of the thrones of his ancestors. The association of Astraea, justice, and a new golden age had the merit of timeliness. While Bianchini composed his cantata, the English painter James Thornhill planned his great painting commemorating the arrival of George I in England. Among the heavenly welcoming committee was Astraea with her scales. In Thornhill's painting she promised the same new golden age as she did in Bianchini's cantata, but under different auspices, a new golden age to match that of the virgin Queen Elizabeth, whom her admirers called Astraea.[69]

Murray perceived an iron age for himself if the charming Clementina became more to James than a source of young pretenders. He tried unvaliantly to prevent her from acquiring a following in the court of which she was to be queen. The only Jacobites he allowed access to her were his sister and the duchess of Mar, and, reluctantly, Wogan and Mrs Misset. He did not know Clementina; the doughty queenlet wanted to know her court and demanded to meet it. Murray gave way to the god-daughter of the pope, and the courtiers made their way to her convent.[70] As these intrigues progressed, Wogan and his musketeers were made citizens of Rome. An impressive ceremony conferred this high honor. The centerpiece was a speech presenting the men and their triumph. Who made it? "Monsignor Bianchini, the pope's domestic prelate, well known in the republic of letters, member of the academies of science of Paris and London." A Te Deum, which Bianchini organized with his fellow canons in the basilica of Santa Maria Maggiore, followed. The greatness of the honor conquered Wogan's composure. "I have been led to the CAPITOL with the applause of thousands and received in the Roman Senate . . . I have been dignifyed in all form, with an hereditary title . . . I am the first of my Country-men that has made a Triumphal Entry into Rome since . . ." Since when? "Since the Emperor Constantine."[71]

Clementina asked Bianchini to thank the Roman people on her behalf for the honor it did her cavaliers. Our monsignor, smitten with the princess from the moment he had brought her Clement's greetings, had lost no time in enrolling in her service. He was soon studying English with her. "I offer to[o] rightly the humble process of my very devoted service and worship," he wrote in approximating the language, "too" referring to a French–English dictionary he was sending "Your Majesty Britannique."[72]

The qualities he saw in Clementina he itemized to her father in a letter describing her first visit to her godfather. In the presence of Clement, the Sacred College, Roman nobles, and ladies and cavaliers of her court, she displayed all the Christian and royal virtues possessed by her grandmother Casimira, but with a "soul and discretion that is so refreshing in her and so rare in people born to rule." The presence of this "treasure of innocent grace" in Rome was little short of miraculous. If the imprudent earl of Mar had been in charge of James's affairs when Clementina was arrested, Bianchini opined, she would still be in prison; but Divine Providence and Wogan had saved the day, and everyone now looked forward to the wedding. Let us hope that, unlike bachelor James, the royal pair would always have servants favored by God. Thus did Clementina's very devoted servant assure her father of her safety and the divine sanction of his sacrifice.[73]

Bianchini did not overlook the cultivation of his fellow servants Wogan and Misset. With the hero he was on joking terms. You think you have had adventures? Let me tell you what happened to me on my way to Albano. I had forgotten my health certificate and so could not enter the city and had to stay with friends in an insalubrious part of the country. A Jesuit, a father Kircher, was staying there, and offered to take me in his carriage to more wholesome air. His bizarre carriage (here Bianchini enclosed a picture) had four large balls in place of wheels. Another Jesuit, a Father Lana, and Fontenelle, the secretary of the Paris Academy of Sciences, who happened to be in the area, completed our traveling party. Kircher lightened the load by pumping the air out of the balls; the carriage soon rose into a purer atmosphere, of which Kircher bottled the best, for our comfort on the way to the moon. Wogan may have recognized the contraption as the vehicle that Athanasius Kircher's protégé Francesco Lana-Terzi had designed for lunar travel.[74]

Bianchini's messages to Mrs Misset dwelt on the sweetness of suffering for the Catholic religion. Despite the persecution of the English, he wrote, the new royal couple could be the happiest of princes; for James, though already sorely tried, was "the most tranquil soul in the world," and Clementina had the makings of a perfect consort. To strengthen her, Bianchini sent through Misset letters he had received from James's mother expressing her religious sentiments and containing some elevating words from James's deceased father, who certainly had suffered (though not always sweetly) for his faith. Besides promoting Clementina's piety and resolve, the letters would disclose to her mother-in-law Bianchini's central role in maintaining a "perfect correspondence" between her betrothed and the pope.[75]

Crossed Paths

It suited His Majesty to marry on 17 September 1719 in Montefiascone, the seat of a bishop friendly to James, about 100 km from Rome. Clementina traveled there together with her security team, Wogan and Mrs Misset, and with Lady Inverness (Marjorie Hay) and several other courtiers. These few, together with Murray, Lord Inverness, and some priests, made up the company for the almost clandestine ceremony.[76] Unusually, a documentary painting was made, in which the most conspicuous actors beside the presiding priest were the Murray siblings, James on the right, superciliously pointing, and his sister, Marjorie Hay, on the left, standing ominously erect behind the queen. The kneeling male trio behind her may be Hay, Wogan, and another Irish officer, O'Brien; a preliminary draft for the painting shows the soldiers standing holding halberds (see Plate 12).[77] Wogan signed the register as a witness and received as a reward for all his services the insignificant honor of a knight-baronetcy. Bianchini oversaw the design and coining of a medal to mark the occasion.[78] The public celebration of the marriage took place nine months later, on midsummer day in 1720. Bianchini wrote a cantata for the festivities that sang of "the perfect harmony of these two heroic souls."[79] The harmony was not to last much longer than the singing.

The cantata's cast comprised the sun and the moon, wandering about the Alban Hills dressed as fake shepherds, and two real rustics, Ascanio and Silvio. Since they have the names of the sons of Aeneas, the Trojan founder of Rome, we are to understand them as representations of Italy. It is almost sunrise. The sun, apparently a Jacobite, observes to the moon that the happy hour was approaching when "I in heaven will color the brightest noon with rays of gold . . . and awaken England to greater hope." Ascanio and Silvio overhear the chatty luminaries and deduce from their royal countenances that they are not local bumpkins. Ascanio proposes a song in their honor; Silvio advises moderate panegyrics. They delve into the husbandry they know for apt metaphors. Ascanio, to the moon: "Ape industre, che soavi | formi i favi, ecco il tuo Rè." Silvio, to the sun: "Ape attenta, industriosa | vuoi la Rosa? Ecco il tuo fior."[80] Ascanio: Why are we singing about bees and flowers when we are supposed to be praising heavenly virtues? Silvio: As the luminaries come to share our humble ways, they will know how to interpret our prattle. Bianchini's modest work may have contributed to a short opera still heard, *Ascanio in Alba*, also composed for a wedding between an almost royal pair, with an Italian libretto and music by Mozart.

Bianchini's cantata ends in a quartet in which the sun sings the intriguing lines: "The sun's ray writes in the heavens and the earth the origin, and the years, and the day, of the royal flowers | and to the celestial genius of the pastors | every star, and every heart, pays homage." The verse suggests that James and Clementina would be written into the floor of Santa Maria degli Angeli. Where else than at Bianchini's historical–liturgical pavement does the sun's ray write royal history on the earth, and every star pay homage to such pastors as Clement? If Bianchini had wanted to commemorate the celebration of 1720 with appropriate medallions, he would have placed them at the sign of Cancer, close to the jubilee ellipses, since the celebration took place at the summer solstice. And behold, two royal medallions do lie there (see Plate 13). Although they have no inscriptions, only the most unfeeling skeptic could doubt that they represent the union of the Stuart and Sobieski lines. Now, in June 1719, Bianchini wrote to Sobieski that he had inscribed the names of the British royal couple in bronze and marble in the basilica, "where many years before Her Majesty the Queen of Poland, your mother, deigned to be present at the placement of the bronze that marks the 12th of September and the celebrated name and victories of her husband His Majesty King John III. Her Majesty the Queen of England deigned to come the day before yesterday to examine both sets." Clementina's seventeenth birthday, on 17 July, was rapidly approaching. Bianchini would have a medal ready for that too.[81]

Another explanation, proffered recently as a discovery, interprets the medallions unromantically as "obliquity meters": the sun's image around noon at the summer solstice no longer falls on them as it did three hundred years ago, and the displacement accurately measures the change, as theory predicts.[82] But this, if intended, is an added virtue. Still, there is a mild mystery. The inscriptions Bianchini described to Sobieski were in place in June 1719. Nothing singular happened to the couple that month; they had not even met; but the summer solstice, when the sun blazes highest in the sky, was a reasonable place to indicate the high hopes Rome placed in the union of Sobieska and Stuart. Perhaps, with that symbolism in mind, the date for the wedding had already been set for the following June. Meanwhile, Bianchini changed his mind and replaced the inscriptions with blank brasses; medallions without names belong to sovereigns without kingdoms. A notable symbol from an expert in symbolism! If it puzzled James, it was not the only enigma with which Bianchini teased him at the time. Soon after the belated ceremony, he applied two famous riddles from Virgil to the royal couple. They ask for identification of lands where "heaven's vault is but three ells wide" and where "flowers [grow] inscribed with royal names." Many elucidations of Virgil's riddles have been proposed over the centuries.[83]

Addressing the industrious bee Clementina, Ascanio and Silvio sing about "offspring" of the sun and the moon. That was a good guess. At the time of the celebration of 1720, the queen was three months pregnant. The Young Pretender, Bonnie Prince Charlie, duly emerged on New Year's Eve. A medallion at Santa Maria degli Angeli celebrating the arrival would have been provocative. Instead, Bianchini played with something a little less dangerous. King James's birthday was a major annual event on the Jacobite calendar. To make it conspicuous in the basilica, Bianchini replaced the dull copper medallion he put down to mark James's twenty-ninth birthday in 1717 with a white marble plaque. That the marble lies where the medallion did appears from its placement 1.7m in front of, and about halfway down, the south wall of the transept, thus a perpendicular distance of 9 m from the line.[84] An easy calculation shows that close to the summer solstice the solar image comes to the marble at 10:30 a.m. local time, which corresponds to 9:45 a.m. London time. According to London clocks, James was born between 9:30 and 10:00 a.m.[85] Since the marble lies on the diurnal path of Arcturus, which the English associated with Arthur, it might also be read as a reference to the legendary king who tried to drive the Saxons (Hanoverians!) out of Britain. Monsignor Bianchini the symbolist favored polyvalent types.[86]

The inscription on the marble reads "IaCobVs III, D.g. reX Magnae brIttanIae, etC."—that is, "James III, by God's grace King of Great Britain, etc." (see Plate 14). The raised letters sum to 1721 if read as roman numerals.[87] Does the marble also commemorate the baptism of James's son, performed by the moribund pope Clement in January 1721 according to the Gregorian calendar?[88] Is it significant that James accepted the Gregorian calendar while Britain remained steadfastly Julian, and so, at the turn of the year, he ordained that 31 December 1720 should be followed by 11 January 1721 in all the kingdoms he did not control? If the 1721 plaque does refer to James's newborn son, the count of characters memorialized at the *meridiana* would include three Sobieski and three Stuarts, counting Clementina as both.

Clement died on 19 March 1721. His successor, Innocent XIII, sixty-five years of age, the Protector of Portugal and former nuncio to Lisbon, achieved his tiara with the critical support of Annibale Albani and the trifling pledge to make Annibale's younger brother, Bianchini's protégé Alessandro, a cardinal. The new pope, who was to last three years, followed Clement closely in governing Rome and losing influence abroad. He retained Bianchini as domestic prelate. The Jacobite cause gained momentum. Innocent urged France and Spain to back James and

Clementina and advertised his support by placing them conspicuously at his inauguration. What was no less useful, papal and French sources briefly increased James's income to around 48,000 scudi, which enabled him to maintain a respectable staff, including twenty-six stablehands. He supplemented this income by sending Clementina's old servants back to Ohlau and denying her a household of her own. He would pay dearly for this economy.[89]

Meanwhile James shared his joy at his heir's birth with the heads of European states. Not all were happy to receive his tidings. Notably the independent Republic of Lucca feared that a letter of congratulations from them might irritate George I and ruin the English market for their olive oil. The city fathers directed their agent in Rome to ask Gualterio to accompany him to the royal presence, where he could express Lucca's delight in the increase of Stuarts. Gualterio declined; he could not show his face because loss of a French subsidy had forced him to make embarrassing economies. "I have, however [the agent wrote], discharged my task in another way having Monsignor Bianchini . . . a great confidant and almost a member of His Majesty's family, accompany me." To Bianchini he explained that the usual formalities would be "of very little if any advantage to [James's] royal house and family, but of grave danger to our Republic." The interview took place on 20 March 1721. James remarked that the man from Lucca had not left a written message, but graciously overlooked it.[90]

A year later the Stuarts presented Lucca with a graver problem. On 22 July 1722 Clementina arrived there under an assumed name and left the next day for the baths, the Bagni di Lucca, above the city. The city fathers, knowing who she was, worried over the level of hospitality they should show her. James then announced that he would join his wife in the baths, which caused further flutter. Should he be given the hospitality owed a king, which would irritate George I, or be treated as a private citizen? He aggravated the question by publicly curing the sick by his royal touch, an old therapy practiced by the Stuarts but then recently abolished by the Hanoverians. In this way he managed to irritate the Vatican as well as the Republic of Lucca.[91]

James had not come to bathe for his health or anyone else's. He was on his way to try again to impose himself on Britain. When the planned invasion collapsed before he could leave Lucca, he decided to turn the cheek and issued, "from our royal court at Lucca," a ludicrous manifesto offering to leave the Hanoverians alone if they left Britain to him. "We declare that provided the Elector of Hanover will deliver quietly to us the possession of our own Kingdoms, we will make no inquisition for any thing that is past." Peaceful transfer would benefit everyone: Britons, who would enjoy a regime devoid of faction, division, discord, or prejudice; the elector, who at last could live in honor and prosperity free from "the Crime and Reproach of Tyranny and Usurpation;" and, perhaps most grateful, those who would not be killed or maimed in further invasions. Perhaps the idea came from the duke of Mar, who had proposed that George abdicate in return for British support for enlarging his domains in Germany. George declined the generous terms and harassed the poor fathers of Lucca for allowing the printing of the manifesto on their territory and terrified them by hinting that England might change its purveyor of olive oil. George soon forgave the frightened republic.[92] James was not so generous. While the Stuart "court" loitered in Lucca, he picked a fight with our pacific monsignor that ended their friendship and Bianchini's visits to the Palazzo del Re.

While James was in Lucca, Bianchini followed his affairs through an intermediary between the royal pair and the city fathers. This was a native Lucchese, Eufrosina di Sardi, who had served as secretary to Maria Casimira in Poland and Italy. Bianchini replied enthusiastically to her account of James's inspiring piety toward the Holy Sacrament and the example he was setting, "truly worthy of the royal houses of his birth and his marriage." He then reported on the healthy state of the royal infant, "this treasure so near to God," whom he visited frequently. But it was the health of the infant's mother that most concerned him:

> You say, Madam, that Her Majesty the Queen seems to you to be thinner since she left Germany. She has suffered a lot and if she did not possess courage like that of her grandfather King John, I doubt she would be alive

today. I pray that God finds a way to console her in the future in proportion to the heroic resignation she has maintained by her perfect obedience.[93]

James interpreted this letter as further evidence not of Bianchini's service to his family, but of his siding with Clementina in her growing dissatisfaction with her treatment by James's favorites and by James himself, which had come down to a struggle over responsibility for the education of infant Charles. Bianchini tried to brighten her life, continued to study English with her, and brought her news and books. James resented this meddling and, in pride and pique, wrote an order to "forbid [Bianchini] to come ever again to my household in Rome." He also ordered the public disgrace of Mrs Misset and the withdrawal of her pension for her part in the imagined conspiracy to alienate the affections of the queen.[94] No doubt quarrels among the leading court Jacobites, in which Mar, who cultivated Clementina, opposed his great enemies Murray and Hays, played a part in Bianchini's dismissal. The worldly monsignor knew how to handle intrigues of the court of Rome, but not how to survive backbiting among Jacobites.

Clementina continued obediently in her place until she gave birth in 1725 to a second son, Henry Benedict Stuart. Her dynastic duties done and her efforts to remove Murray and Hay unsuccessful, she determined to struggle no more.[95] In that Jubilee year she retreated to a convent in Trastevere and turned for consolation to acts of devotion that destroyed her health. Her physical decline can be followed in contemporary paintings; by 1727, when she returned to her husband, she had lost the beauty and spirit that had attracted all Rome eight years earlier.[96] Return did not mean reconciliation. They went their separate ways, he continuing to play a worldly king, she winning the reputation of a saint.[97] Unlike her grandmother Casimira, Clementina did not know how to combine extravagant demonstrations of piety with the enjoyment of ordinary extravagance. Had James known how to develop his teenage queen, he would have had an engaging and energetic partner; as it was, he again failed within reach of success.

The year before Bianchini fell out with James, an antiquarian he had helped to promote began a long career spying on the Jacobite court. He was a German, Baron Philipp von Stosch; his employer, the government of George I; his character, shady. He had come to Rome in 1715 with a letter of recommendation to the Albani from Montfaucon. He quickly established a close friendship with young Alessandro Albani, whose senior he was by only a year; Clement thought him so good a scholar and so fit a companion for the papal nephew as to deserve a pension. Very probably Bianchini also received notice of Stosch's antiquarian promise and advised about the pension, which terminated with the pope's death in 1721. That left Stosch with a hole in his finances, which otherwise consisted mainly of a pension from the great enemy of the Sobieski, Augustus the Strong, king of Poland and elector of Saxony, whom he served as antiquarian *in absentia*. Stosch was thus a willing spy with qualities that

Fig. 30. Antiquaries in action around Baron Stosch (caricatured as number 2); others previously introduced in the text are Ghezzi (3), Valesio (5), Fontanini (6), Marsigli (7), Bianchini (8). Per Leone Ghezzi, "Congresso di migliori antiquari di Roma," 1728. Vatican Library, Ottoboni latini 3116.

interested the English: access, as antiquarian, to prelates and princes, and, as homosexual, atheist, blasphemer, pornographer, and thief, a reputation for risk.[98]

These were not character traits that Bianchini admired. But Stosch was an inspired collector, a consummate expert on antique engraved gems, and relatively honest when dealing with antiquarians he respected. Bianchini allowed himself to believe in Stosch's honesty far enough to lend him at least one book, which is far indeed for a bibliophile; the book described images of famous men inscribed on old stones and coins. Stosch's masterly work on engraved gems came out in 1724, the year he too came out, exposed as a spy.[99] Bianchini still did not drop him (see Fig. 30). In collecting mode the monsignor would deal with the devil.

7

DIGGING INTO HISTORY

After his labor with the column of Antoninus Pius, Aventine explorations, and collecting for the aborted Ecclesiastical Museum, Bianchini had many occasional connections with excavations and their products before he became fully engaged with shovel and inkpot in the 1720s. He oversaw several significant archaeological works, of which the most important took place in Rome before 1708, in Anzio in 1711–12, and at Urbino and elsewhere between 1715 and 1718.[1] We are already familiar with his preservation of the bas-reliefs in Urbino. His analysis of an item from the excavations in Anzio is another example of his imaginative treatment of damaged relics.

The Anzio excavations, undertaken as groundwork for a villa for Alessandro Albani, yielded a prize cracked ancient marble in 1712. The broken stone bears year-by-year lists of Anzio officials, including five or six slaves or freed men from the villas that the emperors maintained there. These footmen and gardeners acted as semi-magistrates primarily for overseeing public entertainments, which Bianchini supposed performed in the space where the stone was found. Later archaeologists accused him of relying more on imagination than on ruins in interpreting the space as a theater; but his reconstruction (in the style of *Istoria universale*) had the merit of incorporating not only the informative stone, but also the information on it and the life of the imperial household at Anzio deducible from it.[2] Still more could be squeezed from the marvelous marble. It presents a calendar for half a year (the remainder being broken off) from AD 36 to AD 51 with dates of some important events that

Bianchini had not been able to fix by other means. He did not get around to publishing these tantalizing results until after his parting from James III gave him time to return to higher things. "Now that Your Excellency is free from attendance on the King of England, you will be able to observe whether the pole is still in its place."[3] And what digging in the mud of Rome was revealing.

The Center of Empire

In April 1720, Francesco Farnese, duke of Parma and Piacenza, obtained permission to excavate his gardens on the Palatine hill. The pope allowed the duke to export whatever he found except for large objects, which by remaining in place would enhance the city of Rome, and hordes of coins and jewels valued at 10,000 scudi or more, which he would have to share with the Holy See. The duke's agents did not find much of high commercial value.[4] Statues, inscriptions, murals, and mosaics, pediments, capitals, and columns, went to enlarge the cultural heritage of Parma or the collections of the duke's friends, patrons, clients, and customers. Sculptors bought pieces of marble for reworking and, among many other works, upgraded a chapel in Ottoboni's church of San Lorenzo in Damaso, where he had decided to be buried. And, of course, contractors bought fragments of stone and loads of bricks from the palaces of the Caesars to use as building material. At the conclusion of the spoliation, Farnese's men reburied most of the site for reuse as gardens and vineyards.[5] The choice items from the dig soon left Parma for Naples via his heir Antonio, a son of Philip V of Spain. When in 1734 the Spanish replaced the Austrians in Naples, Antonio became its king and settled there with Farnese antiquities, which included the antique sphere from which Bianchini had dated the voyage of the argonauts.[6]

Bianchini could do little to arrest spoliation of the Palatine. He was not in charge and, though still President of Antiquities, lacked the power of enforcement. In any case, he recognized that the only way to pay for expensive excavations was to couple the interests of collectors and anti-quarians with those of philanthropists and responsible proprietors. He regarded Farnese as responsible and helped to select and repair objects dispatched to Parma or the ducal villa in Colorno.[7] On visits to Parma during the works on the Palatine he advised about installation and left his mark by inserting a *meridiana* into the ducal palace. Later he made the Farnese a present of some of his archaeological drawings. He became a family friend, not only because he always did with nobility, but also because the duchess was Clementina's aunt, the very same whose visit to Innsbruck had made possible Jeannot's escape from her privy. Bianchini kept the duchess informed about Clementina's self-sequester and papal efforts to entice her from her monastery.[8]

Farnese's diggers uncovered many wonderful things that could not be moved easily and that interested Bianchini far more than the exportable treasures. They made possible his reconstruction of the imperial build-ings on the Palatine built up and burnt down from the time of Augustus to their resurrection under Domitian (reigned AD 81–96). The overlap of palaces, presence of later structures, layered rooms on the hilly site, and the ruinous state of most of it guaranteed that Bianchini's reconstruc-tion would be unsafe. And how could he avoid being misled by unrep-resentative fragments, architects' caprices, and items displaced from their original positions? To minimize these dangers, Bianchini thought archi-tectonically, as he had in a lesser way at Anzio; the builders of the palaces, the great architects of the Vitruvian school, must have worked to a grand plan, "the idea of it all," which he thought he could discover.[9] The fire at which Nero fiddled, and later conflagrations under his successors Ves-pasian and Titus, reduced the task to finding clues in the ruins to the sup-posedly homogeneous design of Domitian's palace. It was completed by AD 96, the year in which the emperor was murdered, for many good rea-sons, on the day and hour foretold by the astrologer he had put to death for telling it.[10]

Monsignor Bianchini acknowledged that he would fail in detail. "I can-not presume to have grasped fully and followed out the idea of the famous architects, and the perfections of this edifice and its every part, and fear that I might have diminished them in trying to represent them."[11] He persisted, however, convinced that a knowledge of the layout and working of the palaces was essential for a correct view of the early Roman Empire. He relied on later generations to correct his errors, which they were only too happy to do. Those who ridiculed him for reconstructing Domitian's domain as a baroque palace in the style of Bernini missed the important point underscored by a long-time student of Roman antiquities and sometime secretary to James III, Andrew Lumisden. "Although this work, which is full of ingenious remarks [Lumisden wrote of Bianchini's *Palazzo de' cesari*], does not give us, perhaps, the real plan of the imperial palace, yet it gives us such an one as is not unworthy of the Roman emperors, and may afford many useful and curious hints to architects."[12] That is the point: Bianchini's reconstructions brought the fair-minded historian insights into the ideas of ancient imperial architects and the lives of the individuals their buildings housed.

In disciplining his imagination, Bianchini relied on a principle similar to the proposition of equal progress in equal times that informed his *Istoria universale*: the fundamentals of good architecture perdure. Domitian's architects, being good, must have designed much as did the best architects of Bianchini's time; in their shared art, Vitruvius and Francesco Nicoletti, the prize-winning young architect who designed an astonishing depiction of Domitian's establishment for Bianchini's *Palazzo de' cesari*, were contemporaries.[13] On this principle, Rome could boast of being not only eternal but also, in its best buildings, unchanging in style and purpose. The fundamental principle of this eternal design was symmetry. From the beginning of his close involvement in the Palatine operations, Bianchini sought the clues from which to infer the expected symmetry and recover the ancient plan.

He began building his palace using the best map he could find of the Palatine hill. Its author, an Augustinian from Verona, Onofrio Panvinio, drew up the map for a book on the games the Romans staged in the Circus Maximus, which runs along the southwest root of the Palatine. Panvinio's

primary concern was to indicate the preferential seating for viewing chariot races afforded by the Theatrum Tauri (see A in Fig. 31). This imperial
press box was a precious guide for Bianchini, as was also the adjacent stadium or hippodrome, FGHK, and the exedra (a semicircular outside apse
for relaxation) halfway down its east side, both conspicuous features of
Domitian's palace.[14] The structure above the line FK centered on the stadium axis is a fancy based on baths built after Domitian and on bits of
the Claudian aqueduct (indicated by the line of square dots to the right
of the stadium) extended to serve the northeast quarter of the hill. Let
us call this miscellany "Panvinio's Fancy". Note also the area between the
theater and the libraries N, O, including the Domus Augustiana, which
were refashioned into the central courts of Domitian's precincts.

In 1721 excavators exposed an elliptical pool to the southwest of the
Domus Augustiana. It turned out to belong to a substantial nymphaeum
(a recreational space containing a fountain) that a prepared imagination

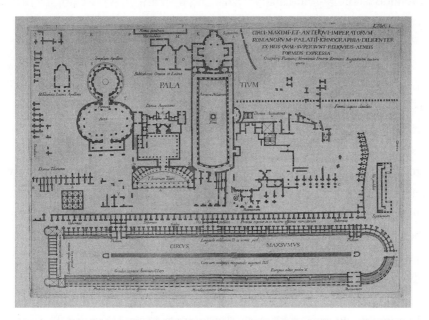

Fig. 31. The best map of the ruins on the Palatine available when Bianchini
began his study of them. North is at the top. Onofrio Panvinio, *De ludis
circensibus* (1600), as reprinted in FB, *Palazzo de' cesari* (1738), tab. I (at end).

could assimilate into an exact geometrical and functional counterpart to the exedra attached to the stadium to the east. Bianchini determined that a line parallel to the axis of the stadium passing through the center of the Theatrum Tauri precisely bisected a line drawn between the centers of the exedra and the nymphaeum. Baroque symmetry then required a stadium where Panvinio had put Apollo's temple, and a "Major Complex" northeast of this hypothetical stadium to reflect Panvinio's Fancy northeast of Domitian's stadium. In 1724, workmen finished uncovering the parallel complex Bianchini had expected.[15]

This complex had three large rooms. The middle one measured 38 m along the axis of the hypothetical hippodrome matching the stadium, and 30 m in breadth, wall to wall, making it bigger than the nave of St Peter's, "in symmetry and magnificence without an equal." Bianchini made it even more imposing by including the thickness of the walls, which brought it up to 150 roman feet or 45 m, which agrees with measurements recorded on modern large-scale maps.[16] This grand chamber, whose domed roof, supported on arches, may have reached over 30 m above the floor, was cleared of mud early in 1726. The sixteen arches, fashioned from diverse rich marbles, separated as many niches, each able to accommodate a colossal statue. Two of these found outside the chamber, a Hercules and a Bacchus, forthwith went off to the Farnese museum in Parma. On either side of the grand chamber was a smaller and lower room whose widths when added to that of the chamber made the northeast façade of the complex 75 m, external measure. With good will, the external dimensions of Panvinio's Fancy agreed with those of the complex. That was enough to prompt Bianchini to draft a symmetrical palace on the resemblances between a nymphaeum and an exedra, and between two sets of approximately matched ruins. These parts are marked "→" and "I" and "Major Complex" and "Panvinio's Fancy" in Fig. 32.[17]

Since Bianchini's architecture required super symmetry, he had to fill in only one quarter of his design to know the rest. He chose the southeast quadrant where rooms and courtyards were relatively plentiful, but

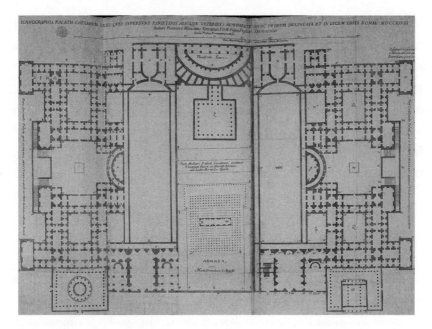

Fig. 32. Bianchini's design of Domitian's palace. North (forum side) is at the bottom. FB, *Palazzo de' cesari* (1738), tab. VIII.

fatally misleading, since most of them date from a century or more after Domitian's death. Reasoning from the exposed ruins, ancient texts, and his architectural instincts, Bianchini imagined the arrangement of rooms and courts depicted in the lower right-hand corner of his plan. He left the space between the stadia mainly empty, on the authority of gardeners who told him that they seldom ran across submerged walls there. Eureka! Domitian's famous gardens were just what Banchini needed to fill up the space between his imagined stadium and Domitian's real one. Folding his plan along the axis of Domitian's stadium, Bianchini had the design of the southwest quadrant. He filled out the northern quadrants by reproducing in them the same arrangements of rooms and courts he had set up in the southeast. They provided abundant accommodation for nobles and servants and stairs and corridors enough to ensure that they did not mix.[18]

While Bianchini worked at his design, more fragments came to light that seemed to strengthen his assimilation of the nymphaeum to the exedra. He duped himself: later work revealed the nymphaeum to have been a large court outside Domitian's very large dining hall.[19] Although a significant mistake, it did not infect Bianchini's measurements of the overall palatial footprint. From top to bottom, from the Theatrum Tauri facing the Circus Maximus to the line of the façade of the chamber complex facing the Forum, Bianchini counted 800 feet (240 m); the frontage facing the circus ran to 400 m; whence the entire palace, exclusive of its main entrance, fit into a rectangle with sides in the agreeable ratio of 3 to 5.[20] Fig. 33 shows the reconstruction of one of the façades.

The entrance to so magnificent a palace had also to be magnificent. It had, therefore, to give access from the Forum to the noblest rooms in the palace, the Major Complex unearthed in 1724. With an eye to Vitruvius, Bianchini designed a fanciful entrance that completed "all the parts of Domitian's palace according to the rule of architecture and roman practice." The distance from the complex to the foot of the Palatine toward the Forum (to the northwest) was almost equal to that of the complex to the foot facing the Circus Maximus. Bianchini made it 225 m down to the Arch of Titus. That was exactly three times the width of the complex. Evidently, the entry way was a huge ramp 75 m broad and 225 m long, anchored on one end by the complex and at the other by a step roughly parallel to it running 75 m from the Arch of Titus northwest to a missing arch, which

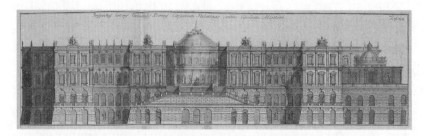

Fig. 33. Reconstruction of the southeast–northwest façade of "the palace of the Caesars." FB, *Palazzo de' cesari* (1738), tab. XII.

Fig. 34. Reconstruction of the pompous stairway and main entry to Domitian's palace. FB, *Palazzo de' cesari* (1738), tab. X. The audience chamber is the structure on the extreme right.

Bianchini supposed celebrated Domitian.[21] It was now a question of filling in this great space plausibly. Bianchini's solution appears in Fig. 34.

The visitor to Domitian's palace (as designed by Bianchini) began the climb from the city level (the Forum) on one of a pair of ramps 9 m wide bordering the sides of a garden terrace. The ramps could accommodate horses. The climber or rider arrived at a semi-circular path around a vestibule that gave access to an atrium, which, in this scheme of Russian dolls, opened to a courtyard that fronted the chamber complex. Bianchini assigned special functions to each of these spaces following descriptions in Vitruvius, Virgil, and Suetonius, and the evidence of fragments unearthed in the Farnese gardens that he associated with the entry way.[22] He placed arches, niches, and statues where he saw fit, furnished posts for guards and passageways for servants, and guided the senatorial visitor through the throng of officials, petitioners, and clients who crowded the avenue to the great audience chamber, the middle room of the complex.[23] There he left him and us, and also his book, for he did not finish the description of the palace he drew up in step with the discoveries of its extent and content.

Believing that his external elevations (see Figs 33 and 34) were reliable, Bianchini did not hold back the manuscript owing to doubt and went so far as to commission and approve the figures, except perhaps for the large mis-en-scène by Nicoletti.[24] But he could not give the project his full attention. As usual, he was busy with other things; also, he had an accident. On 17 August 1725, while measuring a wall in the northwest quadrant of the palace with a tape and two servants, he fell into a big hole (as Ciampini also once did) while heedlessly pursuing his observations. He grabbed a handhold at the brink and swayed in midair while his servants tried to pull him out. They were able to let him down part way with the measuring tape, which was fortunate, as the hole turned out to be 3 m deep. The unfortunate monsignor landed on sand, seriously twisting his leg, but otherwise he escaped damage. The rumor that, despite his injury, he immediately called for a lantern to explore the space he had so painfully discovered may be rejected. Likewise, we may doubt that he died in the accident, "a martyr to antiquity," although it did put him out of action for four months. When he recovered, he had to add to his schedule occasional visits to spas to assuage a residual painful limp.[25]

His nephew and literary executor, Giuseppe Bianchini, an antiquary and Oratorian, decided to make a great book of the incomplete *Palazzo de' cesari*. With a Latin translation on alternate pages and twenty good plates, Giuseppe had a book of respectable size, which could have been more respectable had he not omitted passages he did not understand and items in his uncle's papers that would have clarified some of them.[26] The book was expensive. After much negotiation prosecuted with the help of Maffei and Polignac, the king of France, Louis XV, put up 500 scudi (3,000 livres) toward publication costs. With a royal patron and famous author, the *Palazzo de' cesari* enjoyed some success as the first account of the palaces written in Italian. But soon Bianchini's creative interpretations of the ruins he accurately described attracted the criticism of the learned, and the book, like the Palatine itself, fell out of favor for a hundred years. When serious digging on the hill resumed, his discoveries and inferences had to be made anew.[27]

The wild Aphrodisian had let his imagination loose in playing the architect with Domitian's ruins. That had its merits: he regarded them as prompts to conceive the palace as it would have appeared to a senator proceeding up the imposing stairway to a meeting in the magnificent audience chamber, or to a servant scuttling down unseen passageways bearing food, gifts, or messages to residents. And he thought through the serious problems facing an architect charged with providing air, light, water, open space, accommodation, and mobility to many hundreds, perhaps thousands, of people within a structure vast and grand enough for deified emperors. Like the *Istoria universale*, the *Palazzo de' cesari* is a work of the imagination controlled by thorough knowledge of the relevant texts and artifacts. And, apart from his profligate application of the symmetries that captivated architects of his time and mathematicians of all time, Bianchini's elevations do not differ much in overall impression from modern conjectures.[28]

A similar synthesis of symmetry and mathematics was conspicuous in a field in which Bianchini had had some exercise at the beginning of his career. This was the art of fortification, which emphasized symmetrical layouts such as Bianchini drew for his undercover survey of the fortress of Guastalla (see Fig. 8). But he had no need to look to modern fortifications or baroque palaces for an example of conspicuous symmetry in Roman buildings. Santa Maria della Rotonda enjoys the many symmetries of the Roman Pantheon, which came into use thirty years after the death of Domitian. No doubt it gave Bianchini some ideas. An ignoramus once asked Bianchini if he knew Domitian's palace. "Do I know it?"' he replied. "I made a part of it."[29]

The Servants of Empire

In November 1725, halfway through Bianchini's convalescence, a band of legal bandits dug up an ancient columbarium in a garden along the Appian way owned by one of them. Their purpose was profit, their methods crude, their yield disappointing. Apart from a sarcophagus or two, a few funerary objects, and a mosaic floor they could not move, they found only ashes. Or, rather, urns full of ashes, placed in niches and identified by inscriptions naming their inhabitants. The gang soon fell out over the spoils and referred their dispute to Alessandro Albani, whom they expected, rightly, to be interested in buying them. During negotiations Albani's mentor hobbled to the digs, recognized the importance of the inscriptions, and persuaded Albani, Davia, and Polignac, by then a cardinal and the French ambassador to Rome, to pay for a proper exploration.[30] They did so with Bianchini in charge of operations. In a few weeks he had made drawings of the entire sepulcher and copied some inscriptions, working carefully but hurriedly, fearing that the earlier quick and clumsy methods had imperiled the structure.[31] It soon collapsed.

Bianchini's Christian charity did not extend to pillagers of ancient monuments. The organized despoiling of the Palatine by the educated Farnese duke was one thing; the purely mercenary grab by the Appian Way gang was uncivilized as well as greedy. It had taken the Romans centuries to amass their treasure of nature and art, Bianchini wrote, and centuries for barbarians to destroy most of it. To which of the periods does our age belong? "Looking into the question would excite too many grudges." The collapse of the columbarium was not unprecedented. Foreign visitors often saw famous relics of ancient architecture sink under the ruins. "Observing the destruction of an example of this master art that they had admired a month before . . . they are uncertain whether in Rome they have entered the cultured age that conserves, or the barbarian that destroys."[32]

Bianchini reserved a special Hell for scavengers who removed inscriptions that could support grand historical narratives. He expected that the labels in the columbarium were of such a kind. They dated from the time of Augustus's widow Livia, which coincided with the construction of the earliest of the Palatine palaces whose remains were surfacing while Bianchini directed work at the columbarium. Quite a synergy! From the geography of the inscriptions still in place that described the persons in the urns he hoped to deduce the relative status of the offices they had held. No luck: the distribution of the ashes around the columbarium was disappointingly democratic. Arriving at this result took courage as well as geometry and philology. We might imagine Bianchini in any of the postures of the people in Fig. 35, suspended in space, copying and drawing deep underground, or shoveling mud from a mosaic floor.[33]

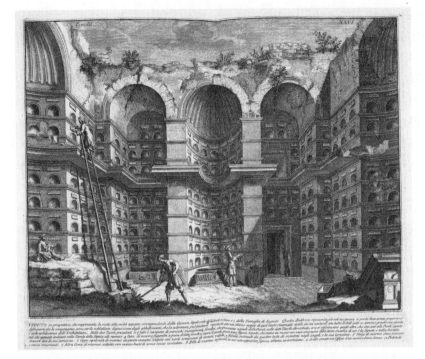

Fig. 35. Columbarium of the slaves and free servants of the Empress Livia. FB, *Camera ed inscrizioni sepulcrali* (1726), plate after p. 16.

Bianchini found many attractive opportunities for meaningful measurement in the columbarium. He measured the size of the tessellations in the mosaics, of the niches, urns, ledges, and arches, and found that the architectonic dimensions, the length and breadth of the floor, the height of the ledge giving access to the upper niches, the size of the topmost pilasters, and so on, followed "the very best taste in architecture," the rules of Vitruvius. Musical intervals abounded: the ratio of the heights to the ledge and pilaster made a major fifth, 3:2; of ledge to topmost arch, a major third, 5:4; of other dimensions, fourths and octaves. For arithmetical exercise, there were niches and urns to count: how many of each, urns per niche, individuals per urn. Bianchini estimated the total of slaves and freed men and women whose ashes filled the urns at three thousand. From other sepulchers of Augustus's and Livia's servants from the same time he reaped a similar number of souls, six thousand in all, "an entire legion of freed and enslaved men and women of the house of Augustus."[34]

Bianchini published 222 of these inscriptions, many of which gave civil status (slave or freed, and if freed by whom) and sort of employment. Most of the jobs performed by the legion presented no surprises: they functioned as personal attendants, secretaries, cleaners, laundresses, cooks, footmen, gardeners, ostlers, just the sorts found at the palaces of Bianchini's great patrons. A modern estimate puts the number of distinct jobs at fifty-five; Bianchini described and analyzed forty-six. The number of servants measured the power and prestige of the householder in the Rome of Clement XI as well as in the age of the Caesars, with the difference, however, that the number of slaves tends to increase, whether wanted or not, by breeding.[35] Bianchini could more easily identify with Livia's household, not only because he was used to servants but also because, as the pope's chamberlain and domestic prelate, he was one himself.

Among the specialized servants required by the empress were a "custodian of her purple"—that is, her imperial dye; an overseer of her summer cloths, another for winter ones, and a third for folding both; trustees for her bracelets, earrings, and pearls, and an *ornatrix*, or rather several, to see them worn properly; doctors, midwives, and apothecaries; a bath staff, water carriers, perfumers, oilers, masseuses; and jewelers, goldsmiths, and pearl setters, to make splendors for her and gifts for her friends. Among Augustus's dead servants we find a *lector* or two to read to him when he could not sleep; a *nomenclator* to remind him of the names of his friends; and a *lanipendio* (wool weigher) and similar officials to protect against swindlers. There were also the *dispensitor,* a trusted slave usually wealthy enough to own slaves himself and have a freed woman as wife, who paid the household bills; and that most useful functionary, the *silentarius,* charged with shutting up chatterboxes.[36]

Bowing to Davia's wish for prompt publication, Biancini printed his description of the columbarium in 1726 before the discovery of an annex adjoining it. That did not leave him time enough to integrate the information from the inscriptions with the views he was developing about the plan of Domitian's palace. Consequently, his *Camera* presents only the raw material for the history he might have written. Had he turned his hand to it, he would have had the help of two other descriptions of the *Camera*, one by Ghezzi, the other by a Florentine antiquary, Francesco Gori, who wrote in detail about its layout and content without having seen it; and also Ghezzi's draft of his account of the annex, uncovered in 1728 during excavations underwritten by Polignac.[37] Ghezzi emphasized art and collector's items; Gori depicted everything and attempted a reconstruction of the columbarium as an institution; Bianchini gave both a clear general view of the monument and an exact transcription of the inscriptions, for which he still receives high marks.[38]

A Plethora of Popes

"The city is glad, the church leaps for joy, the elderly rejoice, virgins in triumph sing God's praises, the poor are enriched, the needy prosper, captives are comforted, the maimed rise up, and the sick are healed." All this happened on 29 September 855, when the Holy Spirit inspired the choice of Benedict III as pope. The Spirit missed the Holy Roman Emperor, however, who imposed his own candidate, one Anastasius, who had the disadvantages of being excommunicate and anathematized; he rushed to Rome and immediately defaced pictorial records of his earlier condemnation and (as collateral damage) some icons of the Savior and his mother. "This he should not have done." The Roman people and clergy threw him out after a week's service as Christ's vicar. From this nadir Anastasius rose to papal secretary under Benedict's successor St Nicholas I. The post gave him another opportunity to eliminate incriminating documents from the archives and to acquire the expertise that lifted him to librarian of the Roman Catholic Church under Nicholas's successor Hadrian II.[39] These preferments did not mend his behavior; he was condemned as an accomplice to murder but bounced back to protect Bulgarian Catholics against Greek Orthodoxy. Like Bianchini, Anastasius had an unusual career for a librarian.[40]

The church champion, sometime murderer, and antipope Anastasius Bibliothecarius is better known as the author or compiler of the *Liber pontificalis* (familiarly the *LP*), the fundamental source of reliable information and doubtful stories about the first 112 popes. Bianchini devoted thousands of hours to an edition of the *LP*, which he published in three stout quartos between 1718 and 1728. This by no means finished the business, however. The first volume contained the full *LP*, according to an earlier edition based on a manuscript in the Vatican, augmented by a few variant readings from other sources; the subsequent volumes contained all sorts of material Bianchini raked together from previous editions, newly discovered manuscripts, coins and medals, ancient inscriptions, astronomical data, liturgy, and ornament. With these scrapings he established the chronology, and illustrated the reigns, of the individual popes, but not all of them. Volume 2 brought him through Silvester I (314–35); volume 3, to Pelagius II (579–90); volume 4, a posthumous work edited by Giuseppe Bianchini and three other savants, reached Paul I (757–67). The four editors together did not have the strength of one monsignor and abandoned the project sixteen popes short of the finish. There were three other editions of the *LP* made in the eighteenth century; but of them all, according to the editor of the definitive modern edition, "the splendid volumes of Bianchini deserve first place."[41] And a closer look.

During the sixteenth century, scholars of the caliber of Panvinio, for whom Bianchini had a high regard, accepted Anastasius as the author of *LP*. By the later seventeenth, "historical studies and criticism having reached maturity," scholars of equal eminence, such as Bianchini's patrons Schelestrate, Ciampini, and Noris, had deprived Anastasius of all his lives save, possibly, those of his patrons Nicholas and Hadrian.[42] Bianchini devoted much of the first volume of his edition of Anastasius to describing the ancient sources of *LP* and evaluating their trustworthiness. Schelestrate et al. had worked over the catalogues of the fourth and sixth centuries that had provided the information about the earlier popes. Were the sources reliable? Yes! The consistent rendering of names, dates of inauguration and death, and reign spans in the style of Roman records guarantees it. "Who can doubt that the manuscripts transcribe entries by officials of the Roman church?" Bianchini offered readings from many documents civil and religious, on stones and in manuscripts, from his inspection of Albani collections, Neapolitan sarcophagi, and Hippolytus's chair, from personal letters, deeds of gift, synod records, calendars, coins, and graffiti, to confirm that the entries in LP have the earmarks of a dedicated bureaucracy.[43] The argument would have been familiar to readers of Istoria universale.

The first volume of Bianchini's Anastasius, published in 1718 as he was creating a place for himself at the Jacobite court, had a political slant. It was an early publication of the Vatican press, which Clement XI established to publish manuscripts from the Vatican Library demonstrating the munificence and rights of the popes and the range and quality of Roman scholarship. The LP answered these objectives perfectly. As to importance, the lives and doings of the popes were "the chief part of ecclesiastical history" for Roman Catholics and Protestants alike. The large quartos, well printed in several fonts and formats and accompanied by instructive illustrations, sufficiently demonstrated munificence. And, as a vehicle for the publication of the miscellaneous material studied, developed, and discussed at the many academic meetings Bianchini attended, nothing could serve better. All sorts of out-of-the-way information could be gathered into notes to the individual lives. In one representative case of erudite overkill, his and others' annotations to a life the LP disposed of in 100 words sum to 6,370. When in good form he could produce a note of 6,370 words himself.[44] There could never be enough notes, "no surviving words or deeds left unexplored." Combining his compulsion for collection with his cosmopolitanism, Bianchini invited scholars of all nations to send him corrections and additions for inclusion in later volumes. "For I have held for a long time that by following experts a company of well-regulated minds can approach accessible truths as closely as humanly possible."[45] There spoke a senator of the Republic of Letters.

The second volume of Anastasius, published in 1723 when its author was busy on the Palatine, is the most interesting of the series for his contributions. In order to fix the dates *anno domini* of the sequence of popes, he had to repair several slips and outright errors in the catalogs of reigns; and, being meticulous, he devised some rules for the job. The first is to give precedence to unmutilated public monuments set out by responsible officials at the time of the historical events they record. The next best evidence is what scholars say, if by chance they agree, but good scholars only, such as Panvinio, Petavius, Noris, and Bianchini himself. Credence should also be given to dates that put games and festivals on days reserved for them in the ancient calendars, and to dates determined by astronomical calculations for eclipses mentioned in histories or inscriptions. Bianchini possessed a valuable key for the application of astronomy in a pettifogging marble then recently exhumed from an old Catholic cemetery. This stone bore the time of day, day of week, and age of the moon when certain consuls took office, data enough to enable him to work out the year of their service and the reign of contemporary popes.[46]

Determining the timing of the Passion from Last Supper to Resurrection required particular care and delicacy. All authorities agreed that it began on 25 March, but differed as to the year: AD 28 or 29? Most evidence, summarized persuasively by Noris, favored 28 and placed Christ's birth in 5 BC. But one datum pointed unequivocally to 29: the names of the consuls during the year of the Passion were known, and records showed that they had held office in AD 29. Bianchini calculated his way through this impasse. Pious tradition claimed a total solar eclipse on the day Augustus died in AD 14; modern astronomy could find no such event but revealed an almost total eclipse for the same calendar date in the year 13. Bianchini reconciled consuls and eclipses by subtracting a year from the Roman reckoning, and, by transforming 29 into 28, removed the consular obstacle to the dating of the Passion. Everything slipped into place if he violated his rules and gave astronomical deductions priority over public monuments drawn up by accredited officials. Astronomy also confirmed a full moon on Friday, 26 March 28, and Bianchini's calculations using Jewish rules established that it was indeed the Paschal moon of Christ's Passion.[47]

There is still no agreement about the year of Christ's death; Bianchini's proposal is perhaps the earliest admissible.

Any competent mathematician could use modern astronomical tables to calculate the dates of ancient eclipses. But only Bianchini thought to apply astronomy's chief tool to increase the number of original sources for *LP*. High up in the nave of the basilica of San Paolo fuori le mura, some 45 m above the ground, ran a line of almost invisible painted medallions of popes, beginning with Peter and extending through the fifth century, each labeled with name and duration of office. In Bianchini's time the inscriptions were hard to read, owing to distance and damage; today they cannot be read at all, because they perished in a fire in 1823. His idea to inspect them through the spy glass he always carried thus not only engineered another source of papal chronology, but also preserved most of what we know about the medallions and inscriptions. The popes were dusty, Bianchini jubilant. He ran to inform his friends of his discovery and to ask Clement, who took a great interest in images of his predecessors, to order workmen up ladders to remove the dirt.[48] Nor, in his enthusiasm, did he disdain to climb up with them; he had often dangled dangerously before, and would again, in his haste to copy inscriptions or take impressions.[49]

Dirt and distance were not the only obstacles to interpretation. The images did not always follow chronological order, and many numerals were painted badly, with an "X" where a "V" should be, or vice versa, or with a superfluous "I" or lacking a needed one. Bianchini unscrambled them by considering the manner of depiction, facial hair, tonsure, and vestments in the portraits, and adjusted the inscriptions to agree among themselves. When properly arranged, the pictorial series and their inscriptions matched the manuscript lists kept by officials of the early church. "I confess that [if the conclusion be doubted] I have no idea what more certain can be said to make a historical or chronological matter credible." Or what could illustrate more clearly the merit of his historical credo: do not ignore known relevant sources however unpromising or difficult of access. "It would be unjustifiable indolence and intolerable carelessness to neglect to publicize these precious relics."[50]

Having brought the evidence from San Paolo into agreement with the fourth-century catalogue, Bianchini gave an exact chronology of the first forty-two popes in synoptic form and again in the abundant notes he devoted to each of them, "desiring to leave nothing unexplored that seemed doubtful."[51] Again he was able to enrich his annotations in synergy with his astronomical investigations. Having nothing to do but work on Anastasius, dig around the Palatine, observe at Santa Maria degli Angeli, and cultivate patrons, Bianchini was engaged, in 1723, in extending the meridian line through Urbino he had laid out when visiting the Jacobite court there. In the Sabine hills, just off the line between Rome and Urbino, lies the old Benedictine abbey of Farfa, a once powerful political and monastic center, with a particularly rich archive, now a national library monument. Bianchini stopped off there to rest from geodesy and enjoy some decaying manuscripts. Three of these, though late (eleventh century) and not entirely reliable, furnished stuffing for notes about the earliest popes. It would be a work of supererogation to enter further into the manuscripts that Bianchini edited and exploited to fill his compendium. His coverage can be indicated sufficiently by mentioning that it includes the itinerary of an unknown visitor to Rome in the seventh or eighth century and an inventory of the holdings of the Christian cemeteries in and around the city. Nothing remotely relevant was irrelevant. Knowing every street, gate, and sarcophagus of the city after thirty-five years of residence and twenty as czar of marbles, Bianchini sought, collected, and protected antiquities "not only from diligence, but truly also from love."[52]

Love brought forth eight pages of chronological correspondences over the interval 50 BC to AD 313 between dates in the Julian calendar and dates by Olympiad and reckonings from the foundation of Rome. Bianchini added a continuous day count, the Julian period, beginning in 4713 BC, some years before Creation, according to Hebrew scripture and the *Istoria universale*; invented by Scaliger to ward off "all danger of any confusion," it is now running around 2,460,000, and in an altered form is still used to announce dates of creation, not of worlds but of canned goods.[53] The curious may want to know that 4,713 derives from 7,980 = 15×19×28, the product of the length of an Indiction (a Roman interval of record keeping) by a Metonic cycle and a leap-year cycle, all of which started at zero in 4713 BC. "If this is not clear, I cannot imagine what can be wanting." Between his chronological correspondences, Bianchini recorded, in separate columns, securely datable events in civil and ecclesiastical history. The gigantic effort resulted in specifications so reliable, according to their creator, as to preclude the need to return to the documents on which he based it. Thanks be to God! It was the culmination of the program that began with the commissioning of the Clementine *meridiana* to "confirm and revise chronology in general."[54] Indeed, it was the work for which Bianchini's study of astronomy and history, his *meridiana* and his *Istoria*, had uniquely and providentially equipped him.[55]

Bianchini's contributions to this second volume of Anastasius won applause among the learned. "Your great Anastasio is something extraordinary," Maffei wrote in 1724. Muratori showed his appreciation by inserting Bianchini's work into his edition of *LP*, through which some of Bianchini's *Anastasius* came to the attention of Edward Gibbon. The stylist of Rome's decline and fall did not like *LP* but had to use it. "The style is barbarous, the narrative partial, the details are trifling; yet it must be read as a curious and authentic record of the times." That is also the way Bianchini read and enriched it. The result confirmed his reputation as "one of the great lights of living literature," as "an embodiment of antiquarian erudition."[56] Bianchini's *Anastasius* was not only a monument to his scholarship but also an exemplar of a lost form of knowledge production, the joint creation of a virtual college of highly placed, well-supported celibate men deeply versed in the sources of antiquity. These men believed, if they believed as Bianchini did, that the knowledge they produced from critically read texts and authentic relics was reliable and, perhaps with adjustments here and there, an answer to sceptics and cynics who doubted that historians could say anything both certain and significant.

The later volumes of *Anastasius* bring nothing new in form, not even the lengthy treatise on papal chronology promised in volume 2, but offer some revealing biographical details, not of popes but of monsignor their chronicler.[57] During his forty-four years in Rome, he writes, he has had the enviable opportunity of assisting at the inhumation of three popes—Innocent XI, Alexander VIII, and Clement XI. The inscriptions on their caskets and tombs followed the style of *LP*, even the legend on the pompous tomb of Alexander erected by grateful grandnephew Cardinal Ottoboni; from which Bianchini confirmed his view that inscriptions concerning the old popes, if written in the perennial form, should be considered good-faith efforts to convey authentic information. Before declaring this doubtful finding, he offered an irrelevant curiosity that, combined with his fascination with sepulchers and epitaphs, suggests a serene interest in last rites. The curiosity: the *praecordia* (heart and viscera) of a pope who dies in the Vatican are preserved under a brief inscription in a chapel there reserved for holy relics; whereas the innards of a pope who dies in the Quirinal rest in the neighboring church of Saints Vincent and Anastasius under a still briefer label.[58] Further to autobiography, Bianchini enters into great detail about the changing forms of papal vestments, reflecting his love of clerical garb in all its colors and distinctions.[59]

The fourth volume of Anastasius, being posthumous, carries on Bianchini's plan as well as a group of only four could manage; but they were not equal to mastering the contextual history of the Germanic kingdoms and the incessant discovery of new relics, to say nothing of astronomical calculations. It is worth recording, however, as a footnote to the acceptance of the Copernican theory in Italy, a remark that one of the editors, Gianfrancesco Baldini, a dear friend of Bianchini and a good mathematician, inserted in the life of Pope Donus (AD 675–8). Commenting on a comet mentioned in the *LP*, Baldini described, unapologetically, the modern theory in which, he said, comets move around the sun, "like planets."[60]

The scrupulous editor of the standard edition of *LP*, Louis Duchesne, gives high marks to Bianchini's performance for his exact transcriptions

210

and publications of manuscripts, shrewd judgments, life-by-life commentaries (the overstuffed notes), and detailed chronologies. To these may be added unscrambling the genealogies of the kings of Aragon, teasing out the temporal dominion of the Holy See and the jurisdiction of popes, and controverting Pietro Giannone's famous anti-curial history of Naples.[61] To be sure, the merits were many, the coverage boundless, the erudition staggering, but—it is Duchesne's opinion—the method fell far short of modern standards. Precisely what Bianchini had said of the efforts of his predecessor Panvinio! Bianchini's work, Duchesne explained, "is interesting and useful despite its disparate assemblage of erudition." But not very useful.

> As far as my notes and dissertations . . . are concerned, I will surprise no one by saying that they owe almost nothing to anything resembling . . . Bianchini's. Historical sciences, especially those devoted to Christian antiquities, have made such progress in the last century and a half that I have to hand infinitely more resources than were available at the beginning of the 18th century. Perhaps also I have given myself . . . a little more trouble than people would have taken then.[62]

Duchesne exemplified the highly professionalized critical history built up during the nineteenth century. Although a Catholic priest, he taught at the École pratique des hautes études, an institute for advanced study, in its section for religious sciences set up by the anti-clerical French Republic in 1886. His judgment was a repudiation of the antiquarian, collector-driven, catch-as-catch-can elements in the practice of the learned ecclesiastics of Bianchini's circle. At the end of his career he acknowledged that their haphazard ways, depending largely on the books and artifacts available in Rome, could not keep up with the pertinent monographs mercilessly multiplying throughout western Europe and even Britain. "How difficult it is just to acquire these books from afar and at great expense! And how great is the time needed to read and reflect upon them!"[63]

8

FROM MARS TO VENUS

War and Peace

For military exploits and literary achievement, Count Luigi Ferdinando Marsigli of Bologna was easily a match for Sir Charles Wogan of Kildare. By 1711, when astronomy again brought him into contact with Bianchini, Marsigli had spent many years with the Turks, first as a gentleman associate of the Venetian agent in Constantinople, Pietro Civran, then as a captive, finally as an opposing general. He knew Turkish languages, customs, religion, government, and military operations, and studied the natural history of the Ottoman lands. The first installment of this natural history, contained in a letter to Queen Christina, whose academy he had frequented when in Rome, describes his observations of the currents and countercurrents in the Bosporus. Montanari had taught him, so he began, that life implied motion, from which it followed that travel was the liveliest of occupations for humans and ceaseless movement its counterpart in water.[1] When he devised this conceit, in 1681, the Turks too were in motion, not yet mobilizing against Vienna, but threatening Venetian stations in the Mediterranean. Civran was recalled. A sensible man, he returned by ship, generously including in his baggage books and specimens collected by Marsigli; while that mobile adventurer returned to Venice overland through territory made hostile by plague and people. After a brief sojourn in Italy, Marsigli determined on a career in the imperial army and set forth for Vienna.

212

To recommend himself to the imperial command, he made maps of French fortifications using Montanari's applied mathematics, just as Bianchini would do at Guastalla. He quickly rose to serjeant in a surveying unit in the Austrian army. When the Turks began their great offensive of 1683, Count Marsigli found himself measuring marshy land south of the Danube equidistant from Vienna and Buda. Tartar horsemen also found him and sold him to a commander in the sultan's army. He slaved at grinding, brewing, and serving coffee until the army reached Vienna; he then watched Sobieski's victory from the losing side, and fled with the rest. After the execution of his master for deserting an untenable post, Marsigli was bought by the proprietor of a coffee bar. The cruelty of his new owner drove him to attempt escape; caught, he was sentenced to death; on the morning of his execution, he was rescued by breakfast. His master, squeezing the last drop of coffee from him, made him serve two thirsty soldiers. Between cups the barrista told the brothers, for such they were, about his plight and the prospect of a good return if they bought him cheap and ransomed him dearly. Done! After many adventures the trio reached the brothers' village, Marsigli contacted Civran, and Civran paid the ransom.[2]

Marsigli returned to imperial service, in which, with his knowledge of surveying, fortification, and Turkish tactics, he rose quickly to general. He was the main engineer of the works for the successful siege of Buda in 1696 and the expert designated to negotiate boundaries with the sultan's representatives at the peace conference at Carlowitz (a castle on an island in the Danube) in 1699. On one of many surveys associated with the negotiations, Marsigli passed close to his rescuers' village. He looked up the brothers, wined and dined them and plied them with coffee, and sent them away with new clothes and a thousand sheep. The ransom money had not brought them prosperity. The local pasha had invented a reason for jailing them, and they had been forced to ransom themselves.[3]

During his extensive travels along the Danube, Marsigli amassed a vast quantity of natural- and human-historical items, natural-history specimens, ancient relics, medieval and later manuscripts, and books, some picked up directly, others through pillage, still more by purchase. Among his agents for books and drawings was Georg Einmart, an etcher in Nuremberg who amused himself with astronomy. Einmart urged on Marsigli the acquisition of up-to-date instruments to take latitudes and longitudes for his surveys; they would prove unexpectedly useful in establishing an academy of science that Marsigli dreamt of setting up in Bologna. The War of the Spanish Succession returned him to Italy almost as discreditably as his repatriation from slavery. He had been sent as second in command to the fortress of Breisach, a poorly equipped Austrian bastion against French incursions into the empire from Alsace.[4] Marsigli spent the winter of 1702 begging field officers and the imperial staff for construction materials, manpower, and weapons. When French troops came to call in August 1703, the Breisach garrison, facing swift and inevitable defeat, capitulated with honor. The general staff whose dilatoriness had made surrender inevitable did not think surrender honorable. It called a court martial, which decided to execute Marsigli's superior and to humiliate and discharge him.[5]

While under arrest Marsigli shipped his collections to his palace in Bologna as an endowment for his academy. Manfredi was there, building an observatory on the palace roof with the continuing advice of Bianchini, who procured instruments for it in Rome. The completion of the observatory toward the end of 1702 freed Manfredi to receive and arrange the material dispatched by Marsigli, who, after coming to terms with his disgrace, followed his boxes to Italy in time to head up the ragtag show army that Pope Clement fielded to protest against the emperor's takeover of Comacchio.[6] He then pressed ahead with the institution that would house, maintain, and study his collections and manuscripts, all of which he intended to give to the university of Bologna. The far-seeing plan immediately aroused short-sighted opposition; the professors, rejecting the implied criticism of their ancient establishment, did not desire to bring their teaching into the eighteenth century. Undeterred, Marsigli pressed his philanthropic campaign with the city's senators. Those who recognized the university's stultification were inclined to accept his gift as the nucleus of an independent institution provided that the pope, who held the ultimate authority over the city's finances, agreed to divert sufficient tax revenue to its upkeep. Since Marsigli expected his academicians to teach, the revenue would have to cover salaries as well as maintenance.[7] To help persuade Pope Clement to pledge the required income, Marsigli and Manfredi turned to their mutual friend, the pope's polymathic chamberlain.

Bianchini was well disposed to help. Not only was he a great champion, assiduous attendee, and occasional founder of academies, but also he trusted and admired Marsigli. They had known one another as alumni of Montanari's school, corresponded about antiquarian matters (then most recently about the ruins of a Roman bridge over the Danube), and held a similar cosmopolitan ideal (like Bianchini, Marsigli opposed Muratori's academy for its Italian inwardness).[8] And, as we know, Bianchini had been very helpful in setting up the observatory in Marsigli's palace. Their strategy in approaching the pope hinged on his previous grant of a charter to an association of painters for whose meetings Marsigli had made space in his palace. The association, known as the Accademia Clementina, followed the Arcadia in rejecting the baroque in favor of relative clarity and realism, which helped to fit it for incorporation into the nascent institute.[9] The projectors therefore stressed the symbiosis of art and science in which Clement had already invested significantly, notably in creating the grand *meridiana* at Santa Maria degli Angeli. At just this time, 1710–11, the pope was considering a proposal, marked with Monsignor's fingerprints, on which Marsigli could hinge his appeal. This was to build a museum of calendrical innovations in the Vatican's Tower of the Winds around a meridian line installed there 125 years earlier in connection with the Gregorian calendar reform. What was needed to make the museum another example of the symbiosis was artwork portraying the modern astronomy that confirmed the reformers' conclusions.[10]

Among Marsigli's accumulations were paintings made from telescopic images by Einmart's daughter Maria Clara, who, like her father, was both artist and astronomer. The cultural consultants Manfredi and Bian-chini thought that the paintings would interest Clement. They were right; the pope asked for copies. An artist stood ready, Donato Creti, a distin-guished member of the Clementine Academy. Creti's designs instanced the cultural value of Marsigli's instruments while appealing to many of Clement's sensibilities. Each telescope, accurately pictured in use by fash-ionable people in an arcadian landscape, would be directed toward the planetary object best suited to it, "so that [Manfredi wrote to Marsigli] all the chief instruments in your observatory will appear in the paintings, and the painter will be able to show his skill in the several postures of the observers." Creti left sufficient sky space for a miniaturist, Raimondo Manzini, to insert Maria Einmart's depictions of telescopic images of the sun, moon, and planets. An eighth canvas showed a comet, not drawn from life, as the other depictions, no doubt because, peace then being in the offing, dire portents did not appear in the heavens.[11]

Three of the results of this unusual collaboration advertise Galileo's spectacular discoveries favoring the Copernican system. Jupiter shows three of his four satellites as first described by Galileo. The Jovian sys-tem offered a complete analogy to a planetary system centered on the sun. Saturn has its ring, which Galileo saw as protuberances or han-dles, suggesting that, like the mountainous moon, planets are not perfect spheres, as the old astronomy taught. The disclosure of the rings required better lenses than Galileo possessed; in Creti's rendition, the observers employ a tubeless telescope with the objective lens mounted on a post and connected to the eyepiece by a string, as in the portable apparatus Bianchini invented (see Fig. 22). Venus is shown as a sickle; her phases as revealed by the telescope afforded a geometrical demonstration that her orbit circles the sun, not the earth. Creti placed a young woman in the foreground attending to her hair while gazing at the goddess of love; she sees the planet as a brilliant disk because the factors deter-mining Venus's brightness—its distance from the earth and illuminated surface—compensate one another over most of its orbit.[12] Bianchini could be confident that the Copernican insinuations in Einmart's drawings would not upset Clement. The pope had not proceeded against a Neapolitan publisher who had reprinted Galileo's condemned *Dialogue* in 1710 nor against a journal controlled by the Arcadia for reviewing the reprint favorably under the common fig leaf, "justly condemned by ecclesiastical censure."[13]

Clement decided to support Marsigli's institute. The city fathers made a large palace available to it with space for the teaching collections, the academicians, and an observatory. This last, developed by Manfredi, came online over a decade later with the injection of 1,500 scudi from Clement's successor, Innocent XII, who wanted it completed by the jubilee of 1725. Davia made it the playful present of an armillary with the sun in the center. Manfredi immediately procured armillaries representing the systems of Ptolemy and Tycho to flank Davia's compromising gift and assure the authorities that, in the matter of world systems, the astronomers of Bologna had no definite opinion. The observatory still exists, as a museum, in central Bologna. The paintings of Creti-Manzini remain in Rome, in the Vatican galleries. And a celebrated set of wax anatomical models made for the institute continues to exemplify the symbiosis of art and science that inspired Marsigli's initiative.

The main practical purpose of the observatory, as Manfredi saw it and as he urged Marsigli to present it to Clement, was to anchor a trigonometric survey of the papal states like the one that Cassini's son Jacques was then beginning in France.[14] Italy desperately needed such a survey, which (so Manfredi wrote to Marsigli) could be accomplished by two astronomers measuring trigonometrically for two years. The initial expense for instruments and other supplies would be a little under 1,000 scudi. "With this capital in hand I offer to make the trip with one companion for only the cost of food and travel, and to ascertain in a couple of years the longitude and latitude of the principal places in Italy." The time was propitious. Preparations could be made while obtaining other items necessary for the institute. No doubt the initial investment was large. "Still I propose it, Your Excellency, so that, if it pleases you, you can suggest it to the Holy Father, under whose protection the work could be essayed; if we lose this opportunity, it is doubtful that anything will be done in the matter."[15] Clement could appreciate the difficulties of the proposed survey from the rocks, rivers, and inconveniently placed trees in the rugged roadless Arcadia in which Creti depicted the sort of heavy equipment required.[16] Marsigli did not support the survey, and Manfredi did not undertake it. The pope's intrepid domestic prelate did.

Bianchini began his survey with measurements he had made at Urbino during his time with the Jacobite court there. The nature of the work appears from Fig. 36.4.[17] The line of longitude (the meridian line) through Urbino (designated V) runs due north to graze the inside of the eastern hill *hg* of Montefiore, while the line of sight grazing the outside of the western hill *ab* strikes the campanile of the church of the Theatines O at Rimini. Carried backwards, OV meets Monte Acuto Y at the southern confines of the duchy. Looking south from Y with the acute vision of Bianchini, an observer could see Assisi and Perugia; carried further, the line passes through Castel Gondolfo on the way to the Mediterranean. Besides the various angles taken with the telescope and quadrant, Bianchini had to measure some distances on the ground to fix the size of the various triangles defined by prominent hills and church towers. He chose the slope GC on Monte Cantiano (see Fig. 36.3), which, besides being gentle, had in G a point on the meridian through V.

From the cathedral in Urbino, Bianchini sighted G and C, conspicuous as the end points of a spinney, and measured the angle that GC subtended ($2°58'$); since angle VGC = $90°$, he had VG, from which, by angular measurements from other sites, he could calculate exact distances between, say, Pesaro or Rimini and Urbino. And, having fixed the latitude and longitude of Urbino by the observations in which James had participated, he could give the geographical coordinates of any place in the duchy. For further enlightenment, Bianchini added orienting images: Fig. 36.1 shows Rimini, Montefiore, and the cathedral of Urbino as seen from Monte Acuto, some 30 km away; Fig. 36.2 offers a similar view from G on Monte Cantiano; and Fig. 36.3 depicts these mountains as seen from Urbino.

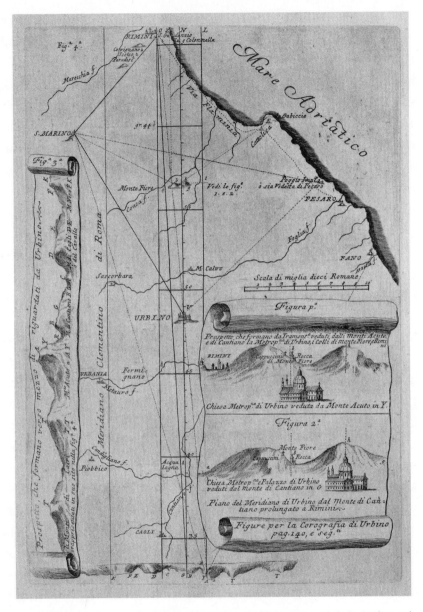

Fig. 36. The northern end of the Clementine meridian (longitude line) through Rome to Rimini. FB, in Bernardini Baldi et al., *Memorie concernenti la città di Urbino* (1724), second plate after p. [148].

In pushing the meridian of Urbino beyond the northern frontier of the duchy at Montefiore, Bianchini had the cooperation of his old friend Davia, who had spent most of his time since his nunciature in Vienna in his bishopric of Rimini. Clement had made him a cardinal in 1712, and, thus dignified, he had returned to cultivating astronomy in collaboration with Manfredi and Galiani. The extension of the chorography of Urbino through to Rimini was but an hors d'oeuvre, however, to the main menu, an arduous trigonometrical survey of the papal states from sea to shining sea, from the Adriatic near Rimini to the Tyrrhenian at Anzio. The main result, the line on the far left of Fig. 36.4, labeled *Meridiano Clementino di Roma*, passes through the *meridiana* of Santa Maria degli Angeli on its way from Ardea (35 km south of Rome) to the Adriatic above Rimini.[18] By the midsummer of 1724, after eight years of sporadic challenging work, much of which he defrayed himself, Bianchini had finished the necessary observations and proceeded to plot the line "for your glory [wrote Maffei] and our country's."[19] Bianchini determined latitude and longitude from these observations and comparable ones made by Manfredi in Bologna and by a new collaborator, a young Italian Jesuit named Giovanni Battista Carbone, whom he met in Rome in 1722 when Carbone was on his way to Lisbon.[20]

Carbone proved an important intermediary because of a rapprochement that began several years earlier between his king, João V, and the Vatican. João V had irritated Clement by opposing France during the Spanish war and later tried to appease him by contributing generously to a fund for defense of the Papal States against a new threat from the Turks, once again, in 1715, at war with Venice. The pope hired Marsigli to advise about defending the Adriatic coast around Rimini, where Bianchini's Clementine line met the sea; the victory achieved in 1717 was so decisive (although not owing to Marsigli) that the pope ordered Bianchini to establish an academy of eloquence to celebrate it. The omnicompetent monsignor resurrected an old academy, the Umoristi, for the purpose, which said its say and died again.[21]

The following year João harvested his investment against the Turks by sending a high-level diplomatic delegation to Rome. As he was rich from Brazilian gold, pious from a Jesuit education, and a show-off by nature, he was able to astound blasé Rome with a mission that filled three hundred carriages. In return for drafts on his wealth, he wanted an upgrade of the titles and powers of some of his bishops and (what he did not get) equalization of the diplomatic status of Lisbon with that of Vienna, Paris, and Madrid. João's generosity extended to knowledgeable patronage of the arts and sciences. He took over an academy of science in Lisbon and almost bought himself the Arcadia by giving it grounds and buildings on the Gianicolo. He also knew something about Bianchini's favorite studies, since the Jesuits had taught him mathematics and astronomy and he kept Carbone as a close advisor, tutor to his daughter, and cultural intermediary with Italy.[22] Through Carbone, Bianchini presented the king with some of his easy-reading books.

One of these was the coffee-table volume on Urbino. Bianchini's contributions to it interested João for possible application to his country and colonies. Indeed, he had intended to send Corbone to measure Brazil before discovering that the astronomer would be more useful at home. Bianchini informed the king that the chorographical work reported in the Urbino book was but a beginning; the end, the account of the Clementine meridian of Rome, was in sight, needing only a few calculations and a Maecenas to cover the cost of publication. As we know, by then Bianchini needed new patrons. Would Your Royal Highness like to pay for it?[23] His Royal Highness would not. But, with the help of Carbone, João agreed to finance a magnificent volume on Bianchini's investigations of the planet Venus.[24]

The Goddess of Love

Bianchini began flirting with Venus the year before he became close to James. He wanted to use the planet to measure the entire solar system by

Cassini's method, mentioned earlier, as improved by himself. It required a prominent planet P and a bright star S close together (angularly speaking) and a clear horizon. The idea is to measure the difference in time between the apparent settings of P and S to the real observer O at the earth's surface and a hypothetical observer C at the earth's center, from which astronomers measured the revolutions of the heavens. The plan may be grasped from the unlikely situation offered in Fig. 37, where the planet P and star S are aligned in the direction CO towards O's zenith at the equator; the polar axis is perpendicular to the plane of the paper. Six hours later S sets simultaneously for both observers O and C: the distance to the star is so great in comparison to the radius of the earth a that O and C can be considered coincident. A short time earlier, P had set for O because a is not negligible in comparison to the distance to P; the planet crosses O's horizon before it runs through an arc of 90°, as seen from C. Provided with a good clock, O can time the difference, which measures the angle χ. As appears from the figure, χ, being very small, is given very closely by a/r, where r is the distance to the planet. In practice, the observed motion takes place at a latitude φ and declination δ different from zero and with a star and planet that do not line up perfectly on the meridian.

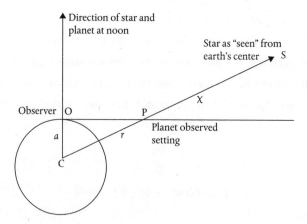

Direction of star and
planet at noon

Star as "seen" from
earth's center   S

Observer O   P
                                χ
                         Planet observed
        a r setting

C

Fig. 37. Diagram of the Cassini–Bianchini method for determining planetary parallax.

These details are easily included with a little trigonometry, but the essential measurement is the same: the difference in time between the settings of the planet and the star. The technique in effect enlists a star as a proxy for an observer at the earth's center.[25]

On 3 July 1716 Venus and Regulus, Cor Leonis, the heart of the lion, came close enough together to make possible a measurement by Cassini's method. Bianchini was ready, in a darkened room in a palace on the Palatine Hill put at his disposal by Pope Clement, with a telescope of 23 palms (5.1 m). It enabled the detection of both the planet and the marker star during daylight. Since Venus's apparent distance from the marker star continually changes owing to her revolution around the sun, Bianchini had the problem of eliminating the consequences of Venus's motion from the time difference Δt between the settings of Venus and Regulus. By following the pair for three whole days, he determined that on average orbiting the sun caused the planet to fall behind the star by 110 beats of his clock in six hours. On 3 July 1716, Δt amounted to 106 beats over six hours; if no other factors intervened, 110–106 = 4 beats had to be attributed to parallax. The clock beat 143 times a minute; converting to angle measure, Bianchini made χ equal to 24 $1/3$ arcseconds; corrected for declination, 23 $2/3$; corrected for latitude, well, Bianchini neglected to do that. With his value for χ, $r = 8000a$.[26]

To promote this number to the great desideratum, the "astronomical unit," the radius of the earth's orbit, he had only to measure the angular separation between the sun and Venus toward sunset on that July day, as indicated in Fig. 38. The observed separation was 44°; the computed angle between Venus and the earth as then seen from the sun, 100°; the astronomical unit, 13,400a, or 86 million km. He took immense pride in obtaining it. "This will solve a most pressing problem in Cosmology, Astronomy and Physics, namely the size of the whole Solar System, which follows as a splendid corollary from the observation of Venus's parallax, and is fixed so finely and accurately by this method that we can hardly expect equal certainty, it seems, from any other observation undertaken hitherto." He was mistaken. Cassini had come 48 million km closer

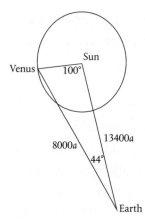

Fig. 38. The Sun–Venus–Earth system according to Bianchini's measurements.

to the modern value (150 million km) by comparing his observations of Mars made in Paris with ones made at the same time by a colleague in Guiana.[27] On reflection, Bianchini must have recognized that he might have put too much faith in his ticks; if he had counted 2.5 rather than 4, he would have duplicated Cassini's result. He postponed publication until he had a chance to confirm his measurement at the next approach of Venus to the heart of Leo. The rendezvous was scheduled for 1724.[28]

At the beginning of his eight-year wait, he received a more earthy assignment. It concerned a hoard of gold and silver objects that builders in Perugia had discovered in 1717. Rules required that the hoard be surrendered to the pope; the pope required an assessment from Bianchini. What did it signify and to whom did it belong? With great learning Bianchini explained that the coins, silver plate, and clasps or buckles (*fibulae*) comprising the treasure were almost certainly a reward for outstanding military service. Whose? The coins fixed the date sometime after the death of Justinus (527), the uncle and predecessor of Justinian; the discovery fixed the place as Perugia; and a long siege of Perugia, which fell to the Goths in 547, offered opportunity for the display of valor. Bianchini pitched on the garrison's commander, the Byzantine general Cyprian, as the hero. Speculating further, Bianchini proposed that the donor of the treasure might have been the pope or the bishop of Ravenna. For the first possibility he adapted information from his *Anastasius*; for the second, from Bacchini's *Agnellus*, "though he bubbles over with anachronisms." He knew he was on slippery ground. In transmitting his account to Clement, he remarked that he had "added a commentary, or if you prefer, a conjecture."[29] While waiting for the stars to line up, the speculative historian again broke through the antiquarian.

He was about to speculate more extravagantly with his redesign of Domitian's palace when, in 1724, he had to prepare himself for the favorable conjunction of Venus and Regulus. Alas! his stars were not favorable. He was denied use of the Palatine palace that the late Pope Clement, "that generous Maecenas of my studies," had put at his disposal, and he could not find or rent a space that provided the necessary unobstructed view to the western horizon. Probably prospective patrons did not want him to punch a hole in the roof of any room allocated to him, as he had done in 1716. The business would have to be postponed to 1732, a date that, as Bianchini predicted correctly, he would not be able to keep.[30] Meanwhile, however, he could employ a more complicated makeshift of two planets and *several* stars. On 19 November 1727 Venus and Saturn, though widely separated in the zodiac, had the same declination. It will not be necessary to follow his refined measurements with several clocks and telescopes and his stellar observations, which resembled those at the Clementine *meridiana*. The outcome was an astronomical unit of 114 million km, exactly what he would have obtained if he had assigned 3 clock ticks to parallax in 1716. Bianchini seems to have had a greater interest in the process than in the product, for, referring to the measurement of 1716, he gives 11,200*a* in one place and 13,400*a* in another.[31]

After Galileo's disclosure of moon mountains and solar spots, astronomers searching the planets for surface features had found the belts and red blot of Jupiter and cracks in the rings of Saturn. Cassini had glimpsed a spot on the fresh face of Venus in 1677 and 1678 using a lens of 23 palms, but no one since had found any flaws. A more exquisite scrutiny began when Cardinal de Polignac, after serving for many years as papal ambassador to France, returned to Rome in 1724. During his earlier stay in the city, he had supported the master optician Giuseppe Campani in developing mountings for very long telescopes and now wanted to try them on astronomical objects. There was reason to believe that they would reveal something of interest, for, even in the damp of Paris, Cassini had discovered two small moons of Saturn using Campani lenses of 150 palms and 205 palms in a tubeless telescope. Since this discovery, made in 1684, nothing of any significance had been detected through super-long lenses.[32] Polignac thus had hope and hardware and easily acquired the needed software, his friend Bianchini, the world's greatest performer on long telescopes and, as Campani's executor, another source of long lenses. In a preliminary test, they tried whether they could spy something on the moon through a Campani lens of 150 palms (36 m) that no one had reported before. Immediate success: on 16 August 1725 Bianchini saw a reddish shaft of light streak across the floor of the large crater Plato. He supposed that it might be sunlight piercing a previously unknown rift in the crater wall or an atmospheric effect—an encouraging observation though transient, and still, in 2022, unexplained; he no doubt saw something, however, as sightings similar in appearance and mystery have been reported from time to time in Plato and other lunar craters.[33]

On to Venus. Bianchini and Polignac planned a first look through a telescope of 100 palms for the week following their discovery of the Platonic gleam. Malign stars again intervened. On the 17th the monsignor, his head still in the moon, had his fall on the Palatine. When he was able to pursue Venus again, the planet was 40° above the horizon at sunset, a very nice target had it not been necessary to anchor the distant objective lens 20 m above the level of the eyepiece. Bianchini therefore sought viewing places beneath a structure, like a bridge, on which he could mount a pole of reasonable height to provide a line of sight from below the bridge to Venus. His familiarity with the palaces of his patrons and the churches of Rome supplied several eligible sites: Palazzo Barberini on the Quirinal, a garden attached to a church on the Esquiline, a villa at Albano built by his old collaborator the artist Carlo Maratti. From these places, in February and March 1726, Bianchini, Polignac, several other dignitaries clerical and lay, and a Scot scholar promisingly named Hope saw, or said they

saw, smudges along the terminator (the border between the dark and illuminated portions of the surface) of Venus.[34] If they had been able to see
further than Cassini had sixty years earlier with a telescope of 23 palms, it
was not by standing on his shoulders but by using longer lenses.[35]

The magnification of 112 made smudges along the terminator as large
as lunar features appeared to the naked eye. Consequently, Bianchini
warned, only observers could be depended on who could make out the
Moon's maria without the aid of a telescope. Some keen-eyed observers
corroborated his deduction from the motions of the markings and the
relative positions of the sun, earth, and Venus that Venus rotated once in
24 1/3 days on an axis inclined 15° to the ecliptic. The axial rotation, which
Cassini had conjectured to be twenty-three hours, seemed anomalous-
ly lethargic, but Bianchini's careful procedure carried conviction. From
these and subsequent observations, he deduced a map of Venus. Contemporaries were nonplussed. Was there nothing Monsignor Bianchini could
not do better than anyone else? Since no one had the vision, technique, or
tools to duplicate the survey, he had to put up with many visitors eager to
see the spots on the "Mirror of Divine Beauty, the brightest of all bodies,
the brilliant star of the morning."[36]

One evening in August 1727 this company of eager visitors included a
nobleman sent by the duke of Parma, an agent of the Holy Office, a marquis, two abbots, and Bianchini's steady supporters, Maraldi and Polignac.
All seemed to agree that they saw what Bianchini saw. On 7 January 1728,
in the face, apparently, of clear-sighted doubters, he collected signed
testimonials from those present, including two priests, a professor of
theology, and three laymen. "Vidi et testo," each said. After careful study
of Bianchini's papers, Manfredi did not doubt that his friend had detected
the marks and thence correctly inferred the rotation of Venus. We must
be grateful, he wrote to a colleague, that Bianchini had been able to put
the finishing touches on his "immortal book on Venus, in which there
are so many excellent discoveries in no way inferior to those that conferred glory on Galileo, Huygens, and Cassini."[37] The book scarcely had

an equal, according to an English reviewer, for "the novelty of the Phaenomena, the great labour and industry required to discover them, the exactness and accuracy of the Observations, or the consequences necessarily
resulting from them, either with respect to the planet Venus in particular, or that of our Solar system in general." All this was more than
enough, so Maraldi wrote Bianchini, "to enroll you among the greatest
of astronomers."[38] Alas! Despite his technical virtuosity and willing witnesses, the whole business was an illusion. Venus preserves her privacy
with a cloud cover impenetrable to ordinary light.

It is not easy to account for the mistaken observations of so skillful
an observer. Among causes suggested are dust on his telescope lens and
unusually persistent cloud formations in the Venusian atmosphere.[39]
There is also the intriguing possibility that optical illusions arising from
working at the limit of resolution of Bianchini's very long telescope were
at fault. In any case, his error allows an estimate of his authority. Maraldi
ascribed his inability to see the features to the weakness of his telescope
and the haziness of the Parisian sky. Carbone and other astronomers
accepted Bianchini's Venus as a true rendering, although they could not
confirm it.[40] As late as 1877 the astronomer Camille Flammarion, well
known for his high-level popularizations of science, reproduced some of
Bianchini's sketches of Venusian features, although he had only twice seen
something like them himself. He supposed that rifts in the cloud cover
during tranquil days on Venus allowed glimpses of the surface. But soon
Giovanni Schiaparelli overcame the authority of his countryman, proved
the impossibility of observing the Venusian surface in ordinary light,
and conjectured that Bianchini had described cloud formations. While
correcting one optical illusion, Schiaparelli inspired another, by calling
the lines he detected on Mars *canali* (channels); which, mistranslated as
"canals," gave rise to the idea of Martian engineers.[41]

In working up his field drawings for publication, Bianchini insensibly
gave the splotches greater prominence and distinction. Clear rendering, one of his strengths, turned into a fault; it reified what at best must
have been fuzzy, kaleidoscopic blotches. The process has attracted the

attention of several recent historians concerned with the stabilization of images in science.[42] Besides practical and pedagogical considerations, the stabilizer's expectations come into play. Bianchini wanted Venus to have lunar features to allow him to make naming gifts, as Galileo had done in calling the satellites of Jupiter the Medici stars and Cassini had done in bestowing a few moons around Saturn on Louis XIV. Bianchini had a larger treasury than either of his predecessors; like the argonauts, whom he regarded as the originators of the practice of naming celestial features after people, he had a pristine globe to populate. Naturally, he awarded the biggest Venusian feature, an ocean, to João V; three lesser ones to João's seafaring forebears, and one each to Columbus and Amerigo Vespucci. Nine other explorers, mostly Portuguese, shared as many straits and promontories. Bianchini awarded the last of the seas (the seventh to anyone counting) in Venus's midriff to "the prince of them all," Galileo, for his inventions useful to sailors and his pioneering investigations of the planet's phases. Galileo's waters ran into the strait of Cassini and skirted promontories dedicated to the academies of Bologna and Paris, which represented academic patronage of astronomy because of the support they had given Cassini. Bianchini thereby disposed of all notable Venusian features apart from two oceans, one at either pole. The northern went to Marco Polo, the southern to Magellan.[43]

There may be more here than repaying patronage. Apart from King John and, perhaps, his relatives, Bianchini had free choice in his bestowals. In picking Galileo, Cassini, and the academies, he acknowledged his own intellectual heritage; and, in giving companies as well as individuals a place on the planet, he further underlined the importance of academic collegiality for him. As for the explorers, they spread "the glory and power of Europeans" east and west, and, more importantly, acted as "collaborators and assistants to ministers of divine grace in procuring eternal salvation for innumerable nations."[44] If we read Catholics for Europeans and Turks and Britons for the nations most immediately in need of salvation, we can translate the arrangements on Venus into a more elaborate version of the message on the floor of Santa Maria degli Angeli. Bianchini had some trouble deciding on the definitive list of the missionary helpers, since most of them were bandits. A pity! But, although we may not like the means, we must acknowledge them: history is not morality.[45]

Bianchini designed an elaborate armillary to bring his Venusian discoveries immediately to the eye and understanding (Fig. 39]. A small globe

Fig. 39. Armillary of the Copernican orbits of the earth and Venus. FB, *Hesperi et phosphori nova phaenomena* (1728), tab. V, at end of book.

V, 1.25 cm in diameter, marked with seas, straits, and promontories by a myopic miniaturist sits in its harness KH with its polar axis LN fixed at 15° above the large horizontal circle XZ representing the zodiac and the plane of the ecliptic. The harness is attached to a semi-circular slip of metal CVE 15.25 cm in radius and free to revolve around the central pillar of the instrument; the slip GΩF, 20.3 cm in radius, carries a similar contrivance for the earth Ω. A candle at A, or, better, a lamp with a lens, acts as the sun. To follow the phases and enjoy the markings of Venus, the observer moves V and Ω into the zodiacal positions for the day of interest and views the miniature Venus from the place of the slightly larger earth. Bianchini gave as an example the situation when Cassini thought he saw a spot on the planet: looking from the model earth, the historian–astronomer would see exactly what Bianchini thought his predecessor should have seen. A similar machine could be made to present the same phenomena on Tycho's system, in which the earth rests, while the sun, carrying all the planets, revolves around it. That would require making the candle-holder move along the line ΘΔΛ (the dotted arc on the base of the armillary); to accommodate its entire orbit together with that of Venus demanded a larger and more complicated machine. That would be no problem for God, if He had wanted to make a Tychonic world, but it made a difficulty for craftsmen in this one.[46]

In return for João's patronage, Bianchini undertook to procure a pair of globes, among the largest and most beautiful available, for the royal library, and to explain the mysteries of Italian time-keeping. In return for John's gift of a Newtonian reflecting telescope, Bianchini sent a model of his Venus system, beautifully worked in silver and gilt.[47] He also had made a few large globes showing the named territories, which he gave to the academies of science of Paris and Bologna, and to the cabinet of King João.[48] What else should the king acquire to bring Portugal into the cultural league of the leading European nations? At his request, Bianchini drew up an account of the most notable rarities to be found in Italy. Of course, he did not try to fulfill this order literally, but took advantage of a trip to a spa in Tuscany to report on the rarities of Florence, Parma, and Marsigli's

Institute in Bologna. Of Florence he says nothing that would not occur to a modern traveler. In Parma he mentions the theater with its superb acoustics (which can still be seen and heard), galleries filled with paintings by Correggio and Parmigianino, and gems, statuettes, inscriptions, and objects from the Palatine excavations; but what he recommended as particularly engaging were waterworks and hydraulic games in the gardens of the Farnese villa at Colorno.[49]

The most pertinent example for encouraging the king "to bring to a still higher level the liberal arts and sciences already protected and promoted by [him]" was the Bologna Institute, founded by a general and supported by cardinals and popes. To make the example easy to follow, Bianchini described the content and purpose of each room in the Institute. He began with Clementine fine arts on the ground floor, with sculpture, painting, and architecture; rose to the *piano nobile* for natural history and the physical sciences and their applications, especially to military matters, fortification, artillery, siege works, everything pertinent to "this noble profession;" and climbed further to Manfredi's new observatory, marvelously equipped with Campani lenses, globes, clocks, and models of world systems. Nothing equaled the Bologna Institute. "Whence it can be reasonably proposed as an example to any protector of the sciences, who wants to establish such a thing." The observatory was reached by 225 steps, ten of which made Bianchini's height of 6 Roman feet (180 cm). From which we learn that he was a tall as well as an energetic man.[50]

It is a cruel but instructive irony that Bianchini's last substantial work in astronomy should have foundered on the very ground he had made uniquely his own. His combination of words and images in astronomical and historical investigations had won him his scientific reputation. It had succored him in his maps of cometary journeys, universal history, description of ancient coins and medals, trigonometrical survey, and early archaeology, but it had also exposed him to serious error, as in his reconstruction of the Palatine palaces. It was particularly risky in his examination of Venus. The wish to find features fit for naming encouraged the rendering of fleeting blotches as permanent markings. Giving a name to even real features does not of course guarantee preservation of the identity of the party honored.[51] The Portuguese heroes have vanished from Venus. In compensation, we have craters on the moon and Mars named after Bianchini, in one case ingloriously identified as "an Italian lawyer and keen amateur astronomer."[52]

The Cosmos

In describing the phenomena from which he deduced the axis and revolution of Venus, Bianchini represented the solar system by a Copernican diagram (see Fig. 40). The armillary he designed for three-dimensional appreciation of the phenomena has the sun at the center encircled by the orbits of Venus and the earth. From neither representation, however,

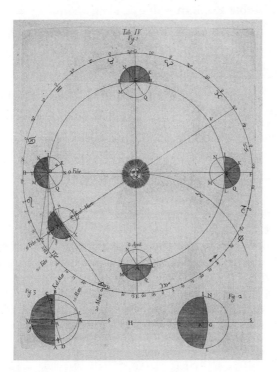

Fig. 40. Planisphere of the Copernican orbits of the earth and Venus. FB, *Hesperi et phosphori nova phaenomena* (1728), tab. IV.

was it safe to infer that the true system is heliocentric. Venus's phases showed that the planet circled the sun, which it does in both the Tychonic and the Copernican system; and the armillary could be made Tychonic by increasing its size at the same scale by 75 percent. "For the sake of economy we prefer to use the smaller diagram of the Copernican system ... I thought it necessary to point this out in case anyone should deem that this view of the phases favored one theory more than the other." Diplomatic as ever, Bianchini no doubt had in mind that Portuguese astronomers taught the Tychonic system and that most instruction in the schools had not advanced that far. But no one could ignore that, for practical purposes, the system championed by Galileo had the edge. "In using the more compact one, we believe that our exposition will be easier to follow." Anyone can work out the phenomena on the Copernican armillary, "at a single glance," whereas a Tychonic one would require "rulers and little circles rotatable about movable centers, and [complicated] systems of numbering." Still, it presented no problem. Double explanations were standard in textbooks, "and indeed everyone knows the conclusions both in the system of Tycho and that of Copernicus."[53]

The strategy of reasoning like Copernicus while allowing the possibility of the Tychonic system had been used to circumvent the condemnation of heliocentrism since the latter part of the seventeenth century. By the 1720s most censors, and notably the reviewer of Bianchini's book on Venus, allowed it. Here is what the reviewer of Bianchini's book on Venus wrote.

> The most erudite and, in matters of astronomy, most expert author prudently put forth all his facts the more truly to be comprehended, so that no one could seek arguments from them in support of [either] of the two most widely held systems of the world rather than of the other, and he shows clearly how the phenomena can equally well be explained according to either system. Whence [the reviewer somehow deduced] astronomers, of whose discipline the Church stands greatly in need for computing Catholic feast days ... will rejoice that heaven appears day by day more revealed.

The compliant author of this opinion was Bianchini's friend Gianfrancesco Baldini, who assisted in observations at Santa Maria degli Angeli and would help edit the fourth volume of Bianchini's *Anastasius*.[54]

Bianchini thought in Copernican terms and, in unguarded moments, wrote that way as well. In an early essay on chronology, he spoke of the "grand ellipse that Saturn describes around the sun," which can be construed, though tortuously, in the manner of Tycho; whereas his casual statement in Nova phaenomena some thirty years later that he had to break off observations, "because the Earth's rotation carried the planet to a part of the sky obscured by the actual palace of Barber[in]i," must be read as Copernicus's diurnal motion.[55] And, although he liked Cassini's method of obtaining planetary parallax in part because it was independent of the choice of world system, in deducing from it the distances of the planets Bianchini referred to their rotations around the sun.[56] Moreover, as Manfredi observed, making Venus moonlike helped to elevate the earth to a planet via Galileo's discovery that the moon is earthlike.[57] All astronomers agreed, however, that the strongest evidence in favor of the Copernican system would be detection of an annual parallax of the "fixed" stars. Bianchini began a search for it around 1716. And he found something close to what he sought.

He found that certain stars in the constellations Capella and Lyra appeared to move in tiny circles as the earth ran its course around the sun. But, since the directions of the earth's motion and the star's rotation that he detected did not fit that expected for stellar parallax, he did not publish his results. A grave misfortune! His friend in Genoa, Salvago, recognized their importance, "for the eternal glory of your name and [the excitement of] jealousy among the oltremontani." An Englishman, James Bradley, independently observed this "aberration" and attributed it to differences in the direction and magnitude of the velocities of light and of the earth around the sun. Manfredi examined Bianchini's journal of observations and determined that he had seen what Bradley saw. "Thus Bianchini's observations support Bradley's laws of aberration with new, very important, and entirely unexpected evidence; for Bianchini recognized in 1719 the extent of the motions in Capella and Lyra, that is, eight years before Bradley devised his ingenious hypothesis for explaining the motions of the stars."[58] This hypothesis, Manfredi pointed out, requires the annual motion of the earth; and although not, perhaps, as persuasive as stellar parallax, it seemed to many hesitant cosmologists to settle the question of the world system.

Why did Bianchini not continue his exploration of the annual motion of the stars? He knew that he had touched an important issue. With a change in stars and instruments he might have accumulated enough data to prompt the discovery of aberration, though not, perhaps, of its cause. Here a modern astronomer might diagnose a woeful lack of inspiration and, perhaps, a reluctance to commit himself to the requisite motion of the earth against his church's official position. In his *éloge* of Bianchini for the Académie des sciences, Fontenelle took him gently to task for "his great care always to indicate how everything can be made to accord with Tycho."[59] That was to ignore the spirit and harmony of Bianchini's life work. His models in science were his teachers Montanari and Cassini; the one for his Galilean methods and willingness to suspend judgment, the other for his *meridiana* and discoveries of planetary features.[60] Bianchini admired exact description, insightful inference from careful observation and experiment, and phenomenological laws; although interested in every detail of the natural history of the heavens, he did not, like the Cartesian Fontenelle, concern himself overmuch with the truth about the world system. "Nature is more fertile in invention than we can comprehend."[61]

Bianchini had tried to follow the directions for philosophizing properly that he had given to the Aletofili forty years earlier. The message may be recalled: we make mental pictures to represent our knowledge; the pictures cannot penetrate to the bottom of things; they are largely free choices, although we must be able to demonstrate their verisimilitude publicly using evident reasons and be willing to change them when experience requires it. "[Our task is] to reduce the causes of all sense experience to the few, clear principles of quantity and motion that nature presents to everyone as unquestionable and that mathematicians suppose without needing proof."[62] This was the philosophy that Montanari had instilled so thoroughly in his students that Marsigli reminded Bianchini of it late in the Venus investigations. On Christmas Eve 1726, the general advised the monsignor that "the modern method of observation is the right one, but it is still in its infancy, and we must not be so rash as to expect instantly to deduce systems from observations."[63] We will almost certainly never know the essence of matter, or the shapes, sizes, and numbers of the ultimate particles. But that should not bother anyone who strives to make a faithful portrait of carefully culled observation and experiment—provided undetected hopes and expectations do not lead the portraitist astray.

Bianchini's approach to history also followed the epistemology he presented to the Aletofili. As a historian, he interested himself primarily in facts, in exact dates supported where possible by astronomical events, in hard objects such as coins and medals. He rejoiced in finding parallel stories in gentile histories and Jewish scripture that confirmed capital events like the universal flood. His religion taught him how history began and how it would end. His task as historian was to document the journey, not to explain its meanderings. He did not feel obliged to devise or defend a moral history any more than he felt constrained to assert a world system. He saw no opposition between his science and his faith, and no danger to believers in following up his investigations. "[My methods] can be repeated at will by whoever loves to revere and contemplate the workings of Divine Wisdom in the disposition, immensity, and motions of the heavenly bodies."[64]

The frontispiece of *Nova phaenomena* represents much of what a censor of Bianchini's posthumous essays called his "lofty learning combined with a singular piety and a very well-regulated mind."[65] Under the angel trumpeting something new sits a Minerva pointing toward Venus against a background of the ancient monuments among which Bianchini mined antiquities and settled his long telescopes (see Fig. 41). On the left, an Atlas carries the Farnese globe shown with the Centaur who invented the constellations and taught astronomy to the argonauts turned toward the observer.[66] Under the Atlas a putto offers a simulacrum of Venus, shown with her newly discovered seas. On the ground lie symbols of

Fig. 41. João V contemplating a world without a center. FB, *Hesperi et phosphori nova phaenomena* (1728), frontispiece.

geometry and astronomy and a puzzling piece of arithmetic. The multiplication may relate to a cycle of Venus's appearances; in five of its sidereal periods there are as many days (1,120) as contained in three Julian years plus one Venusian revolution of 24 $^1/_3$ days, the sort of numerical coincidence in which Bianchini delighted. At the center, on the pedestal, sits the image of João V, to whom an attractive assistant presents an armillary of the Venusian system with an unoccupied center. She is too modest to insert the body that would indicate which system of the world she favors. Like Bianchini, she prefers living quietly to grubbing publicly for the unreachable truths of natural and human history.

9

EXEUNT

The Others

Of four protagonists, Maria Casimira, Dowager Queen of Poland, was the first to depart this life. First she left Rome, in 1714, pressed by debt; she had lived well beyond her means and retreated to France, where she died in 1716.[1] Consequently, she was spared watching the supererogatory fasting and other gratuitous mortifications that destroyed her granddaughter's health and charm. In 1735, at the age of thirty-three, Clementina collected her eternal reward. The reigning pope, a twelfth Clement, had her buried with all possible pomp. Everyone turned out to watch her cortège pass from the church of the Holy Apostles, across the street from the Palazzo del Re, to the Vatican; a contemporary likened it to a triumph of the Caesars, except that their acclaim depended on the spoils they had seized, whereas hers was owing to virtue alone.[2]

To commemorate the first anniversary of Clementina's death, the Congregation for the Propagation of the Faith published a handsome booklet with many celebratory poems in strange languages and scripts and an admonition to "the Prince of Wales," then sixteen, to take up the great cause. A specimen of this unseemly piece of papal Jacobitism runs as follows: "Why the delay? Wake up! Relieve the parched Anglican kingdom and your Britannic subjects ... Your father expects it ... Your mother hopes for it and brings strength from the high heavens, where she reigns happily and readies honor and glory for you." A footnote explains that, on the prince's birthday, a very bright star appeared not seen before, "to the great wonder of everyone."[3] Not quite everyone; lynx-eyed Bianchini missed it.

In 1743 Clementina, who had wanted to be queen of England, took up permanent residence in St Peter's across the nave from Christina, who had not wanted to be queen of Sweden. Moving in had its little difficulty. A bust of Clementina, in full color mosaic, enlivens her tomb. It scandalized the rich imagination of the administrator of the Vatican Cathedral, who insisted that she appear as a penitent. James preferred that posterity see her as he first did. The pope agreed and the mosaic, by Pietro Paolo Cristofori, shows her in fashionable décolleté, a little older than at her wedding. Clementina gazes beyond the viewer to a monument to her husband and her sons begun four score years after hers. Intended originally to commemorate only her younger son, who, at his death in 1807, was the Dean of the College of Cardinals and, to lovers of lost causes, King Henry IX of England, Scotland, etc., its scope increased in a design proposed by the famous sculptor Antonio Canova. Lack of funds prevented its realization until the prince regent of England, acting for his mad brother King George III, picked up the bill. Thus did the Hanoverians remember their predecessors, and celebrate the end of the Stuart line, with a monument conveniently located 2,000 km from London.[4]

While in St Peter's, we should cast an eye at the magnificent tomb of Bianchini's patron Alexander VIII Ottoboni. It matured over the interval between the jubilees of 1700 and 1725. The casual visitor will notice it for the contrast between the pope, presented in dark bronze, and two large white angels, and for its rich assortment of colored marbles. But what claims our attention is the frieze below the main action (see Plate 15). It celebrates Alexander's canonization of five saints in 1690. In gratitude, three men offer Alexander gifts. The first of these magi carries a lamp in antique style designed by Bianchini. The tall young man standing behind the gift-bearer and gesturing toward the lamp is almost certainly Bianchini himself, acknowledging the patronage that gave him his start in Rome.[5]

Clement XI does not have a monument in St Peter's nave. Instead, at his request, he lies under a slab in the ornate chapel of its choir. His memorialists gave him high marks for character and stewardship. One of these appreciations, a full-length biography, came to Bianchini as reviewer. He rated the work excellent for bringing out Clement's "virtue, science, abstinence, patience, piety, love, in a word, the charity of God," qualities rarely found together except in obituaries, but which he, Bianchini, had observed in Clement "in more than thirty years, publicly and in private, daily and intimately."[6]

Although Clement does not reside in St Peter's nave, his guiding hand may be seen there on the monument he raised to his predecessor a few removed, Innocent XI. Clement admired Innocent for bringing about the Holy League that enabled Sobieski's victory at Vienna, for disciplining fast-living prelates, and for raising charitable works for other Christians. Later popes did not think the case for Innocent's beatification compelling. In 1956 his corpse migrated to a transparent sarcophagus under the altar of St Sebastian, where his further activity could be monitored. It is said that the Vatican intended to sanctify him in 2003, to recall the old Holy League in the aftermath of 9/11, and that an accusation in a historical novel canceled his canonization. The novelists charged that Innocent had favored the ouster of James II to advance the banking interests of his family. Perhaps for this reason the Vatican turned Innocent out of his sarcophagus to make room for the blessed John Paul II.[7] The story of the aborted canonization, which the Vatican denies, suggest that the Turkish–Stuart policies of Clement XI and his predecessors have not lost all relevance.

Clement's candidate for Sobieski's counterpart in the west, the Old Pretender, tried twice again in the 1720s to seize Britain by force. The earlier attempt, in 1722, took him no further than Lucca; the second, in 1727, intended to upset the accession of George II, got no further than Lorraine. James then returned to Rome and another war he could not win. The disaffection of his queen was diminishing him financially as well as politically, for Benedict XIII, Innocent XIII's successor, had assigned a substantial part of his papal pension (then 16,000 scudi) to Clementina during her stay in her convent. James conceded much to obtain peace. Among the few things the dysfunctional royal pair could agree about easily was that their elder son should be raised a Catholic and taught to think himself a king. Bonnie Prince Charlie had no chance of a restoration, but in 1745, true to the spirit and illogic of his house, he tried armed intervention.[8] His followers paid dearly for his fantasies. Charlie returned in defeat to live and die in Rome. His father began his everlasting life on the first day of 1766. Following Clementina's example, he lay first with the Santi Apostoli and thence proceeded, in a grandiose procession, to the Vatican. An orator at his funeral extolled him as "the honor, ornament, and exemplar of Rome and the entire Christian world." He had suffered grievously by preferring exile "to betraying his conscience and perishing forever."[9]

The gallant group that rescued Clementina dispersed in the early 1720s. The Missets and Jeannot accompanied Wogan to Spain, where Misset died in 1733 and Jeannot in 1739. Mrs Misset lived into the 1750s and perhaps later. Gaydon and O'Toole returned to Dillon's Regiment. Sir Charles Wogan distinguished himself in the celebrated battles to recover Oran, seized by the Moors while Europe disputed the Spanish Succession. On the eve of a fight in February 1732, he composed the anonymous letter to Swift mentioned earlier. He had been "one of the merriest fellows in Europe," he wrote by way of introduction, but had sunk into sullenness during exile, and now "lash[es] the world with indignation and grief." On the morrow he would have to cut off the heads of some Saracens "because they will not become Christians." Not that he had the least objection to their religion. "With all my spleen and vexation of spirit . . . I would not shed one ounce of blood in anger or enmity, or wrong any man of a cracked sixpence, to make all the world Catholic; yet I am as staunch a one myself as any pope in the universe."[10]

During the siege of Oran in the summer of 1732, Wogan was ordered to protect a convoy of ammunition and provisions. Almost hit by canon shot, "so covered with Dust, that he could hardly see," he managed to secure the convoy before a bullet forced his retirement from the field. He recovered from his wound to become governor of Don Quixote's homeland of La Mancha, whence he helped agitate for the quixotic '45.[11] In that year he wrote his memoir of his rescue of Clementina, which he dedicated to the queen of France, the third queen (including the Spanish) with whom he had struck up a friendship. He would wait to publish it, he said, until Clementina began to move toward beatification and James toward restoration; the conditions were not met, and Wogan's memoir slept undisturbed for 150 years.[12]

In addition to his very long confessional letter, Wogan sent Swift the poems and verse commentary on the penitential psalms mentioned earlier. He hoped that some of it might be publishable after improvement by Swift's trenchant criticism; he had not been able to profit from the comments of his unlettered comrades at arms, since none of them had "genius or freedom of thought enough to comprehend these notions."[13] This was not quite true; in Spain he met Philip, duke of Wharton, like him a Jacobite man of letters in the Spanish army. They had also in common an admiration of Swift and a requited love for a woman, luckily not the same one, serving the Spanish queen. Wharton wrote an appreciation in rhyme to accompany Wogan's blank verse commentary on the penitential psalms; "Ev'n SATYR shall grow dumb, and owne it true | That MILTONS Spirit still survives in you." The appreciation ends lamenting their common fate:

> These lines, MY FRIEND (LATE and IN EXILE found)
> Test of our Union, must with yours be bound
> That both may share a happyer Fate than Ours
> Wellcome at home, whoever guard the doors.

But Wogan's accomplishments gave grounds for hope.

> Your LAWRELLS, as your BAYS, be ever green
> GREAT in your VERSE, as on the MARTIAL SCENE
> WHOSE ESSAY WAS TO FREE A CAPTIVE QUEEN.[14]

From such clues in the manuscripts, Swift worked out the identity of his correspondent and replied, with compliments. He thought Wogan's style remarkably good for an exile, a soldier, and an Irishman.[15]

Swift's literary friends, whom he entertained with Wogan's "poetical history in prose of your own life and actions," agreed. Wogan was only too good: Swift, Pope, and Gray were just rhymers, trying to make folks "merry and wise, whereas your Genius runs wholly into the grave and sublime." Consequently, and because the Irish do not read, Wogan's hope to make some money from his writings to give to charity would not work in Ireland. Nor would his history of government under the Stuarts have done well in England. It is easy to see why. Wogan blamed parliament for their fall. James I, having squandered the lands left by Elizabeth, had to negotiate with it, "[a] forlorn and desperate situation." It refused to keep Charles I and Charles II solvent; the first turned beggar, the second, rake. James II had some income and inclination to rule, but parliament canceled him when the Tories agreed to drive out Catholics. The Tories remained supine, the exiled Stuarts hopeless, the Hanoverians impregnable. The only hope was an independent Ireland, standing to England as Portugal did to Spain.[16] Such an outcome may no longer be farfetched.

In return for Swift's attention, Sir Charles sent him a kilderkin of good Spanish wine (some 16–18 gallons), which was quickly consumed at literary sessions of the kind at which Wogan's work was read. Swift asked for recommendations for a good Spanish substitute for claret, "for my disorders, with the help of years, make wine absolutely necessary to support me."[17] Wogan had time to supply another kilderkin or two. He died, unpublished, in 1754, twenty years after his first approach to Swift.

The Monsignor

Bianchini retained his positions under Innocent XIII, who resembled his predecessor Clement in character and policy. The new Innocent contributed to war chests to fight the Turks and conquer Britain, affirmed *Unigenitus*, and tried to restrain the Jesuits. He added to Bianchini's burden of honors and income by appointing him a *Referendarius utriusque Signaturae*, a title awarded to a prelate slightly below a bishop distinguished for responsibility, experience, and trustworthiness. Many *referendarii* served as administrators for absentee bishops and as judges or advisors in the "Signatures" (bureaus) of Grace and Justice. These last dealt with rare and delicate cases of simony, non-residency, matrimony, impediments to ordination, and so on, cases requiring judges with good sense as well as a knowledge of canon law. The appointment, which, owing to the number of his colleagues, probably did not exercise Bianchini's talents often, gave him the right to wear still another distinctive purple costume.[18]

Bianchini described himself as *referendarius* (and as a canon and domestic prelate) on the title page of an oration he delivered to the College of Cardinals as they prepared to choose a successor to Innocent, whose short reign ended in 1724. His words, "delivered from the pulpit with his wonted energy to universal applause," advised the conclave to await the guidance of the Holy Spirit, and, while waiting, look around for someone better prepared than most of them to carry the weight of the Holy See. This Atlas should also be a man of prudence, gravity, rectitude, and zeal. The Holy Spirit told the Albani to put forward Paolucci; the emperor told his cardinals to veto him; the conclave settled on a pious old man, a Dominican, who vetoed himself. Bianchini had anticipated the possibility. He had urged that the cardinals deal with refusal as their predecessors had in 1700 and bring forth a pope as good as Clement XI. After two days of browbeating, the Dominican yielded and called himself Benedict XIII. He was no Clement. Weighed down by superfluous rereading of the twelve huge volumes of church annals compiled by Cardinal Baronio over a century earlier, he had no idea how to rule and allowed his favorites to pilfer the treasury that Innocent XIII had left in surplus. All he could do for Bianchini was to appoint him archivist of Santa Maria Maggiore and order him to collect all decrees and constitutions about the basilica.[19]

One of Benedict's first impulses was to call a synod for the jubilee year 1725 to discuss the problems of clerical discipline and doctrine afflicting his diocese (the "province") of Rome. He appointed Bianchini as its historiographer. Nothing could have been more appropriate were it not that almost nobody wanted a synod. Although the Council of Trent had mandated that bishops hold synods with their clergy, few dioceses held them with any regularity; the popes, though bishops of Rome, did not call them at all; and cardinals without Roman jurisdictions were not eligible to attend. Nonetheless, most prelates in Rome went to the opening session of the synod on 15 April 1725, which gave its historiographer something to write about.[20]

Bianchini did not undertake under Innocent XIII or Benedict XIII ambassadorial duties of the kind Clement XI had assigned him. His last such duty, performed late in 1720, was another hat delivery, this time to a Venetian in Brescia, Giovanni Francesco Barbarigo. Why not a higher-ranking prelate? "We are sending you our beloved son Francesco Bianchini, our *cameriere d'onore*, who is very dear to us for the rectitude of his life and the preeminence of his mind and capacious learning," qualities already known to the new cardinal. Bianchini had for some time enjoyed a pension from the church of Verona for his physical maintenance and for "nourishing my spirit with the books I need," which Barbarigo, as the town's bishop before his translation to Brescia in 1714, had seen paid.[21]

Bianchini passed through Verona on the way to Brescia to borrow some lackeys from his relatives to fill out his retinue of four servants, a secretary, and a chaplain, announced his arrival in Brescia by special messenger, donned the appropriate peacock vestment (*veste pavonezza*), and entered the town as a returning hero. All the local worthies and many Venetian officials paid their respects. Music, parades, flags, salutes greeted the monsignor, as if he were the party receiving the distinction.[22] As usual with the princes he served, Bianchini established friendly relations with Barbarigo. One of the projects that he had in hand when he died was the canonization of Barbarigo's uncle Gregorio, who had been a strong candidate at the

conclave that elected Alexander VIII in 1689.[23] Had Barbarigo won, Bianchini might not have written his universal history and almost certainly
would not have laid down his *meridiana*.

During his triumphal march through the Veneto, Bianchini visited
Montanari's widow in Venice and a factory outside Brescia he had inspected some thirty-two years earlier together with his patron Correr. The
factory made instruments of war. Our gentle monsignor remembered
seeing a great wheel used to polish steel; he had returned to measure its
speed. He found that a point on its circumference traveled a mile in two
minutes, about as fast, he said, as a horse he once clocked racing from
the Piazza del Popolo to the Piazza Venezia, a mile and a third in three
minutes and a half.[24] The equation falters arithmetically ($16 \neq 21$) but not
psychologically. As we know, Bianchini used measurement to master his
experiences.

After his accident on the Palatine, he visited several spas to ease spasms
in his leg. He became an expert on their curative waters and latitudes
and other matters of physical interest that he investigated with instruments he took with him whenever he traveled. Despite the waters, his
strength waned. He completed his exposé of Venus, his account of Livia's
columbarium, and the third volume of his *Anastasius*; but he had to leave
the fourth volume as well as his dissertations on the Farnese sphere and
ancient musical instruments, his record of a quarter-century of observations at the Clementine *meridiana*, and many smaller pieces to his friends
to publish. There were so many other things to do: before noon, duties
as canon at Santa Maria Maggiore, "so that I cannot devote the most precious span of my time, my mornings, to my studies;" after noon, favors
to run for lesser gods, princes, cardinals, ambassadors, friends.[25]

A violent fever attacked him on 22 January 1729. A month later, on
19 February, he watched the little miracle at Santa Maria degli Angeli for
the last time. On 1 March he bought some antiques, although he could not
leave his bed without help. Great men came to visit him, but not James
III, who did no more than send an emissary to ask about the state of the

man who had served him so generously.[26] On 2 March 1729 Bianchini attended church, received extreme unction, paid his creditors, "and, having distributed various things to his household, he sat down at a small table with a crucifix and a spiritual book and without the least disturbance ... tranquilly passed to another life." His journey could not have been comfortable, however, as he was wearing a hair shirt. It had been his secret practice for some time.[27]

Following his wishes, Bianchini's fellow canons buried him in Santa Maria Maggiore. A plaque put up by them points out his unusual blend of erudition, purity, and modesty. His modesty will not be offended by pilgrimage to his remains, as they were jumbled together with those of other deserving prelates during repairs to the basilica. In the cathedral of Verona he has done better. The handsome monument there (see Plate 16) adds to the merits recorded in Rome his national and international eminence and his Protestant work ethic. "He was preparing new monuments of rare scholarship to the day he died." In his remarks on native authors included in his guide to Verona, Maffei adds to the familiar character traits "an insatiable desire to please." So great a man could have aimed much higher in the church, could have been another Cardinal Noris; "but his natural modesty and piety did not let him consider it."[28]

In fact, he did much better than Noris. The two greatest contemporary physico-mathematicians, who could agree on almost nothing else, touted their friend Bianchini: Leibniz, as "a most learned astronomer;" Newton, as "a candid seeker of truth." The profound political analyst Montesquieu ranked him among the greatest men in Italy. The master of Maurist historiography, Montfaucon, dwelt on Bianchini's great skill as an antiquarian.[29] The philosopher of history Vico and the art historian Winckelmann, both better known now than the author of the *Istoria universale*, studied it assiduously, the one for its insights into pre-history, the other for its use of images.[30] It is a great book, no doubt, but not for everyone, in the considered opinion of the hard-to-please historiographer Langlet du Fresnoy, though, or perhaps because, it exhibits "an erudition only found in the most illustrious savants of Italy."[31]

This erudition had its practical uses. Among them was to inflate the pride, and symbolize the strivings, of those who would raise Italy to a nation. Count Giovan Battista Carlo Giuliari, canon of the Cathedral of Verona and head of the Biblioteca Capitolare that houses Binachini's manuscripts, chipped out nuggets from this motherlode to stimulate the patriotism of his countrymen. He published Bianchini's account of the fortress of Guastalla, bits of his travel diaries, and as much of his historical card game as the Verona archives contained. The excerpts from the diaries, in which Bianchini appears as an equal to the greatest savants in Europe, seemed to Giuliari a suitable wedding gift for his niece; the idea caught on, and a cardinal and a senator commissioned Bianchini bits, "something more enduring than poetry," to improve their nephews' marriages. The card game carried a more serious message: it bore the imprint, and demanded the recognition, of Italian genius. But foreigners had pinched and adapted it, and servile Italians had reimported and esteemed it, not knowing that it was a native product.[32]

The jubilee of 1725, marked by the second of Bianchini's ellipses, engaged several of our departing players. Bianchini wrote a guidebook to Roman monuments, which he made available to the high-placed pilgrims whose cicerone he was.[33] Pietro Ottoboni became a priest just before, and perhaps for, the occasion, and opened the porta sancta on Christmas Eve, 1724. James and Clementina had special seats for the event.[34] Clement had made some plans to assist the progress and enhance the admiration of the jubilee pilgrims. Chief among his planned amenities was a set of stairs connecting the Piazza d'Espagna with the monastery of the French Minims at Trinità de' Monti. Despite disputes with France over finance and supervision, Roman builders managed to finish the Spanish steps in time. The hope to ready a new Trevi fountain, however, did not materialize.[35]

The jubilee of 1750 saw the prettification of Santa Maria degli Angeli alluded to earlier. As reported, Luigi Vanvitelli, the most prominent Italian architect of the mid-eighteenth century, reworked chapels, put up walls, modified windows, replaced the brick floor with marble, and relocated the main entrance to what had been a side door. Few critics reckoned the result an improvement, and astronomers worried that it would impair the scientific value of the *meridiana*. In the 1730s, before the refurbishment, Sir Martin Folkes set his watch at it and estimated that local noon could be determined there to within two seconds of time. The abbé Jean Antoine Nollet, preceptor to the royal family of France, visited the *meridiana* during the repairs and feared that it could no longer be trusted. Soon it did not matter.[36] While Vanvitelli was compromising the architecture of Bianchini's *meridiana*, opticians and mechanics were rendering it obsolete. They invented achromatic lenses that made long telescopes unnecessary and temperature-compensated mounts that made a sturdy wall as good as a cathedral for securing a stable reference frame. Big churches were no longer scientifically useful observatories. The Clementine *meridiana* declined to a noon mark to check timepieces and convert Italian to European time.[37]

The beauty and mystery of the zodiacal marbles, jubilee ellipses, and historical medallions remain. Only one more of these marks was added after Bianchini's death. It again had to do with Clement's policy toward Poland. Augustus II, the elector of Saxony who succeeded Jan Sobieski, sent his son, Friedrich August, to Rome over the protests of his Protestant mother. With the help of the Jesuits, young Friedrich August discovered that he was a Catholic; his conversion in 1712 put him in line to marry into the Habsburgs and ascend the Polish throne. He did so, as Augustus III, in 1734.[38] Had James III only been as flexible! The last medallion at Santa Maria degli Angeli commemorates the visit of the son of this king of Poland on 6 December 1738, and so can be found at the sign of Sagittarius. The son, Friedrich Christian, proved to be a good Enlightened elector, although a brief one, as he reigned for only seventy-four days. If the medallion was intended as a talisman to effect his succession to his father, or to save Poland from dismemberment, it did not work. The fresh page of history opened by Bianchini on the pavement surrounding his *meridiana* is a graveyard of the unfulfilled hopes of its royal visitors.

The tribe of patrons Bianchini cultivated no longer exists. Electors disappeared with the Holy Roman Empire. Serene Highnesses are now scarce. The kings and queens of France, Spain, and Portugal have gone their way, and what remains of royal prerogative in England would not have tempted James III to leave Rome. The pope's territorial dominions have dwindled to some scattered acres; the Apostolic Palace and Vatican City occupy an area slightly smaller than the US capitol building and its grounds. The bulk of his spiritual dominion now resides outside Europe. Where are the learned cardinals who spent their evenings together balancing between two worlds, raking over early church history while seeking its implications for governance in their time? The Vatican boasts a subtle public monument to them in the realization of a part of the Museo Ecclesiastico designed by Bianchini for Clement XI. This partial realization is the Museo Sacro established by Pope Benedict XIV in 1757 in a corridor leading to the Vatican Library.

While practicing as a devil's advocate, the future Benedict XIV often met Bianchini at the learned academies both frequented and gave him some help in the archives of Santa Maria Maggiore. Soon after his election in 1740, Benedict set up a group to study ecclesiastical history under the direction of Bianchini's nephew Giuseppe.[39] The nephew reworked the uncle's seventeen sketches for the content and layout of the Museo Ecclesiastico into large plates, one of which was described earlier (see Fig. 3.7). Bianchini's plan provided for relics sacred, profane, and Jewish, emphasized emperors and martyrs, integrated texts and objects—in short, proposed to realize in three dimensions and Christian piety the pedagogical methods developed in the *Istoria universale*. Bianchini's insistence that the museum be located next to a library further emphasized the integration of texts and monuments.[40]

To help persuade Benedict to realize the plan, Giuseppe could point to the Biblioteca Capitolare in Verona, which had mounted a miniature version incorporating items from Francesco's collections and (texts count!) most of his books and manuscripts, arranged in an order he had prescribed. Among the many items on history and antiquities may be noticed a copy of Panvinio's *De ludis circensibus*, which had oriented Bianchini on the Palatine Hill.[41] The artifacts, books, and manuscripts were intended, like the Museo Ecclesiastico, "for instruction in the holy Catholic religion and the sciences most fitting the life of ecclesiastics."[42] And the installation at the Biblioteca Capitolare was not the only precedent. Over many years Bianchini had supplied Maffei with objects for his famous Museo lapidario in Verona, and with competent artists for illustrating them. Benedict undertook to bring the portraits of popes in San Paolo fuori le Mura, which Bianchini had discovered in working on Anastasius's *Liber pontificalis,* up to date. That made a project as ambitious as the Museo Sacro, for the most recent of the existing portraits dated from the eighth century.[43]

Bianchini's museum design divided the display horizontally into centuries and vertically into sacred and profane. Strict chronology was to reign, established by annual lists of Roman magistrates and kept before the eye by busts of emperors from originals, copies, or coins, and of popes from the portraits in San Paolo. The objects ordered by this coordinate system were supposed to convey the character of antiquity and the heroism of the church at a glance.[44] In its present form, revised from Benedict's by Pope Pius XI in the twentieth century, the museum is too heterogeneous in time and space to follow the order, or inculcate the lessons, prescribed by Bianchini. But there are some points of which he would have approved. The museum abuts a great library. It offers the viewer coins, catacomb relics, and other items Bianchini recommended and probably some things from his own collections.[45] And so, although Maffei's expectation that the Museo Ecclesiastico would be a perfect counterpart to Bianchini's "most learned Storia universale" was disappointed, enough remained of it in Benedict's Museo Sacro to make it (to use the image on the *Istoria*'s title page) the omega to Bianchini's historical writings.

The Arcadians shall put the omega to this story. In their meeting place on the Gianicolo, the Bosco Parrasio (the Arcadian Wood) completed in 1726 with a grant of 4,000 scudi from João V, there stood, or would have stood if the plan had been perfectly implemented, statues representing intelligence, eloquence, playfulness, and chastity. That is, representing

Bianchini. The stream that nourished the wood was fancifully assimilated to the Castalian and Hippocrene springs to which Greek mythology assigned the sources of truth and poetry; prosaically, it derived from the Acqua Paola, waters for which Bianchini had been responsible.[46] To these symbolic references to Selvaggio Afrodisio the pastors added his portrait and an apt epigram.

> Everything that lies open on our globe he understood
> And everything that lies hidden in the vast universe.
> This, however, was not the perfection of his studies
> The greater glory was to pursue them with decency. [47]

NOTES

The following abbreviations are used:

AB	Anastasius Bibliothecarius, *Vitae romanorum Pontificum ... cura Anastasii S.R.E. Bibliothecarii* (1718–35)
BL	British Library, London
CR	Luca Ciancio and Gian Paolo Romagnani (eds), *Unità del sapere* (2010)
DBI	*Dizionario biografico degli italiani*
DNB	*Oxford Dictionary of National Biography*
FB	Francesco Bianchini
FBC	Francesco Bianchini Papers, Biblioteca Capitolare, Verona
FBV	Francesco Bianchini Papers, Biblioteca Vallicelliana, Rome
HAS	Académie des Sciences, Paris, *Histoire et mémoires*
HP	Francesco Bianchini, *Hesperi et phosphori nova phaenomena* (1728)
HS	J. L. Heilbron, *The Sun in the Church* (1999)
IU	Francesco Bianchini, *Istoria universale* (1697)
IU (1747)	Francesco Bianchini, *Istoria universale* (1747)
IU (1825)	Francesco Bianchini, *Istoria universale* (1825–7)
KS	Valentin Kockel and Brigitte Sölch (eds), *Francesco Bianchini* (2005)
MV	Alessandro Mazzoleni, *Vita di monsignor Francesco Bianchini* (1725)
NP	Francesco Bianchini, *New Phenomena of Hesperus and Phosphorus* (1996)
PT	RSL, *Philosophical Transactions*
RSL	Royal Society of London
SP	Historical Manuscripts Commission, *Calendar of the Stuart Papers* (1902–23)
Uglietti	Francesco Uglietti, *Un erudito veronese* (1986)

Prologue

1. Foscolo, *Opere*, ii (1981), 1917 (quote); Marini, *Iscrizioni* (1785), p. viii, quoted by Carini, *Il muratori*, 1/4 (1892), 145.
2. Noris, in *HS* 154.
3. Cf. Johns, in *KS* 46–9; Sölch, in *KS* 179.
4. *MV* 4; Giuliari, in FB, *Relazione delle cose* (1882), 7.

Chapter 1

1. *MV* 1; Favaretto, in *KS* 28; Romanin, *Storia* (1858), 302–8.
2. *DBI*, s.v. "Bianchini;" Menniti Ippolito, *Fortuna* (1996), 22–5, 169–74.

3. Uglietti, 23–4.

4. Maffei, *Verona* (1771), pt 1, 126–31, 135–41, 169–70, 174–6; the earliest edition is 1731.

5. Maffei, *Verona* (1771), pt 2, 5–11 (ancient Rome), 16–17, 24–6 (Duomo), 60–2 (*Museo lapidario*), 73–6 (Scaligeri), 101–24 (fortifications); for good paintings, 30–4, 43–8, 71–2, 78–9, 86–8.

6. MV 2–3; Federici, *Elogi* (1818–19), iii. 33.

7. *Two Gentlemen from Verona*, I.i, line 6.

8. MV 2.

9. Masini, *Bologna* (1666), 4, 7, 109, 112.

10. De Brosses, *Lettres* (1931), ii. 37–9; FB, in FBC ccccxxx: ii, fo. 57.

11. Masini, *Bologna* (1666), 203.

12. Masini, *Bologna* (1666), 656–7, s.v., "Collegio del B. Luigi."

13. Grendler, *Jesuits* (2017), 305–12; Lorenzo Magalotti to Cardinal Francesco de' Medici, 20 April 1686, in Targioni Tozzetti, *Notizie* (1780), i. 219 ("rarity").

14. Dal Prete, in CR 210–11.

15. Ferroni to Viviani, 12 May 1692, in Torrini, *Physis*, 15/4 (1973), 420.

16. Ferroni, *Dialogo* (1680), "Al lettore;" Targioni Tozzetti, *Notizie* (1780), i. 220–4, for the syllabus to Ferroni's unpublished course in physics.

17. Ferroni, *Dialogo* (1680), 6–11. Cf. Torrini, *Physis*, 15/4 (1973), 412–13.

18. Ferroni, *Dialogo* (1680), 13–14.

19. Ferroni, *Dialogo* (1680), 15–16; Remmert, in O'Malley et al., *Jesuits II* (2006), 296, 303–4; Ribouillault, in Fischer et al., *Gardens* (2016), 115, 123

20. Ferroni, *Dialogo* (1680), 16.

21. Heilbron, in McMullin, *Church* (2005), 288–91.

22. G. A. Borelli to Alessandro Marchetti, 24 February 1659, in Derenzini, *Physis*, 1 (1959), 227.

23. Maffei, in *IU* (1747), a2^v; Carini, *Il muratori*, 1/4 (1892), 149.

24. FBC ccccxxxviiic: xv, fos 254–5; MV 3.

25. MV †2r, 117; Ferrone, *Giornale critico della filosofia italiana*, 61 (1982), 4.

26. FB, *Opuscula varia* (1754), ii. 8–9.

27. FB, "Oratio" (1685), in BL 1572/868 (1) (quotes), and *NP* 19.

28. Tiraboschi, *Biblioteca modenese*, iii (1783), 255–7.

29. Rotta, *Miscellenea seicento*, 2 (1971), 67–8, 72, 143; Montanari to Antonio Magliabechi, 16 March 1677, in Rotta, *Miscellenea seicento*, 2 (1971), 143 n. 23; Middleton, *Experimenters* (1971), 154, 360, 366–8; Targioni-Tozzetti, *Notizie* (1780), ii. 652, 657–61.

30. Rotta, *Miscellenea seicento*, 2 (1971), 72–5; Montanari to Marchese Bonifazio Rangoni, 21 March 1672, in Campori, "Notizie" (1877), 71–2, for the instrument business.

31. Cavazza, *Settecento* (1990), 44–6, 135; Rotta, *Miscellenea seicento*, 2 (1971), 98–9; Montanari, *Pensieri* (1667), 6 (quote); Altieri Biagi and Basile, *Scienziati* (1980), 513n.

32. Montanari, *Pensieri* (1667), 20–40; cf. Galileo, *New Sciences* (1974), 21–8.

33. Goméz Lopéz, *Passioni* (1997), 104–7; Oldenburg to Malpighi, 22 December 1668, in Oldenburg, *Correspondence*, v. 279, and to Montanari, 19 November 1670, in

Oldenburg, *Correspondence*, vii. 269 (quote); Malpighi to Oldenburg, 15 April 1670, in Oldenburg, *Correspondence*, vi. 628–9; Tiraboschi, *Biblioteca modenese*, iii (1783), 266–70.

34. Oldenburg to Malpighi, 15 January, and to Ercole Grani, 21 January 1669/70; to Montanari, 19 November 1670, and to Malpighi, 15 March 1670/1, in Oldenburg, *Correspondence*, v. 430, 439; vii. 269–70, 517.

35. Porzio, *Sorgimento* (1667), 64, 72–5, 84.

36. Montanari to Francesco Redi, 3 March 1675, in Rotta, *Miscellenea seicento*, 2 (1971), 140 n. 12, and to Magliabecchi, 22 September 1626, in Campori, "Notizie" (1877), 76 (learning English); Gómez Lopéz, in RSL, *Notes & Records*, 51/1 (1997), 35–40. Boyle's books were prized by Italian experimentalists; Borelli to Magliabecchi, 22 November 1663, in Galluzzi, *Physis*, 12 (1970), 287; Oldenburg to Montanari, 15 March 1670/1, in Oldenburg, *Correspondence*, vii. 517–18.

37. Brodsky et al., RSL, *Notes & Records*, 41 (1986), 1–3, 6–17; Goméz Lopéz, *Passionei* (1997), 164–74.

38. Montanari, *Speculazioni* (1671), 2–12, 17 (quote), 18–20, 27–32.

39. Montanari, *Speculazioni* (1671), 34, 42–64; Montanari to Cassini, March–April and August 1669, in Barbieri and Cattelani Degani, *Nuncius*, 12/2 (1997), 436–8; Cassini to Montanari, 27 January 1673, in Cavazza, *Settecento* (1990), 138; RSL, Letter Book IV, 53, under date 27 October 1670.

40. Montanari, *Cometes* (1665), 6, 11–12, 27–30.

41. Montanari, in Accademia de' Gelati, *Prose* (1671), 370, 371, 379–80.

42. Montanari, in Accademia de' Gelati, *Prose* (1671), 375, 387.

43. Accolades from, resp., Ponzi, *Observationes* (1685), 23; Justel, Auzout, Cassini, and Oldenburg, in Oldenburg, *Correspondence*, v. 77, 79; ix. 23; vii. 516.

44. Phipson, *Meteorites* (1867), 14–15 (quote); Olson and Pasachoff, *Fire* (1998), 63–75; Burke, *Debris* (1986), 18–19.

45. Montanari, *Fiamma* (1676), 4, 17, 43–4.

46. Montanari, *Fiamma* (1676), 50–3; Rotta, *Miscellenea seicento*, 2 (1971), 90–2, 159 n. 146, 160 n. 147. Cf. Campori, "Notizie" (1877), 66.

47. Montanari, *Fiamma* (1676), 56–7.

48. Montanari, *Fiamma* (1676), 58–61, 62 (quote), 64–5, 68, 73 (quote).

49. Tiraboschi, *Biblioteca modenese*, iii (1783), 259–60 (salary). To the examples of appreciation of the *meridiana* in San Petronio given in *HS* (4–5, 23, 94, 112, 138, 152, 329 n. 4) can be added the evidence of two spittoons supplied to meet the needs of the many visitors coming to see it. Smith, *Sketch* (1807), ii. 169, 369.

50. Soppelsa, *Genesi* (1974), 129–30; Rotta, *Miscellenea seicento*, 2 (1971), 131–3; Montanari, *Copia* (1681), 2, 4. The installation included a *meridiana*.

51. Montanari, *Manueletto* (1680), fo. A.3, 112, 120–2, and *La zecca* (1683), in Graziani, *Economisti* (1913), 239–40, 295–6, 364–79; *HS* 150.

52. Montanari, *Astrologia* (1685); Montanari to Magliabechi, 22 September 1676, in Campori, "Notizie" (1877), 75; Rotta, *Miscellenea seicento*, 2 (1971), 106–7, 118–31; Cavazza, *Studi e memorie per la storia dell'Università di Bologna*, 3 (1981), 436–7, 445.

53. Montanari to Cassini, 5 September 1681, in Cattelani Degani and Lugli, *Nuncius*, 19/1 (2004), 222.

54. Montanari, "Discorso," in Montanari, *Forze* (1694), 277–80, 284 (Guarini), and in Altieri Biagi and Basile, *Scienziati* (1980), 513, 517–20, 522.

55. Montanari, "Discorso," in Montanari, *Forze* (1694), 293–7, 298 (Descartes), and in Altieri Biagi and Basile, *Scienziati* (1980), 525, 528–9.

56. Montanari, "Discorso," in Montanari, *Forze* (1694), 309–11, and in Altieri Biagi and Basile, *Scienziati* (1980), 530–2, 535.

57. Cf. Montanari, *Astrologia* (1685), 6–7: "I believe with Galileo that the open confession of not knowing something is equally worthy of a true philosopher."

58. FB, MS 2833, fos 159–161ʳ, Biblioteca Civica, Verona, transcribed in Rotta, *Miscellenea seicento*, 2 (1971), 197–201.

59. FB, MS 2833, fo. 177, Biblioteca Civica, Verona.

60. HS 149–50.

61. Montanari, *Della natura* (1682–3), in Altieri Biagi and Basile, *Scienziati* (1980), 541–3.

62. Altieri Biagi and Basile, *Scienziati* (1980), 551–2; pinched by Edmund Halley, *PT* 17/194 (1691), 540–2.

63. FB, MS 2833, fo. 117ᵛ on Galileo, Biblioteca Civica, Verona.

64. FB, MS 2833, fos 112–13, 133, 138ᵛ, 139ʳ, 175–81, and plates 1–3, Biblioteca Civica, Verona; Uglietti, 29–30.

65. Aristotle, *Meteorologica*, 1.7; Cavazza, *Studi e memorie per la storia dell'Università di Bologna*, 3 (1981), 443–5, 449–51, 460–4; FBC ccclxxxvii-ccclxxxviii, on the comet of 1683; Rotta, *Miscellenea seicento*, 2 (1971), 201, 207 n. 7; Soppelsa, *Genesi* (1974), 124.

66. Schizzi, *Obs.* (1902), 4, 7–8, 47 (1st quote), and plate XII; Montanari, in Accademia de' Gelati, Bologna, *Prose* (1671), 381 (2nd quote); Tinazzi, Fondazione Giorgio Ronchi, *Atti*, 59/3 (2004), 408–9, 422–40.

67. Rotta, in *Miscellenea seicento*, 2 (1971), 80–2 (quote from a letter of 1685), 85–8.

68. MV 12; Favaretto, in KS 29–31.

69. MV 7; Patin, *Introduzione* (1673), fo. a8ʳ (quote), and *Histoire* (1695), "Préface."

70. DBI, s.v. "Corner, Elena;" Pighetti, *Vuoto* (2005), 26–36, 80, 88–91; Brugueres, in *Pompe funebri* (1686), 13–24.

71. FB, in *Pompe funebri* (1686), 169–70 ("All Olympus is home to you").

72. Grendler, *Jesuits* (2017), 325–7, 331.

73. Platania, *Europa* (2005), 245–50, 257–63, 279–82.

74. Crescimbeni, *Notizie* (1720–1), iii. 361–5; Donato, *Accademie romane* (2000), 72–3. The *Vite* stresses the "ideal of a scholar in perfect equilibrium among erudition, good taste, sociability, and moderate religiosity."

75. Albani, *Discorso* (1687), 6–7, 12, 18, 19–22 (quotes); Albani's speech on James's accession, in Crescimbeni, *Notizie* (1720–1), iii. 364.

76. Zaccagni, *Rivista abbruzzese*, 14 (1899), 3, 6–7, 252–4.

77. Uglietti, 26–7; Santorio, in Ciampini, *Opera* (1747), p. xix; Vacant and Mangenot, *Dictionnaire* (1935), iii/2: 1869–70.

78. Leonio, in Crescimbeni, *Vite*, ii (1710), 208–11, 226–7, 238, 246–8, 250.

79. Middleton, *British Journal for the History of Science*, 8 (1975), 138–54.

80. Rotta, in Di Palma, *Cristina* (1990), 102–3, 108–9; Bignami Odier and Partini, *Physis*, 25 (1983), 254, 257; Graziosi, *L'Arcadia* (1991), 69–72; Renaldo, *Bartoli* (1979), 125–32.

81. Leonio, in Crescimbeni, *Vite*, ii (1710), 215–18; Montalto, *Studi romani*, 10 (1962), 661–4; Fabiani, *Merito* (1694), "A chi legge," and p. 10; Ottoboni, in Fabiani, *Merito* (1694), 67–8.

82. Ciampini to Croone, 20 August 1684 (quote), and FB to Flamsteed, 10 February 1685, in Flamsteed, *Correspondence* (1995–2002), ii. 191, and in Rotta, in Di Palma, *Cristina* (1990), 179–80, resp.; RSL, RGO 1/42, for FB's correspondence with Flamsteed.

83. Rotta, in Di Palma, *Cristina* (1990), 124–5. The secretary, Agostino Fabbri, had assisted Montanari in experiments on Prince Rupert's drops; Fabbri, in Montanari, *Speculazioni* (1671), fo. *5^{r-v}.

84. Eschinardi, *Raguagli* (1680), 3–6, repeated in Eschinardi, *De impetu* (1684), 77–9; the correction concerns the first figure in the second day of Galileo's *Discorsi* (1638), in Galilei, *Two New Sciences* (1974), 114.

85. Eschinardi, *Raguagli* (1680), 24–5, and *De impetu* (1684), 21–4, which offers improvements. "In order to have Galileo's complete teaching on falling bodies all together, I've drawn up the following discourse [pp. 26–9], staying close to what he writes in various places, especially in his Dialogues [*The two new sciences*], and adding some better explanation where necessary."

86. Eschinardi, *Raguagli* (1680), 25–7.

87. Eschinardi, *Raguagli* (1680), 30–5.

88. Eschinardi, *Raguagli* (1680), 12–19, 25–7, 29–33.

89. Eschinardi, *Raguagli* (1680), 35–6, 43–9, 57–8, 61–4.

90. Eschinardi, *Raguagli* (1680), 60–1.

91. Eschinardi, *Lettere* (1681), 12–13, 14 (quote), 34.

92. Eschinardi, *Cursus* (1689), "Ad lectorem," 106–20.

93. Eschinardi, *Cursus* (1689), pp. x–xi, xxiii; Ciampini, *Opera* (1747), iii. 216–24.

94. Berkel, in Smith and Findlen, *Merchants* (2002), 279–81, 288; Berveglieri, *Studi veneziani*, 10 (1985), 82; Hoogewerff, *Oud-Holland*, 38 (1920), 85–6, 102.

95. Onorati, *Apologia* (1698), 43–4.

96. Meijer, *Arte* (1685), [pt 1]. The book is not paginated.

97. Meijer, *Arte* (1685), [pt 2], sundials, [pt 3], navigation.

98. Meijer, *Nuovi ritrovamenti* (1696), [pt 2]; Ribouillault, in Fischer et al., *Gardens* (2016), 104–5, 126–8.

99. Jervis, *Furniture History*, 21 (1985), 4–6; Connors, in Acioglu and Sherman, *Practices* (2015), 51–60.

100. FBC ccccxxii, fos 5, 7, 9–11, 16, and ccccxxxiii; Marzocchi, *Civiltà veronese*, 3rd ser. (1999), 44; Affò, *Istoria* (1785–7), iii. 196–7, 202–6.

101. FB to "Serenissimo Principe" (the Doge?), 11 June 1696, FBV, U.22, fos 101–3; FBC ccccxvi: 1, esp. 21^v; FB, *Relazione* (1885), 9–14; Dal Prete, in CR 217–20.

102. Condren, *International History Review*, 37/4 (2015), 702, 706–9; Rowlands, *English Historical Review*, 115 (2000), 540, 566–7.

103. Benedetti, in *Accademia* (1979), 223–6; Dal Prete, "Scienza" (2004), 147–50, 159–61, 167, and in CR 220–33; Rotta, in Di Palma, *Cristina* (1990), 151–4; MV 6.

104. Cavazza, in CR 101–20; Mamiani, in Cremante and Tega, *Scienza* (1984), 376. Cf. Dal Prete, "Scienza" (2004), 153–6.

105. FB, *Nuova raccolta*, 41 (1785), 4–5, 21; Dal Prete, "Scienza" (2004), 161 n. 709.

106. FB, *Nuova raccolta*, 41 (1785), 8–9.

107. FB, *Nuova raccolta*, 41 (1785), 14–15, 18.

108. FB, *Nuova raccolta*, 41 (1785), 28–9, 35 (quote).

109. FB, *Nuova raccolta*, 41 (1785), 36–7.

110. FB, in Montanari, *Forze* (1694), fos b.3–b.4.

111. Benedetti, in *Accademia* (1979), 223–6; Dal Prete, "Scienza" (2004), 147–50, 159–61, 167; Rotta, in Di Palma, *Cristina* (1990), 151–4.

112. FB, *De emblemate* (1687), as excerpted in Gasperoni, *Scipione Mattei* (1955), 28–9.

113. Baldini, in *Vite*, v (1751), 120; MV 14, 34; Federici, *Elogi* (1818–19), iii. 39.

114. FB to Ciampoli, 23 August 1686, in Beinecke Library, Yale, General MSS 113; 2 October, 19 December 1686, and 30 January and 16 September 1687, in Carini, *Il muratori*, 1/4 (1892), 161–3, 167; Volpato, *Atti e memorie dell'Accademia di agricultura, scienze e lettere di Verona*, 181 (2004–5), 453–9.

115. FB to Ciampini, October 1688, in Carini, *Il muratori*, 1/4 (1892), 170.

116. FB to Ciampini, 19 August, 28 October 1688, in Carini, *Il muratori*, 1/4 (1892), 167, 172–3; Uglietti, 31.

117. MV 15; Alexander VIII to FB, 8 August 1690, in FBV, U.19, fo. 5.

118. *DBI*, s.v. "Ottoboni;" Olszewski, *Römisches Jahrbuch der Bibliotheca Hertziana*, 32 (1997–8), 536–9, 546; Matitti, *Storia dell'arte*, 90 (1997), 202, 231 n. 7.

119. Baroni, *Conformista* (1969), 21, 55, 58–9, 73–5, 83–92, 104–6, 145–7; the lady was a cousin, Margherita Pio Zeno di Savoie.

120. Matitti, *Storia dell'arte*, 84 (1995), 158–60

121. FB, "Studium bibliotechae," BL 1572/868(2).

122. Correspondence of March–June 1691, in FBV, U.15, fos 132–6.

123. Volpato, *Atti e memorie dell'Accademia di agricultura, scienze e lettere di Verona*, 181 (2004–5), 460, 464–70; Rosa, in Bots and Waquet, *Commercium* (1994), 84–92.

124. FB, in Crescimbeni, *Vite*, i (1708), 199 and 201 (quotes, resp.), 202, 206, 210, 212.

125. FB, in Crescimbeni, *Vite*, i (1708), 206–9; Rosa, in Bots and Waquet, *Commercium* (1994), 93–4.

Chapter 2

1. FBC cccliv: viii, fos 177^r–205^v.

2. Montanari, *Forze* (1694), e.g., 192–3, 265–6.

3. HS 96–7, 217–18; Donato, *Nuncius*, 18/1 (2003), 84; Ferrone, *Roots* (1995), 49.

4. Noris to Magliabechi, 24 and 28 June 1692, in Valery, *Correspondance* (1846–7), ii. 338.

5. Marzocchi, *Civiltà veronese*, 3rd ser. (1999), 45; Volpato, *Atti e memorie dell'Accademia di agricultura, scienze e lettere di Verona*, 181 (2004–5), 462–3, 474.

6. Donato, *Accademie romane* (2000), 39, 42–3.

7. Vignoles, *Chronologie* (1738), i, "Preface," fos b.3^v–b.4^r, c.1^v, c.4^v, d.1–6. The longest estimate of the time between Creation and Christ's birth, 6,894 years, was almost twice that of the shortest, 3,483.

8. Klempt, *Säkularisierung* (1960), 93–4, 109–14; Vignoles, *Chronologie* (1738), i. 33, 153, 595–615.

9. Grafton, *Defenders* (1991), 204–13; Popkin, *History* (1979), ch. 11, and *La Peyrère* (1987), 10–17, 21–3, 46–8, 51–2.

10. Simon, *Histoire critique du Vieux Testament* (1678); Popkin, *La Peyrère* (1987), 17–19, 48–9, 87–8; Klempt, *Säkularisierung* (1960), 100, 104–7, 115–17; Reventlow, in Schwaiger, *Kritik* (1980), 23–8; Barret-Kriegel, *Défaite* (1988), 246–53.

11. Martianay, *Défense* (1689). Cf. Lequien, *Defense* (1690), "Epistre," on corruption of the Hebrew text.

12. Vignoles (*Chronologie* (1738), ii. 625–9), allowing 360 days to a year, reduces Babylonian history to 1,200 years. *IU* 68 reports the claims of the Chaldeans to 470,000 years of astronomical observations, and of the Egyptians to 48,800.

13. *IU* 64–5.

14. *IU* 59–60, 61 (quote).

15. *IU* 59, 246–7, 248 (1st quote), 249 (2nd quote), 250–1; Paragallo, *Istoria* (1705), 35, 118–25; Pucci, in CR 265–7; Morello, in CR 196–200. Bianchini received the facts about the deposit, discovered by Francesco Antonio Picchetto, from the professor of mathematics at Rome, Adriano Ariano.

16. Ussher, *Annals* (1658), "Epistle to the Reader" ("from the evening ushering [!] in the first day of the World, to the midnight that began the first day of the Christian aera, there were 4,003 years, seventy days, and six temporarie hours"); Calvisius, *Opus chronologicum* (1685), *passim* (4043); Petavius, *Abrégé* (1682), i, intro., *ad fin.* (4043); *IU* 7.

17. Martianay, *Défense* (1689), 63–4; Vignoles, *Chronologie* (1738), i, fo. c.1.

18. *IU* 17–19.

19. Calvisius, *Opus chronologicum* (1685), 169, 1050, and app., 71 (Scaliger's opinion); Petavius, *Abrégé* (1682), i. 3–9 (Creation to Flood), 60–1, 74–5, 466–9 (savants active between Caesar's death and Christ's birth).

20. *IU* 5, 6 (quote), 9.

21. *IU* 1–2; Chiarlo, in CR 247–9.

22. *IU* 20, 23–4, 51; FB, *Carte* (1871); Chiarlo, in Hübner-Wojciechowski, *Eredità* (1992), 167–8, 178–80; Ricuperati, *Rivista storica italiana*, 117 (2005), 906–7; Dixon, in KS 85–7, 94. The BL has an imperfect copy of the cards; tracts 1572, 868, 1685–1724.

23. *IU* 8–10.

24. *IU* 2.

25. MV 21; Dal Prete, *Nuncius*, 20/1 (2005), 128–30; Bartoli et al., in *Terme di Tito* (n.d.), plates 2, 4–7, and in *Veteres arcus Augustorum triumphis insignes* (1690); Ricuperati, *Rivista storica italiana*, 117 (2005), 914–15.

26. Merolla, in *Letteratura italiana* (1988), ii. 1073.

27. La Mothe le Vayer, *Œuvres*, xiii (1669), 418 (quote), 437–8, 445; Popkin, *La Peyrère* (1987), 5; Chevreau, *Chevraeana* (1700), i. 217–18.

28. *DBI*, s.v. "Fabretti, Raffaello."

29. *IU* 20, 21 (quotes). Cf. Momigliano, *Contributo* (1955), 67, 85–9; Chiarlo, in Hübner-Wojciechowski, *Eredità* (1992), 181–6; Kolendo, in FB, *Camera* (1991), pp. xvi–xxi; Gallo, in Bots and Waquet, *Commercium* (1994), 192–6; Grafton, in Waquet, *Mapping* (2000), 162–3, 173.

30. Barret-Kriegel, *Mabillon* (1988), 25–59, and *Défaite* (1988), 145–75.

31. Thompson, *History* (1942), ii. 15 (1st quote), 18; Barret-Kriegel, *Mabillon* (1988), [9] (2nd quote); Weitlauff, in Schwaiger, *Kritik* (1980), 169–78, 185–90.

32. Barret-Kriegel, *Mabillon* (1988), 67–86; Gross, *Rome* (1990), 264–5; Paravia, in *IU* (1825), i, p. xviii.

33. Mauguin, *Étude* (1909), 91–3, 109; Bertelli, *Erudizione* (1960), 322–3. Cf. Momigliano, *Terzo contributo* (1966), i. 137–40, and Gallo, in Bots and Waquet, *Commercium* (1994), 197–9.

34. Davillé, *Leibniz historien* (1909), 86, 136, 337–9, 355, 373, 376–93, 398–9, 466–7; Robinet, *Iter italicum* (1988), 43–51, 54–62, 87–92; Casini, in CR 18–26.

35. *IU* 31, 32 (quote), 36–7, 39 (quote); Bots and Waquet, *Commercium* (1994), 202–3.

36. *IU* dedication to Cardinal Ottoboni; Claude Estiennot to Mabillon, 23 September 1698, in Valery, *Correspondance* (1846–7), iii. 42; Montfaucon, *Travels* (1712), pp. xiii–xiv, 293. Cf. Rosa, in Bots and Waquet, *Commercium* (1994), 84–94.

37. Cf. *IU* 43.

38. *IU* 67–8, 71 (quote), 72–6; Ademollo, *Spettacoli* (1837), 32 (on carceri).

39. *IU* 78–9, 83–92, and "Al discreto lettore," fo. b.3^r, p. v (quote); Cicero, *De natura deorum*, i, paras 27, 29, in *Brutus*, trans. Poteat (1950), 205–8; Chiarlo, in CR 251–4, on the Eden figure.

40. *IU* 99–106.

41. *IU* 115–16; figures on 93, 114–15.

42. *IU* 117–23, 121 (quote).

43. *IU* 153–65.

44. *IU* 167–85.

45. *IU* 186–201, esp. 187–91.

46. *IU* 216–17, 227. Cf. Cicero, *De natura deorum*, i, paras 27, 29, in *Brutus*, trans. Poteat (1950), 205–8.

47. *IU* 219–23, 227–8, elucidated with the help of St Augustine, *City of God* (1952), bk 5, chs 4–7.

48. Cicero, *De natura deorum*, ii, para. 28; iii, paras 2, 4, in *Brutus*, trans. Poteat (1950), 293–6. Bianchini illustrated the three stages by objects in a three-tier curio cabinet: a Mithraic bull, a triangle, and an image of the Roman pantheon (*IU* 216, 228–9).

49. *IU* 243–4, 245 (quote).

50. *IU* 236–7, 255–6, 262, 264–5.

51. *IU* 367–71, 377–8.
52. *IU* 373–4, 379–80.
53. FB, *Demonstratio* (1752), i, pp. cclxxxi, cclxxxiii, ccxc–cccxv.
54. FB, *Demonstratio* (1752), i, pp. cclxxxiv–cclxxxix, cccxx–cccxxii.
55. *IU* 374–5.
56. *IU* 463–5.
57. *IU* 380–6, 387–8.
58. Federici, *Elogi* (1818–19), iii. 11.
59. *IU* 300–4, 385 (quote).
60. *IU* 413.
61. *IU* 417.
62. *IU* 418–43, 474 (genealogical table).
63. *IU* 477–8, 479 (quote).
64. *IU* 479.
65. *IU* 480.
66. *IU* 481; Needham and Ling, *Science*, iii (1970), 174, 177, 290–1, 296–7.
67. *IU* 483.
68. Schiavo, *Palazzo* (1964), 104.
69. FB, *De kalendario* (1703), 91–3; Fabricius, in Hippolytus, *Opera* (1716–18); Baldovin, *Theological Studies*, 64 (2003), 522–6, after Brent, in Brent, *Hippolytus* (1995), 3–50.
70. FB, AB i, "Praefatio," fo. b3^v. The lady is also construed as truth unveiling herself; Dixon, in KS 102; Sölch, in KS 46, 51–2. Cf. Borgi and Ricuperati, *Rivista storica italiana*, 117 (2005), 947–50.
71. *IU* 42, 369, 571–2.
72. FB to Muratori, 2 May 1699, 13 March and 30 October 1700, in Muratori, *Carteggi*, vii (2014), 402–4.
73. Desimoni, *Giornale ligustico*, 2 (1875), 469, 484; Salvago to FB, 11 May, 8 June, 10 August 1715, and other letters, 1704–23, in FBV, U.18, transcribed at http://uranialigustica.altervista.org/salvago/lettere/bianchini_lettere.htm by Riccardo Balestrieri.
74. Donato, *Accademie romane* (2000), 68–9; Morei, *Memorie* (1761), 1 (quote), 10, 16–21. Giorgetti Vichi, *Arcadi* (1977), gives the real and Arcadian names of all members from 1690 to 1800.
75. Bonjour to FB, 27 September 1705, FBV, U.15, fo. 262.
76. Muratori to Magliabechi, 10 September 1698, in Muratori, *Epistolario* (1901–22), i. 338; Uglietti, 47; Bertelli, *Erudizione* (1960), 374–5; Cochrane, *Journal for the History of Ideas*, 19 (1958), 47–8, and *Catholic Historical Review*, 51 (1965), 166–7; Momigliani, *Terzo contributo* (1966), i. 143–6.
77. Bertelli, *Erudizione* (1960), 375–6; Raimondi, *Giornale storico della letteratura italiana*, 128 (1951), 430–1, 433, 442, 444–5, 452–3.
78. *IU* (1747), "Al cortese lettore," and 452; *IU* (1825), i, pp. xvi, xxi–xxii (quote); Ricuperati, *Rivista storica italiana*, 117 (2005), 884n. (edition of 1699), 925–6, 940; Dixon, *Art History*, 22/2 (1999), 192, 196–9, 204–7; Giuliari, in FB, *Carte* (1871), 15–17.

The sheets Giuliari found were printed by the press of the Camera Apostolica in 1695.

79. Fontenelle, *Eloges* (1740), ii. 379–82.
80. Foscolo, *Opere* (1981), ii. 1260–1 (1st quote), 1918 (2nd quote).
81. De Sanctis, *Storia* (1870), ii. 309.
82. Stark, *Handbuch* (1880), i. 32, 109.
83. Pucci, in CR 261–3; Croce, *Conversazioni* (1924), ii. 102, 107–8.
84. FB, *De lapide* (1698), 8–9, 12.
85. Pastor, *Popes*, xxxiii (1941), 505 (quote).

Chapter 3

1. Bossuet, in Bossuet, *Œuvres* (1815–19), vi. 601, 612, 614 (quote).
2. Platania, *Ultimi Sobieski* (1990), 12, 55–8, and *Polonia* (2011), 99–109; D'Elci, *Present State* (1721), 386–90.
3. Ricuperati, *Rivista storica italiana*, 117 (2005), 883; Platania, *Ultimi Sobieski* (1990), 23–4, 93, 110, 112n., 200–1, and *Studia italo-polonica*, 3 (1987), 88–9; Komaszynski, in Kanceff and Lewanski, *Viaggiatori* (1988), 157–8; Scano, *Strenna*, 25 (1964), 455 (quote).
4. Platania, in Istituto nazionale di studi romani, *Viaggio* (1995), 11–16, and *Ultimi Sobieski* (1990), 66, 69, 73, 88–92, 107–8; Komaszynski, in Kanceff and Lewanski, *Viaggiatori* (1988), 154–5; Palumbo, *Giubileo* (1999), 376–8, 427, 493.
5. Castelli, *Anni santi* (1949), 129; Cattabiani, *Breve storia* (1999), 140–5.
6. D'Elci, *Present State* (1706), 348 (text of 1700).
7. D'Elci, *Present State* (1706), 253–5, 260 (quote).
8. Anon., in D'Elci, *Present State* (1706), pp. viii–x, referring to 1700.
9. D'Elci, *Present State* (1706), 196–200.
10. D'Elci, *Present State* (1706), 342–7.
11. Anon., in *Present State* (1706), pp. xxx–xl; Platania, in Istituto nazionale di studi romani, *Viaggio* (1995), 28–30, 33–8, 46–7, and *Ultimi Sobieski* (1990), 101, 113–20, 125–7; Pometti, *Archivio della Società romana di storia patria*, 21 (1898), 302–4, 307.
12. Scano, *Strenna*, 25 (1964), 452–4.
13. D'Elci, *Present State* (1706), 204–7.
14. Castelli, *Anni santi* (1949), 131–4; Cattabiani, *Breve storia* (1999), 153; Clement XI, *Opera* (1729), *Bullarium*, cols 1–6.
15. Bossuet, in Bossuet, *Œuvres* (1815–19), vi. 549–51; Reboulet, *Histoire* (1752), i. 32–3, 52–3.
16. FB to Gio. Battista Dumeotti, 22 January 1701, in FBV, U.23, fo. 6; MV 25–7.
17. FB to Gio. Giacomo Costanzo (in Brescia), in FBV, U.23, fo. 13.
18. Valesio, *Diario*, i. 299; Uglietti, 28–9.
19. MV 31; Sölch, *Bianchini* (2007), 32 n. 115; the Quirinal gardens had several astronomical decorations; Ribouillault, in Fischer et al., *Gardens* (2016), 108–13.

20. Rossi, *Horarium* (1637), 36; Virgil, *Georgics*, iv. 155; Kircher, *Ars magna* (1646), 493–4, with illustration, pl. viii, fig. 1, and frontispiece, suggesting dials in the Quirinal gardens.

21. Fèa, *Storia* (1832), i. 45, 76, 225, 244–5.

22. Rinne, *Waters* (2010), 39–40, 52–5, 140–54; Fèa, *Storia* (1832), 219; Karmon, *Waters of Rome*, 3 (2005); FB, *Opuscula varia* (1754), i. 35 ff.

23. Mastai Ferretti, *Notizie* (1792), 19.

24. Baldini, in Cremonte and Tega, *Scienza* (1984), 532 n. 11, 534 n. 15, 535 n. 17; Johns, *Zeitschrift für Kunstgeschichte*, 55 (1992), 579.

25. Valesio, *Diario*, i. 512–13 (October 1701).

26. FB to Muratori, 31 December 1701, in Muratori, *Carteggi*, vii (2014), 404; Muratori to Montfaucon, 19 January 1702, in Muratori, *Epistolario* (1901–22), ii. 558; Valesio, *Diario*, i. 563; anon., *Giornale de' letterati d'Italia*, 4 (1710), 67–71.

27. *HS* 89–93 (quote); Bianchini to Flamsteed, 10 February 1685, in Rotta, *Montanari* (2021), 293–6.

28. *HS* 102–9.

29. *HS* 137–8.

30. *HS* 152–3; Tinazzi, Fondazione Giorgio Ronchi, *Atti*, 59/3 (2004), 410.

31. FB, *Observationes selectae* (1737), 267 (text of 1683), 1–6 (comet of 1684), 7–13 (Cassini's method); FB, *Acta eruditorum* (1685), 189, 241, 470.

32. FB, *Observationes selectae* (1737), 11, 13 (quote).

33. *IU* fo. b.2^r.

34. FB, *Observationes selectae* (1737), 106; *HS* 248–54; FBC ccccxx: 5, on tubeless telescopes (figure in Spagnolo, *Manoscritti* (1996), 408).

35. FB, *Opuscula varia* (1754), ii. 57, 64–7.

36. FB, *Opuscula varia* (1754), ii. 69–74, 81.

37. FB, *Opuscula varia* (1754), ii. 75–7.

38. FB, *Opuscula varia* (1754), ii. 137; *HAS* (1701), 127; and (1702), 105. Cassini identified the error that had thrown off the Easter of 1700; Bianchini had the task of investigating its source. Noris to Bianchini, 1702, in Celani, *Archivio veneto*, 36 (1888), 174–5.

39. Reboulet, *Histoire* (1752), 105; Polidori, *Vita* (1727), 100.

40. *DBI*, s.v. "Maratti."

41. Merrifield, *Art of Fresco* (1846), 124–7, quoting Bellori, *Descrizione delle immagini dipinti da Raffaello di Urbino* (1751); Olszweski, *Ottoboni* (2004), 10, 33, 60–1.

42. Paradisi's bill, in FBV, S.83, VIII.2, 469; Scipione Maffei to FB, 12 June 1724, in Romagnani, *"Sotto la bandiera"* (1999), 43.

43. FB, *De nummo* (1703), 70.

44. Bellori, in *Vite*, iii (1821), 189–90, 221–35; Greco, *Ric. stor. arte*, 118 (2016), 89, 92.

45. Heinz, *Thermen* (1983), 112–14.

46. Pastor, *Popes*, xvi (1928), 443–5; Siebenhühner, *Münchner Jahrbuch der bildenden Kunst*, 6 (1955), 187–203.

47. Karmon, *Annali di architettura*, 20 (2008), 143, 145 n. 19, 153 n. 63.

48. FB, *De nummo* (1703), 5–6.

49. FB, *De nummo* (1703), 14–18.

50. FB, *De nummo* (1703), 27, 70.

51. FB, *De nummo* (1703), 3, 27 (quote).

52. Cucco, *Albani* (2001), 185–6. The drawings are at the Istituto di storia dell'arte of the Università di Urbino.

53. FB, *De nummo* (1703), 26–8; *HS* 163–4.

54. Valesio, *Diario*, i. 514–15 (6 October 1701).

55. FB, *De nummo* (1703), 2, 22–3, 32; Valesio, *Diario*, ii. 255, 300; Schiavo, *Meridiana* (1993), after p. 24.

56. MV 38; Federici, *Elogi* (1818–19), iii. 39, 40 (quote).

57. Anon., *Giornale de' letterati d'Italia*, 4 (1710), 64 (quote), 66, 78, 80, 83.

58. FB, *Observationes selectae* (1737), 37; Schiavo, *Meridiana* (1993), 151–2.

59. Miller, *Wife* (1965), 7, 57; Crescimbeni, *Notizie* (1720–1), iii. 369 (quote).

60. Schiavo, *Meridiana* (1993), 82–3.

61. FB, *De nummo* (1703), 25–6, 29, 37–8 44–5, 51–2 (quote).

62. FB, *De nummo* (1703), 59.

63. *HS* 164–6.

64. *HS* 241; *NP* 109; Barbieri and Cattelani Degani, *Nuncius*, 12/2 (1997), 440, on the obliquity at the Bologna *meridiana*.

65. Valesio, *Diario*, ii. 700 (23 September 1703); anon. to Alessandro Albani, 25 September 1706, in FBV, U.23, fos 43–4.

66. Baldini, in *Vite*, v (1751), 119.

67. FB to Manfredi, correspondence of 1704–5, in FBV, U.20, fos 49–52.

68. Scipione Maffei to Ottolino Ottilini, 15 September 1711, in Maffei, *Epistolario* (1955), i. 83.

69. FB, *Observationes selectae* (1737), 37, 46–52, 55, 260.

70. Thus Johns, *Papal Art* (1993), 220 n. 58, citing Filippo Titi, *Nuovo studio di pittura* (1721) and Mariano Vasi, *Roma del settecento* (1763).

71. Salvatore Rotta, *DBI*, s.v. "Bianchini," on 189.

72. FB, in Crescimbeni, *Vite*, i (1708), 215–17, 219, and 218 (quotes, resp.).

73. Sigismondi, in Incerti, *Mensura* (2010), 242–4, 247, and *Gebertus*, 7 (2014), 58–9, 62–3, 70–1, and 12 (2019), 2, 5.

74. FB to Manfredi, 9 April and 26 July 1704, in FBV, U.20, fos 44^{v}–45^{v}, 47^{r}.

75. Cassini, *Journal de Trévoux*, 2 (1702), 152–3, 161; Bonjour, *Calendarium romanum* (1701), and letter to FB, in FBV, U.15, fos 250–62.

76. Celani, *Archivio veneto*, 36 (1888), 164–5; FB, *Opuscula varia* (1754), ii. 140 (text of 1702).

77. *HS* 145–7.

78. FB, *De kalendario* (1703), 140.

79. FB, *De kalendario* (1703), 96, 102–3.

80. FB, *De kalendario* (1703), 91, 94–5, 100. The 99 months comprise 48 of 29 days and 51 of 30 = 2,922 days = 8 years of 365.242 days.

81. FB, *De kalendario* (1703), 138–9; Spagnolo, Accademia di agricultura, scienze e lettere, Verona, *Atti e memorie*, 74/2 (1898), 103 n. 5; MV 36.

82. FB, *Opuscula varia* (1754), ii. 83–7, 108–11, 121–2 (texts of 1703–4); Carini, *L'Arcadia* (1891), 81.

83. Aulìsio to FB, 29 March 1704, in FBV, U.15, fo. 160; Manfredi to Maraldi, 13 August 1704, in Giustino, *Bollettino di storia delle scienze matematiche*, 21 (2001), 86; *DBI*, s.v. "Aulìsio."

84. Le Clerc, *Bibliothèque choisie*, 27 (1713), 193–4 (quote); Lenglet du Fresnoy, *Méthode* (1729), iii. 82.

85. Baldini, in *Vite*, v (1751), 120; Sölch, *Bianchini* (2007), 361 (decree of 30 September 1704).

86. Vogel, *Column* (1973), 5–6.

87. Matitti, *Storia dell'arte*, 90 (1997), 208, 237–8 nn. 64–71.

88. FB, *Considerazioni* (1704), 3–7, 10–14, 20–1, 34–49.

89. Vogel, *Column* (1973), 9; FB to Muratori, 4 and 5 January 1705, in Muratori, *Carteggi*, vii (2014), 406–8.

90. FB to Manfredi, 10 December 1703, in FBV, U.20, fos 43–4.

91. FB, *De kalendario* (1703), 72–85.

92. FB, *IU* 367, 369; Tabarroni, *Coelum*, 24 (1956), 172–4.

93. MV 39–40.

94. FB to Gualterio, 3 January 1710, BL, Add. MS 20549, fos 62–5; Sölch, in KS 181, 183, 193, and *Bianchini* (2007), 9–10, 15–16, 18, 78–83; Liverani, in KS 210, 216–17.

95. Gallo, in Boutier et al., *Naples* (2005), 263–4; Fehl, in Magnusson et al., *Ultra* (1997), 102.

96. Huelsen, *Bulletino della Commissione archeologica comunale di Roma*, 18 (1890), 261–2; Sölch, *Bianchini* (2007), 345–6, 367.

97. *HAS* (1708), 110–11; Fröhner, *Notice* (1878), 15–24.

98. Liebewein, in Beck et al., *Antikensammlungen* (1981), 75–6; FBC cccliv: xvii, fos 1–7.

99. MV 42–4.

Chapter 4

1. Chirico, in Maccavino, *Serenata* (2007), ii. 397–9, 411, 425–40.

2. MV 29; Federici, *Elogi* (1818–19), iii. 37.

3. FB, *De tribus generibus* (1742), pp. viii–xi; Previdi, *Fonti musicali italiane*, 12 (2007), 47–57, 66–9; Johns, *Papal Art* (1993), 77–91.

4. *IU* 124–6; Montfaucon, *Supplément* (1724), iii. 185–97; Blažeković, in Baldassarre, *Musik* (2012), 548–56.

5. FB to Gualterio, 16 August 1710, in BL, Add. MS 20549, fo. 39; Polignac, in CR 279–80.

6. Donato, *Accademie romane* (2000), 44; Finocchiaro, in CR 330–4; Johns, *Papal Art* (1993), 204–6; Gallo, in Boutier et al., *Naples* (2005), 263, 684; Sölch, *Bianchini* (2007), 80.

7. FB, In FBC ccccxxxviii^C: xiii.5, fos 217–18; cf. anon, in D'Elci, *Present State* (1706), p. xxxv.

8. Letter of 10 March [1700], in Bertelli, *Erudizione* (1960), 76–7.

9. Paschini, Atti della Pontificia Accademia romana di archeologia, *Rendiconti*, 11 (1935), 97–8, 101.

10. Donato, *Accademie romane* (2000), 16–25, and in Boutier et al., *Naples* (2005), 245–9; Metzler, *Euntes docete*, 36 (1983), 240–4.

11. Leonio, in Crescimbeni, *Vite*, ii (1710), 212–13, 249–50; Finocchiaro, in CR 330.

12. Cipriani, *Atti e memorie dell'Arcadia*, 5/2–3 (1971), 102, 111–12.

13. Donato, *Accademie romane* (2000), 68–9; Morei, *Memorie* (1761), 16–21.

14. Quondam, *Quaderni storici*, 8/2 (1973), 392–3, 403, 408–13; Morei, *Memorie* (1761), 189–97.

15. Morei, *Memorie* (1761), 10.

16. FB, in Crescimbeni, *Prose* (1718), iii. 69.

17. Donato, *Accademie romane* (2000), 71–2.

18. FB, *Opuscula varia* (1754), ii. 176; Donato, *Accademie romane* (2000), 60.

19. Quondam, *Quaderni storici*, 8/2 (1973), 395, 402, 405, 419 (quote). Cf. Donato, *Accademie romane* (2000), 73–4.

20. Morei, *Memorie* (1761), 75; Giorgetti Vichi, *Arcadi* (1977), 229 ("academico umorista"); FB, *Opuscula varia* (1754), ii. 83–7, 108–11, 121–2 (texts of 1703–4).

21. FB, in Crescimbeni, *Prose* (1718), iii. 65–7, after Niccolò Negroni, in Crescimbeni, *Prose* (1718), iii. 58–62.

22. Muratori, *Opere* (1964), i. 179–80, text of 1703; Waquet, *Modèle* (1989), for French views of Italian letters (and vice versa) around 1700.

23. Perizonius, *Aegyptorum originum ... investigatio* (1711), 251–9, 260 (1st quote); Hendrik Brenkmann (a Dutch jurist studying in Florence) to Perizonius, 1711, in Momigliano, *Terzo contributo* (1966), i. 189 (2nd quote).

24. Muratori to Magliabechi, 19 February, and Giusti Fontanini (Rome), 20 September 1699, and to Montfaucon, 19 January 1702, in Muratori, *Epistolario* (1901–22), ii. 375, 408, 558.

25. Bertelli, *Erudizione* (1960), 16–18, 86–7 (quote). Those to whom it is still a puzzle may like to know that the molecules of air (nitrogen and oxygen) are heavier than those of water vapor.

26. Momigliano, *Terzo contributo* (1966), i. 139–43; DBI v (1963), 126; Vasina, *Lineamenti* (1978), 130–48.

27. Bertelli, *Erudizione* (1960), 81–91.

28. Muratori, *Buon gusto* (1708), I.4, 228–30; I.5, 230–1, 233 (quote); I.10, 251.

29. Muratori, *Buon gusto* (1708), I.6, 236–40. Cf. *Buon gusto* (1715), II.2, 256–9; II.3, 261–3.

30. Bertelli, *Società*, 11/3 (1955), 438–9; Muratori, *Delle forze* (1748), 72–3, re Descartes; Andreoli, *Benedictina*, 6 (1952), 77.

31. Muratori, *Buon gusto* (1708), I.6, 241.

32. Quoted in Marchi, in Romagnani, *Maffei* (1998), 367.

33. Sorbelli, *Benedictina*, 6 (1952), 88–9, 90 (letter of 13 December 1702, quote).

34. Correspondence with Manfredi, 1703, in FBV, U.20, fos 40–1.
35. Burlini, *Accademie e cultura* (1979), 75–6, 80–3, 93; Vecchi, *Accademie e cultura* (1979), 65–7; Soli Muratori, *Vita* (1767), 23–7.
36. Muratori, *Opere* (1964), i. 182–4.
37. Muratori, *Opere* (1964), i. 186–8.
38. Muratori, *Opere* (1964), i. 188–9.
39. Muratori, *Opere* (1964), i. 191–3.
40. Muratori, *Opere* (1964), i. 195–7.
41. Sorbelli, *Benedictina*, 6 (1952), 86–8; Andreoli, *Benedictina*, 6 (1952), 77–80.
42. Sorbelli, *Benedictina*, 6 (1952), 93–5.
43. Sorbelli, *Benedictina*, 6 (1952), 95–8; Muratori-Pritanio to Maffei, 1704, in Muratori, *Epistolario* (1901–22), ii. 739.
44. FB to Muratori, 7 February 1705, in Muratori, *Carteggi*, vii (2014), 409–11; Dupront, *Muratori* (1976), 20–31.
45. Muratori to FB, 20 December 1704, in Muratori, *Carteggi*, vii (2014), 405.
46. FB to Muratori, 7 February 1705, in Muratori, *Carteggi*, vii (2014), 409–11. Cf. Bertelli, *Erudizione* (1960), 81–2; FBC ccclx: ix, fos 107–8; Caracciolo, *Passionei* (1968), 199–201.
47. Muratori's correspondence of February and March 1705, in Muratori, *Epistolario* (1901–22), ii. 745–9, 751.
48. "Lettera apologetica," in Soli Muratori, *Vita* (1767), 242–4, 246 (quote); Waquet, *Modèle* (1989), 378.
49. Soli Muratori, *Vita* (1767), 240, 244.
50. Bonjour to FB, 15 February 1705, in FBV, U.15, fo. 261, and reply, n.d., in Caracciolo, *Passionei* (1968), 47.
51. FB to Manfredi, 7 February 1705, in FBV, U.20, fo. 53b.
52. Cf. FB to Casati (Parma), 31 February 1705, in FBV, U.23, fos. 33^r–34^r.
53. FB, *HAS* iv (1702), 118–20 (cometary observations), analyzed by Cassini, *HAS* (1702), 121–30; FB, *HAS* vi (1704), 142–5 (calendars). Cf. Chiara Curci and Patrizia Devilla, editorial comment, in Muratori, *Carteggi*, vii (2014), 398–9.
54. Waquet, *Modèle* (1989), 48–9, 54–6
55. Thus Andreoli, *Nel mondo* (1972), 153–4.
56. FB to ?, 16 February 1706, FBV, U.23, fos 37^r–38^v.
57. Soli Muratori, *Vita* (1767), 27–8; Trevisan to Pritanio, 14 March 1705, in Soli Muratori, *Vita* (1767), 238–40; Burlini, *Accademie e cultura* (1979), 86–92.
58. Vecchi, *Accademie e cultura* (1979), 55, 58–9; Vasina, *Lineamenti* (1977), 134, 138–9; Golinelli, *Atti e memorie della Deputazione di storia patria per le antiche provincie modenesi*, 11 (1976), 156, 158.
59. FBC cccliv: ix. Cf. Waquet, *Modèle* (1989), 380–1.
60. Andreoli, *Nel mondo* (1972), 143–52; Dupront, *Muratori* (1976), 20–7.
61. Maugain, *Étude* (1909), 120; Waquet, in *Europäische Gelehrtenrepublik* (2001), 94–6, 103; Soli Muratori, *Vita* (1767), ch. ix.

62. Andreoli, *Benedictina*, 6 (1952), 60–3; Vecchi, *Accademie e cultura* (1979), 65; Burlini, *Accademie e cultura* (1979), 91.

63. Muratori to Gatti, 16 April 1705, in Muratori, *Epistolario* (1901–22), ii. 757; "Lettera apologetica," in Soli Muratori, *Vita* (1767), 247.

64. Caracciolo, *Passionei* (1968), 37–43, 49–51; FB, *De tribus generibus* (1742), 49–50.

65. *DBI*, s.v. "Fontanini" and "Imperiali."

66. Caracciolo, *Passionei* (1968), 53–5; Laderchi, *Lettera* (1711), esp. 22–31.

67. Vasina, *Lineamenti* (1977), 145–6.

68. Caracciolo, *Passionei* (1968), 53–4. The censor praised Bacchini's store of learning and found nothing in his revised *Agnello* against sound doctrine and good morals.

69. Bacchini, in Muratori, *Scriptores*, ii (1723), p. 8, col. 1.

70. Bacchini, in Muratori, *Scriptores*, ii (1723), p. 8, col. 2, p. 9, col. 1; Nauerth, in Agnellus, *Liber* (1996), i. 21–7, 30, 196 (quote), 174.

71. FB, *De aureis* (1717), 26.

72. FBC ccccxxx: v, fo. 158 (12 December 1716), referring to Fontanini's *La vita della . . . serva di Dio . . . C. Orsini Borghese* (Rome, 1717); Boncompagni, *Vita* (1931); Spagnolo, *Atti e memorie dell'Accademia di agricultura, scienze e lettere di Verona*, 74/2 (1898), 119–20.

73. Volpato, *Atti e memorie dell'Accademia di agricultura, scienze e lettere di Verona*, 181 (2004–5), 490.

74. FBC ccclx: xxii, fo. 71, with profuse notes and corrections; fos 73–82, on Mabillon's *Epistola di Eustachio Romano de cultu sanctorum ignotorum* (1705); FBC ccclx: xxii, fo. 378^v, on Bayle.

75. FBC ccccxxxviii^C: xv, fos 232–7, 4 February 1721.

76. FBC ccccxx: v, fo. 147.

77. Cf. Vasina, *Lineamenti* (1977), 147–8.

78. Muratori, *Buon gusto* (1715), II.1, 255–6 (moderation); II.9, 277–8 (paternoster); II.12, 284–5 (truth).

79. Heilbron, in Messbarger et al., *Benedict XIV* (2016), 189 (quote); Muratori to Maffei, 20 January 1750, in Marchi, in Romagnani, *Maffei* (1998), 271.

80. Vecchi, *Accademie e cultura* (1979), 69, 70.

81. Waquet, *Modèle* (1989), 381, and in *Europäische Gelehrtenrepublik* (2001), 98–9; Fehl, in Magnusson et al., *Ultra* (1957), 96, 106 n. 94; Soli Muratori, *Vita* (1767), 33–4 (quote); Platania, *Politica* (1993), 32.

82. Golinelli, *Atti e memorie della Deputazione di storia patria per le antiche provincie modenesi*, 11 (1976), 163–4.

83. Waquet, *Modèle* (1989), 382–6.

Chapter 5

1. FB, *De nummo* (1703), 2, 71.

2. D'Elci, *Present State* (1706), 114, 117–20 (cardinal and mistress), 270–2 (Paolucci); *DBI*, s.v. "Davia."

3. Pometti, *Archivio della Società romana di storia patria*, 21 (1898), 313–14, 325–6, 333–5.

4. Legrelle, *Diplomatie* (1895–9), iii. 322, 392; Haile, *James* (1907), 62–3; Pastor, *Popes*, xxxiii (1941), 25, 58.

5. Aquino, *Sacra* (1702), 6–7; Haile, *James* (1907), 79–80.

6. Pometti, *Archivio della Società romana di storia patria*, 21 (1898), 322–32, 346–52, 365, 375, 379–80.

7. FB, *Lettera* (1702), fos 10^v–20^v, 22^v, 32^v–33^r, 36^v, 42^v, 47^v, 54^v.

8. Andreoli, *Nel mondo* (1972), 145; Bulifon, *Journal* (1704), 179–85. FB's position in the procession is inferred from Bulifon's description of the placement of the legate's senior staff.

9. FB, *Lettera* (1702), fos 2^{r-v}, 44^v, 45^v, 58^r, 61^v–62^r, 63^v–64^r (quote); Miranda, *Bianchini* (2000), 25–6.

10. This is the unfriendly anonymous editor of D'Elci, *Present State* (1721), pp. xxxvii–xxxviii (quote), xxxix–xli.

11. Valesio, *Diario*, ii. 392–3 (20 June 1705).

12. Valesio, *Diario*, iii. 480, 517.

13. Valesio, *Diario*, iii. 689 (2 November 1706).

14. Pometti, *Archivio della Società romana di storia patria*, 21 (1898), 390–7.

15. Pometti, *Archivio della Società romana di storia patria*, 21 (1898), 398–412; Papa, *Gregoriana*, 36 (1955), 628–9, 660–1.

16. Legg, *Prior* (1921), 148–50, 165–6, 173–6; Hill, *Historical Journal*, 16 (1973), 251, 257–8; Holmes, *British Politics* (1987), 78–80, 196–201, 211, 245.

17. Bély, *Espions* (1990), 113, 241, 513; Legrelle, *Diplomatie* (1875), vi. 77–8; Stanhope, *History* (1872), 272–4.

18. Hill, *Historical Journal*, 16 (1973), 253–6; Fieldhouse, *English Historical Review*, 52 (1937), 448–55; Biddle, *Bolingbroke and Harley* (1974), 219–39.

19. Legrelle, *Diplomatie* (1895), v. 454, 465, following Torcy, *Mémoires* (1757), i. 345–50; Pastor, *Popes*, xxxiii (1941), 97, 109.

20. Béchu, *Cardinal* (2013), 15–25.

21. "Istruzione," in FBV, S.82, fos 670–2^r.

22. Gualterio to Rohan (1st quote) and Cardinal Rannuzio Pallavicino to Torcy (2nd quote), both 1 June 1712, in FBV, S.82, fos 659, 661; *DBI*, s.v. "Gualtieri (Gualterio)."

23. *DBI*, s.v. "Paolucci."

24. FB to Baldini?, 2 September 1713, in Beinecke Library, Yale, General MSS 113; to Gualterio, 10 October 1712, in BL, Add. MS 20,549; Antonio Baldini to FB, 22 September 1713, in FBV, U.15, fos 182, 184, and Claude (a bookseller?) to FB, 4 February 1713, in FBV, U.19, fo. 719.

25. MV 46; Waquet, *Modèle* (1989), 76; Baldini to FB, 22 September 1713, in FBV, U.13, fo. 184, mentions twenty-two copies of *De kalendario* that Bianchini had left with a bookseller in Paris.

26. Passionei to Paolucci, 20 December 1708, in Caracciolo, *Passionei* (1968), 81; FB to Paolucci, 28–9 August 1712, in FBV, S.82, fo. 579; HS 154, 254.

27. Bély, *Espions* (1990), 333.

28. *MV* 46–51.
29. Andreoli, *Nel mondo* (1972), 147, quoting Bianchini's travel diary; Maffei to Angelo Maria Quirini, 2 December 1712, in Maffei, *Epistolario* (1955), i. 99; Heilbron, in Arabatzis et al., *Relocating* (2015), 99–100.
30. FB to Paolucci, 8 August, and to Cardinal Rohan, 24 August 1712, in FBV, S.82, fos 542, 576; *MV* 52–3.
31. FB to Paolucci, 8 August 1712, in FBV, S.82, fo. 545; cf. Waquet, *Modèle* (1989), 115.
32. FB to Paolucci, 28 August 1712, in *MV* 62–5; Baroni, *Conformista* (1969), 142–3.
33. Réaumur, *HAS* (1713), 299–306.
34. *MV* 68; Waquet, *Modèle* (1989), 98, 124, 128–9; FBV, S.82, T.46.
35. *MV* 54–61; FBC cccliv: x.
36. FB to Gualterio, 1 and 10 August 1712, in BL, Add. MS 20549, fos 235–6, 242–3 (quote).
37. FB to Gualterio, 10 August 1712, in BL, Add. MS 20549, fos 243^v–245^r.
38. Torcy to Bolingbroke, 29 August and 8 September 1712, and answers, 10 and 26 September, in Bolingbroke, *Letters* (1798), 3, 41, 44, 49–50, 94; Prior to Bolingbroke, 17/28 September and 29 December 1712, in ibid., 98 (quote), 251–4; Bély, *Espions* (1990), 247–9; Fieldhouse, *English Historical Review*, 52 (1937), 290–1.
39. FB to Sacra Reale Maestà, December 1714, in FBV, U.22, fo. 105.
40. FB to Paolucci, 23 August 1712, in FBV, S.82, fos 574–5; *MV* 69–70.
41. FB to Paolucci, 28–9 August 1712, in FBV, S.82, fos 582^v–583.
42. FB to Albani, *c.*30 August 1712, in FBV, S.82, fos 588^v, 590.
43. FB to Paolucci, 11 December 1712, in FBC cccxcviii, fo. 82.
44. FB to Clement XI, n.d., in FBV, U.22, fos 3^v, 9–15^r, 16^r, 18.
45. FB to Clement XI, n.d., in FBV, U.22, fos 39–40, 45–8; Baroni, *Conformista* (1969), 142–3.
46. Baroni, *Conformista* (1969), 132, 137 n. 38; Piperno, in Pirotta and Zijno, *Händel* (1987), 350–3; *DBI*, s.v. "Ottoboni;" Matitti, *Storia dell'arte*, 90 (1997), 210–13.
47. Ottoboni to FB, 17 September 1712, in FBV, U.19, fos 40–1, and reply, 26 September 1712, in FBV, S.82, fo. 625^r; and 8 October and 10 December 1712, in FBV, U.19, fos 43–5; *MV* 201 (the serpent dove).
48. Ottoboni to FB, 11 February 1713, in FBV, U.19, fos 46–7.
49. Ottoboni to FB, 17 September and 10 December 1712, 11 February 1713, in FBV, U.19, fos 41, 44, 47.
50. Platania, *Ultimi Sobieski* (1990), 38–41, 167–82.
51. FB to Paolucci, 28–9 August 1712, in FBV, S.82, fo. 579; Legg, *Prior* (1921), 187, 199–200, 203–4; Hill, *Historical Journal*, 16 (1973), 258.
52. FB to "Serenissima Altezza," 25 October 1712, in FBV, S.83, VIII.3, fo. 430; FB, in Paris, to Gualterio, 10 October and 17 November 1712, in BL, Add. MS 20549, fos 276^r, 275, 295^v (quote).
53. FB to Gualterio, 17 November 1712, in BL, Add. MS 20549, fos 295^v (quote), 294^r, 296^r.
54. *MV* 75–6.

55. Baldini to FB, 12 August 1712, in FBV, S.82, fo. 657; MV 78–82; Sölch, *Bianchini* (2007), 366.

56. MV 79.

57. FB to Paolucci, 8 December 1712, in FBV, S.82, fo. 5^r; MV 83–5.

58. FB to Paolucci, 27 October–9 November 1712, in Caracciolo, *Passionei* (1978), 84n.; Andreoli, *Nel mondo* (1972), 150.

59. FB to "Sig. C. d. I.," from Nieuwpoort, 15 December 1712, in FBV, S.82, fos 19–21; FB to Paolucci, 26 December 1712, in FBV, S.82, fo. 34.

60. FB to a cardinal, Nieuwpoort, 15 December 1712, in FB, *Dei viaggi* (1877), 16–18; to Sig. C. P. de Beaumont, 28? December 1712, in FBV, S.82, fos 27–29^r, and to Gualterio, 8 January 1713, in BL, Add. MS 20549, fo. 321^r.

61. FB, *Iter* (1966), 31–3; FBC ccclxxxiv, fo. 43.

62. FB to Paolucci, 16 January 1713, in FB, *Dei viaggi* (1877), 20–1, and *Iter* (1966), 34–6, 55.

63. FB to Paolucci, 31 January 1713, in FB, *Dei viaggi* (1877), 20, 24 (quote).

64. FB, *Iter* (1966), 31, 35–9, 66–7; Ricuperati, *Rivista storica italiana*, 117 (2005), 910n.

65. FB, *Iter* (1966), 39, 42, 45–6 (quote), 49; Guerrini, *Journal of British Studies*, 25 (1986), 305–6, 309; Kahn, *Codebreakers* (1967), 169–70. A bill for 7/5s. for entertainment and a present to FB, "as recommended by the Chancellor," appears in Gardiner's accounts for 1712–13; Oxford University Archives, WPbeta/21/6, courtesy of the Archive's keeper, Simon Bailey.

66. Fontaine to Hudson, 11 January 12/13, in Faccini, in Albarello and Zivelonghi, *Per Alberto* (1998), 132 (quote), 135.

67. FB, *Iter* (1966), 43, 44 (quote), 45, 47.

68. Hearne, 14–15 January (the crucifix), 28 March (the old man), 13 March 1713 (the faults), in Hearne, *Remarks* (1885–1921), iv. 48–9, 135, 101–2, resp.

69. FB, *Iter* (1966), 35, 53; FB to Hudson, 24 January 1713, in FBV, U.23, fos 61–2.

70. e.g., in letters to Hudson in 1714, in FBV, U.23, fos 63–5, 67–9, 71–2.

71. FB, *Iter* (1966), 33–6, 53–5, 57, 69, 72; FB to Paolucci, 16 January 1713, in FB, *Dei viaggi* (1877), 20–1.

72. FB, *Iter* (1966), 53, 60–1 (Sloane).

73. John Conduitt (Newton's executor), "Miscellanea," fo. 4^r, item THEM00168, newtonproject@ox.ac.uk.

74. Conduitt, in Cambridge, Kings College, Keynes MS 130.5, fos 4^v–5^r (on Pretender); Fransen, *Leibniz* (1933), 143–5; Westfall, *Never at Rest* (1980), 265 (anagram), 323 (idolators), 514–20, 775–7.

75. Newton to J. B. Meinke, 1724, draft, in Newton, *Correspondence* (1959–77), vii. 255.

76. Guerrini, *Journal of British Studies*, 25 (1986), 301–2, 310, and Rowlinson, RSL, *Notes & Records*, 61 (2007), 110, 113–15, about the circle of Tory Newtonians of which Keill was a member.

77. FB, optical experiments, in FBV, S.83, VIII:1, fos 260–2, and in *Iter* (1966), 39; Gross, *Rome* (1990), 252–3; Ferrone, *Roots* (1995), 11–13, 27–9, 102–3; Pighetti, *Influsso* (1988), 176n.; *DBI*, s.v. "Galiani."

78. Harrison, *Library* (2008), 101.

79. FB, *Iter* (1966), 39, 57–8, 63; FB to Newton, January 1713, in Hall, RSL, *Notes & Records*, 37 (1982–3), 18–19.

80. Baldini to Conte di Rivano, Parma's agent in Paris, 12 August 1712, in FBV, S.82, fo. 657, and to FB, 22 September 1713, in FBV, U.15, fo. 182, and 12 January, 13 April, and 29 June 1714, in FBV, U.15, fos 187–8.

81. FB, *Iter* (1966), 67–9; RSL, "Journal Books," x. 445, 448 (22 and 29 January 1712–13).

82. Flamsteed, *PT* 15/177 (December 1685), 1215–17, and *Correspondence* (1995–2002), ii. 191, 261; Whiston, *Praelectiones* (1726), 357.

83. Valerio, *Globusfreund*, 35–7 (1987), 98, 106–7 n. 10.

84. IU 367–80.

85. FB, *Demonstratio* (1752), i, pp. cclxxxi–ccxcv, cccxx–cccxxii; Tinazzi, *Società italiana di archeoastronomia, V Congresso* (2005), 69–85. At 71 y/°, 20° 4' amounts to 1,425 years from Ptolemy or 1275 BC. Although it scarcely matters, integrity requires admission that Ptolemy made an error of a degree in placing the vernal equinox in his time, which effectively made his sky correspond to circumstances about AD 75 and compromised Bianchini's calculations of time intervals. Valerio, *Globusfreunde*, 35–7 (1987), 101–3; Toomer, *Almagest* (1984), 1–26.

86. IU 300–4, 385; FB, *Iter* (1966), 42–3. FB remarked that the marble had deteriorated in its exposed position in the Sheldonian Theater.

87. Flamsteed to Abraham Sharp, in Flamsteed, *Correspondence* (1995–2002), iii. 665.

88. FB records the gifts in FB, *Iter* (1966), 59, under 3 February 1713, in the past tense ("donavi").

89. Petavius, *Uranologion* (1630), and *Opus de doctrina temporum auctius* (1703); Harrison, *Library* (2008), 214.

90. Manuel, *Newton Historian* (1963), 98–100, 119–20.

91. Buchwald and Feingold, *Newton* (2013), 187, 193, 215–16.

92. Buchwald and Feingold, *Newton* (2013), 210–12, 233–5.

93. Manuel, *Newton Historian* (1963), 86, 98, 101, 119, 195–6; Buchwald and Feingold, *Newton* (2013), 208, 216–20, 231; 2 Chronicles 12:1–22; 2 Samuel 8:14.

94. Buchwald and Feingold, *Newton* (2013), 201–2, 254, 281–4, 287, 292–3.

95. Buchwald and Feingold, *Newton* (2013), 203–5, 226–7; IU (1697), 451, 489, 530–1.

96. IU (1697), 408–10, 444, 464–5, 484, 511.

97. IU (1697), 320–1. Current scholarship considers Sesostris an exaggeration of Senusret III (*c*.1900 BC), identifies Sesac with Shishak (*c*.900 BC), and accepts most later Parian dates.

98. Buchwald and Feingold, *Newton* (2013), 223–4, 245–6, 269–81; Newton, *Opticks* (1952), 405–6; Gascoigne, in Gaukroger, *Uses* (1991), 185–90.

99. FB, *Demonstratio* (1752), i, pp. cccxvii–cccxviii.

100. Alex Cunningham (a strong Whig historian and diplomat) to Newton, 20 Apr 1716, in Newton, *Correspondence* (1959–77), vi. 331.

101. FB, *Dei viaggi* (1877), 17–24; Rotta, *Bianchini in Inghilterra* (1966), 39, 58, 63, 64, 67; Westfall, *Never at rest* (1980), 807–9, 812.

102. FB, *Demonstratio* (1752), i, pp. cccxxii–cccxxiv; Manuel, *Newton* (1963), 78–88; Gascoigne, in Gaukroger, *Uses* (1991), 185–7; Trompf, in Gaukroger, *Uses* (1919), 216–18, 221, 226–30, 233–4.

103. Bougainville, in Fréret, *Défense* (1758), pp. vi–vii (quotes), xi–xii, xlvi–xlvii, xlxix (devastating).

104. Anon., *Present State*, 2 (December 1728), 453; Valerio, *Globusfreund*, 35–7 (1987), 107 n. 16, pointing to *IU* 253 (Orion), 284 (Boötes), 296 (Aquarius), 309 (Argo), 343 (Aries, Taurus, Perseus); Lippincott, *Journal of the Warburg and Courtauld Institutes*, 74 (2011), 287–90, 298; Manilius, *Astronomicon* (1739), plate after p. xvi, for the planisphere.

105. Schaefer, *Journal for the History of Astronomy*, 36 (2005), 171–82, refuted by Duke, *Journal for the History of Astronomy*, 37 (2006), 87–97.

106. Schaefer, *Journal for the History of Astronomy*, 35 (2004), 161–3, 185.

107. Desimoni, *Giornale ligustico*, 3 (1876), 43; Salvago to FB, 25 November 1715, in Balestrieri, "Delizie."

108. FB to Newton, 17 April 1714, in Hall, RSL, *Notes & Records*, 37 (1982–3), 21.

109. Galiani, letter of 18 December 1706, quoted in Henry Newton to Newton, in Newton, *Correspondence* (1959–77), iv. 506; Ferrone, *Roots* (1995), 9–13; Rotta, *DBI*, s.v., "Bianchini."

110. Ferrone, *Roots* (1995), 20–2.

111. Ferrone, *Roots* (1995), 24, 27–9, and *Giornale critico della filosofia italiana*, 61 (1982), 14–16.

112. Galiani to Giovanni Battari, 16 May 1714, in Ferrone, *Giornale critico della filosofia italiana*, 61 (1982), 18 (quotes); Guido Grandi (Pisa) to Galiani, 30 July 1714, in Ferrone, *Roots* (1995), 33; see also pp. 56–7.

113. Gross, *Rome* (1990), 254; Ferrone, *Roots* (1995), 30; Caffiero, *Archivio della società romana di storia patria*, 101 (1978), 381–2.

114. Baldini, in *Vite*, v (1751), 121; Ferrone, *Roots* (1995), 31–3, 290 n. 51; FBC ccclxxxiv, fo. 43; Dal Prete, "Scienza" (2004), 152 n. 665; Ferrone, *Roots* (1995), 39, 287 n. 15, 293 n. 18.

115. Hauksbee, *Esperienze fisico-mecaniche* (Florence, 1716). Pighetti, *Influsso* (1988), 225–61, printed the translation in Bianchini's papers (FBC cccliv) as if the experiments were his, and dated them to 1687 (p. 178). Cf. Heilbron, *Electricity* (1979), 229–34, 237–9, and Casini, *Newton* (1983), 182, 192–4.

116. Todi, *Giornale critico della filosofia italiana*, 79 (2000), 421–30. Since several pages are out of order in the translation in Bianchini's manuscripts, it must have been reassembled at some point, perhaps after being lent to Dereham.

117. Newton, *Principia*, bk III, Gen. Sch. (*Principles*, ed. Cajori, 546–7).

118. Galiani to Vincenzo Santini, 12 January 1715, in Ferrone, *Roots* (1995), 35; see also pp. 24, 28–34, 63–71.

119. Ferrone, *Roots* (1995), 57–9, 124, 317 n. 6; FBC ccccxxx: v, fos 146–71 (Bianchini's reports to the censors).

120. Galiani to Antonio Leprotti, 26 May 1714, in Caffieri, *Archivio della società romana di storia patria*, 101 (1978), 334–5; Ferrone, *Roots* (1995), 26–7.
121. Galiani to Grandi, 2 March 1715, in Ferrone, *Roots* (1995), 11.
122. Ferrone, *Roots* (1995), 123–4; *DBI*, s.v. "Galiani."
123. FB to François de Campa, abbé de Ségny, 25 November and 8 December 1716, in BPV, U.23, fos 77–83; Béchu, *Cardinal* (2013), 45–50.
124. FB to James III, 9 February 1715, in BPV, U.22, fos 109–10.

Chapter 6

1. Fieldhouse, *English Historical Review*, 52 (1937), 293–6; Bély, *Espions* (1990), 106–7; Saunders, in Frey and Frey, *Treaties* (1995), 179–80; Legg, *Prior* (1921), 445–7.
2. *DNB* 163–8; Ormonde, *Mémoires* (1737), ii. 47–8 (quote), 51; Winton, in Miller, *Augustan Milieu* (1970), 139, 152.
3. Ormonde, *Letter* (2 August 1714).
4. Anon., *At a . . . Meeting* (1714); Gardiner addressed George I on 24 September 1714; *DNB* xxi. 413–14.
5. Anon., *At a . . . Meeting* (1714), quotes on 9, 11, 35, resp., 16, 37–41.
6. Defoe, *Hannibal* (1714), 15, 20–1, 26–8, 34–5, 44 (1st quote), 46–7, 48 (2nd quote); cf. Holmes, *British Politics* (1987), 64, 69, 87, 93–4
7. Seton, *Scottish Historical Review*, 21 (1924), 253–9, 266 (quote).
8. Seward, *King* (2021), 159, 163–73.
9. Haile, *James* (1907), 223, 228, 231; correspondence from 4 November 1716 to 16 January 1717 in Gyllenborg, *Letters* (1717), 12, 15, 22, 26–7, 36–42.
10. *DNB*, s.v. "Stuart, James Francis Edward;" Genet-Rouffiac, in Barnard and Fenlon, *Dukes* (2000), 195–7; Seward, *King* (2021), 175 (quote).
11. Haile, *James* (1907), 221–3, 228, 231; Bodleian Library, Carte MS ccviii, fos 338–59.
12. Clement XI, *Homily* (1717), 21, 23.
13. Corp, *Jacobites* (2009), 22–3, 28, 93–4.
14. FB to Max Emanuel, 18 April; to Charles Albert's uncle, the Archbishop of Cologne (quote), 18 April; to his aunt, the Grand Duchess of Tuscany, 5 June; and to him, 18 July and 3 October, all 1716, in FBV, U.22, fos 113–21; FB, *De nobilissimo hospite* (1716), pp. vi–xii; MV 89–90, 92.
15. MV 90.
16. Elector of Bavaria to FB, 4 June 1717, in FBV, U.15, fo. 238, and reply, U.22, fos 85–6.
17. Alessandro Albani to FB, 7 August 1715, in FBV, U.15, fo. 101. Cf. FB to Queen Mother (Mary of Modena), 23 March and 17 May 1717, on benefices for James's supporters, in FBV, U.22, fos 133–4, 141–3.
18. FB to elector of Bavaria, 3 and 10 April 1717, in FBV, U.22, fos 134, 137–8, and to Altesse Royale, 12 November 1717, in Beinecke Library, Yale, General MSS 113, fo. 127 (quote).
19. FB to "Sire," in FBV, U.22, fo. 87^r, and to Queen Mother, 10 October 1717, in FBV, U.20, fos 145–6 (would like to spend his life in James's service).
20. FB to Queen Mother, 22 June 1717, in FBV, U.22, fos 125–6.

21. Nairne, "Journal," in Bodleian Library, Carte MS ccviii, fos 338^v–339^r, 349^v–350^r (quote); Shield and Lang, *King* (1907), 286.

22. FB to James III, 1 June 1717, in FBV, U.22, fos 129–30; Bodleian Library, Carte MS xxviii, fos 344^v, 345^v, 346^v, 348^r, 350^r, 354^r, 355^v, 356^v.

23. Baldini to FB, August 1717, in FBV, U.15, fos 199, 201; Bodleian Library, Carte MS cclviii, fos 47–8 (9 March 1717–18); Corp, *Stuarts in Italy* (2011), 9–10.

24. FBV, S.82, fos 674–5.

25. Corp, *Jacobites* (2009), 43–5, 51–2, 56, 59; *DNB*, s.v. "Mar." Mar put his own interests above James's, but may not have betrayed him; Bruce, Royal Historical Society, *Transactions*, 20 (1937), 62–8, 81–2.

26. Corp, *Jacobites* (2009), 47, 69, 73; FB to Nairne, 19 March, and reply, 25 March 1718, in BL, Add. MS 20, 312, fos 227–9.

27. FB to Queen Mother, 3 September 1717, in FBV, U.20, fo. 149.

28. Baldini to FB, 22 March 1717, in FBV, U.15, fos 197–8, and 26 March 1719, in FBV, S.83:2, fo. 480^v.

29. Salvago to FB, 31 July 1717 and 31 December 1718, in Balestrieri, "Delizie;" FB, *Observationes selectae* (1737), 158–9, and in Baldi, *Memorie* (1724), 133–4; Bold, *Music and Letters*, 81/3 (2000), 355–8.

30. Baldini to FB, 11 October 1717, in FBV, U.15, fos 201–2; FB in Baldi, *Memorie* (1724), 135–7, 140–5; Baldi, *Memorie* (1724), 58–61.

31. Gualtieri to FB, 10 August 1718, in BL, Add. MS 20550, fos 12–13; FBC ccccxxx: ii; Shield and Lang, *King* (1907), 294.

32. Alessandro Albani to FB, 4, 11, and 20 September, and 16 October 1717, and 21 September 1718, in FBV, U.15, fos 102–3, 106, 114–16, 118; Pezzini, *Fregio* (1985), 19–20, 24, 27, 30–1.

33. FB, in Baldi, *Memorie* (1724), 79–85, 86–9; Molari and Molari, *Trionfo* (2006), 8–9, 19, 27–8.

34. Mar to James III, 6 and 13 April (quote), and James to Mar, 10 April 1718, in *SP* vi. 264, 289, 309.

35. On news, e.g., letters of 1718 in FBV, U.15, fos 151–8; on overconfidence, Gualterio to Nairne, 24 November 1717, quoted in Corp, *Nairne* (2018), 337 n. 47: "he gets irritated at the least opposition," a judgment in opposition to the usual praise of Bianchini's ready complaisance.

36. MV 90–1 (the portrait); Nairne to Gualterio, 21 October 1717 and 4 August 1718, quoted in Corp, *Nairne* (2018), 291, 337, from BL, Add. MSS 31260 and 31261; Alessandro Albani to FB, 21 September 1718, in FBV, U.15, fo. 118.

37. Antonio Baldini to FB, 23 January and 7 September 1716, in FBV, U.15, fos 167, 169; MV 29, and FBC ccclxxxii: xviii (hunting).

38. Lewis, *Connoisseurs* (1961), 31–5; Corp, *Jacobites* (2009), 102–3, and *Nairne* (2018), 353; Nairne, memo to file, 25 September 1718, and to Mar, 2 November 1718, in *SP* vii. 322, 488 (quote); Bodleian Library, Carte MS cclviii, fos 148–59; FB to Gualterio, 16 November 1718, in BL, Add. MS 20550.

39. Corp, *Jacobites* (2009), 109–12, and in Monod et al., *Loyalty* (2010), 180–205.

40. Szechi, in Corp, *Stuart Court* (2003), 55 (quoting a letter of 28 February 1718), 69–70; Gregg, in Corp, *Stuart Court* (2003), 65–83, and *Studies in History and Politics*, 4 (1985), 44; Haile, *James* (1907), 135; Corp, *Stuarts in Italy* (2011), 18–22.

41. Platania, *Politica* (1993), 3, 10, 14, 18, 29 n. 121, 34.

42. Gregg, *Studies in History and Politics*, 4 (1985), 45–9; Miller, *Wife* (1965), 29–30.

43. Wogan, "Manuscript," i (obscenity), iii–vi, vii (doggerel), 42–88 (psalms), Galway/Kilfenora Diocesan Archives.

44. Wogan, in Swift, *Miscellaneous Pieces* (1789), 48, 89; Sherburn, *Early Career* (1934), 46, 51.

45. Wogan, in Swift, *Miscellaneous Pieces* (1789), 39, 50–3, 80–2; Murtagh, *Irish Sword*, 2 (1954–6), 43–53.

46. Wogan, in Swift, *Miscellaneous Pieces* (1789), 68–71, 87–8 (quotes).

47. Maher, in Maher, *Irish* (2021), 76–9.

48. Clarke de Dromantin, *Refugiés* (2005), 183–8, 497–506; Genet-Rouffiac, in Monod et al., *Loyalty* (2010), 206–7, 211–12, 217–18; O'Callaghan, *History* (1870), 310.

49. Wogan, *Preston Prisoners* (1716), verse ix (quote); Maher, *History Ireland*, 25/2 (2017), 20–3. "Lord Derwentwater Lights," an unusual auroral show occurred on 6 March 1715–16, when the lord's corpse arrived at his home in Northumberland; Jankovic, *Journal of British Studies*, 41/4 (2002), 429–31, 435–6.

50. Flood, *Life* (1922), 22–6; Sankey, *Prisoners* (2005), 158–64; Maher, in Maher, *Irish* (2021), 77–9.

51. FB to James Sobieski, 15 June 1719, in FBC cccxcv, fo. 73[r].

52. Gregg, *Studies in History and Politics*, 4 (1985), 49 (quotes); Tayler, in Tayler, *Court* (1938), 5–7; Wogan, in *SP* iv. 95–6, 388–90.

53. Gregg, *Studies in History and Politics*, 4 (1985), 50–1; James to Murray, 4 June 1718, in Tayler, in Tayler, *Court* (1938), 23–5.

54. Miller, *Wife* (1965), 41–4.

55. Wogan, in Gilbert, *Narratives* (1894), 401; anon, in Gilbert, *Narratives* (1894b), 164–6; Tayler, in Tayler, *Court* (1938), 7–9.

56. Tayler, in Tayler, *Court* (1938), 26–7; Frati, *Nuova antologia*, 220 (1908), 422–3; FB to Gualterio, 9 November 1718, in BL, Add. MS 20550, fo. 24.

57. Boylan, in Gilbert, *Narratives* (1894), 2–6.

58. Miller, *Wife* (1965), 68 ("Adieu"); Wogan, in Gilbert, *Narratives* (1894), 43–5, 56–8, 59 (1st quote), and in "Manuscript," 9 (2nd), Galway/Kilfenora Diocesan Archives.

59. Boylan, in Gilbert, *Narratives* (1894), 8–9; Wogan, in Gilbert, *Narratives* (1894), 61–5; O'Callaghan, *History* (1870), 309; Miller, *Wife* (1965), 91.

60. Boylan, in Gilbert, *Narratives* (1894), 1–16; Wogan, in Gilbert, *Narratives* (1894), 74–8.

61. Hamel de Breuil, *Revue d'histoire diplomatique*, 9 (1895), 82–4, 87–90, 97.

62. O'Callaghan, *History* (1870), 312–13; Boylan, in Gilbert, *Narratives* (1894), 18–23; Wogan, in Gilbert, *Narratives* (1894), 86–91; Miller, *Wife* (1965), 123–4.

63. Boylan, in Gilbert, *Narratives* (1894), 23–5; Wogan, in Gilbert, *Narratives* (1894), 82, 96–102.

64. Gregg, in Cruickshanks, *Ideology* (1982), 186–8.

65. Haile, *James* (1907), 255–62, 272–7; Shield and Lang, *King* (1907), 299, 306, 311–14, 322–36; Corp, *Jacobites* (2009), 122–30; anon. to FB?, 26 March 1719, in FBV, U.19.

66. Wogan, in Gilbert, *Narratives* (1894), 92–5, 102; Frati, *Nuova antologia*, 220 (1908), 420, 424; Platania, *Politica* (1993), 29–30, 35; Maher, in Maher, *Irish* (2021), 81–4.

67. Frati, *Nuova antologia*, 220 (1908), 427; Platania, *Politica* (1993), 26; *Historical Register*, 4 (1719), 189, cited by Gilbert, *Narratives* (1894), p. vii.

68. Wogan?, *Female Fortitude* (1722), 2–3; anon, *Remarks* (1722), pp. iv–vi, 2–10, 20; Frati, *Nuova antologia*, 220 (1908), 427 (the newsmonger's decapitation, 3 February 1720); Bruckmann, *Eighteenth Century Life*, 27/3 (2003), 35–7, 41; Miller, *Wife* (1965), 143 ("adorable Queen").

69. Markuszewska, in Maher, *Irish* (2021), 150–2; Yates, *Journal of the Warburg and Courtauld Institutes*, 10 (1947), 56–75. The painting is on the west wall in the Old Royal Naval College in Greenwich.

70. Corp, *Jacobites* (2009), 131, 134–5.

71. Wogan, in Gilbert, *Narratives* (1894), 103, and in "Manuscript," 22–3, Galway/Kilfenora Diocesan Archives.

72. FB, speech to Conservatori, 26 June 1719, and letter to Clementina, 10 September 1719, in FBC cccxcv, fos 74^v, 87^v.

73. FB to Sobieski, 15 June 1719, in FBC cccxcv, fos 71^r, 73^r, 74^r, 75^r; and, on the visit to the pope, FB to ?, 17 May 1719, in FBC cccxcv, fos 58^r, 59^r, 60^r.

74. FB to Wogan, summer 1721, in FBC cccxciv, fo. 40^{2r}; on Lana, Heilbron, *Electricity* (1979), 189–92.

75. FB to Eleanor Misset, 8, 11, 16 June, in FBC cccxcv, fos 69^r–71^v.

76. Corp, *Jacobites* (2009), 13. The other witnesses were another Irish officer, Hay, Murray, and James's confessor; Gittins, in Maher, *Irish* (2021), 169–72.

77. Identifications suggested by Corp, in Maher, *Irish* (2021), 133; draft in Morét, in Ebert-Schifferer et al., *Maratti* (2016), 214–18.

78. FB to James III, 20 September 1719, in FBV, S.82, fo. 668, and U.20, fo. 151.

79. FB, letter of 9 September 1719, Beinecke Library, Yale, General MSS 113.

80. "Busy bee who makes sweet combs of honey, behold your king | Bee diligent and industrious, do you wish a rose? Here is your flower." FB, *Giorno* (1720).

81. FBC cccxcv, fo. 75^v.

82. Sigismondi, *Gebertus*, 11 (2018), 21–3.

83. FB to James III, 27 July 1720, FBC ccccxxxviiib, fos 392–3; Virgil, *Eclogues*, 3, lines 104, 106 (trans. H. R. Fairclough, 1916); Dix, *Classical Philology*, 90/3 (1995), 256–62.

84. Distances deduced from Schiavo, *Meridiana* (1993), 68–9, 156–60, and measurements by the author.

85. Luttrell, *Relation* (1857), i. 442 ("between 9 and 10 in the morning, 15 minutes before 10"); Shield and Lang, *King* (1907), 15 (ten o'clock). The calculation amounts to taking the height h of the hole admitting the sun's rays (20.34 m) as the perpendicular of a gnomon whose origin lies $hctn\phi$ = 22.67 m along the line extended south of the vertex (ϕ = latitude = 41.9°). The hour line through James's marble

makes an angle β = 16° with the *meridiana* extended. This corresponds to an hour angle $\alpha = \tan^{-1}[\tan\beta \; / \; \sin\phi] = 23.5°$, or a little over an hour and a half before noon.

86. Schiavo, *Meridiana* (1993), 161.
87. FBV, S.83, VIII:2, fo. 568^v; Schiavo, *Meridiana* (1993), 68–9, 156–60. Geneva, *Astrology* (1995), 40, gives other examples of extracting dates from royal titles.
88. Platania, *Studia italo-polonica*, 3 (1987), 104–5.
89. *DBI*, s.v. "Innocento XIII;" Corp, *Stuarts in Italy* (2011), 19–22, 25, 121–3, 127, 133–4.
90. Letter to FB, n.d., in FBV, U.19, fo. 34 (2nd quote); Sabbatini, *Mediterranea*, 15 (2018), 100–2, 102n. (1st quote).
91. Antonelli, in Platania, *Da est* (2006), 232–44; Gregg, in Cruickshanks, *Ideology* (1982), 185.
92. Sabattini, *Mediterranea*, 15 (2018), 105–12, 542, 546–7, 558–60, 563–3 (quotes).
93. FB to Contessa di Sardi, 22 August 1722, mentioned in Corp, *Stuarts in Italy* (2011), 146, who very kindly supplied a copy of the letter; Daugnon, *Gli italiani* (1905), 161, 264–7; Antonelli, in Platania, *Da est* (2006), 247, 251–61.
94. Corp, *Stuarts in Italy* (2011), 146–7, 157–8, 165; Faccini, in Albarello and Zivelonghi, *Per Alberto* (1998), 138–9; FB to Clementina, in FBV, U.22, fos 99–100.
95. Corp, *Jacobites* (2009), 145–6; Tayler in Tayler, *Court* (1938), 28, 38; Gregg, in Cruickshanks, *Ideology* (1982), 183, 193.
96. Corp, in Maher, *Irish* (2021), 134–44.
97. Corp, *Stuarts in Italy* (2011), 28–30; Maffei, *Verona* (1771), pt 2, 167.
98. Justi, *Zeitschrift für bildende Kunst*, 7 (1872), 291–9; Lewis, *Connoisseurs* (1961), 39, 49–50 (homosexual, etc.), 54–6, 67–8.
99. Stosch, *Gemmae* (1724), 58–9; the book was Fulvio Orsini, *Imagines et elogium virorum illustrium* (Rome, 1570). Stosch had seven books written by Bianchini; [Stosch], *Bibliotheca stoschiana* (1759), pt 1, 15, 63, 165 (including *Anastasius*), and pt 2, 79–80 (including *De kalendario*).

Chapter 7

1. Lanciani, *Storia* (2000), 6, 11–12, 15–16, 21–2, 25–6, 37, 39, 44–8, 52–3, 59, 68, 70, 79, and Miranda, *Bianchini* (2000), 64–70, 83–8, for indications of FB's extensive *ex officio* participation in excavations.
2. Winckelmann, *Nachrichten* (1997), 18.
3. AB ii, pp. ccxxi–ccxxiii, and FB, *Camera* (1727), 73–9, tabs VI and VIII; Kolendo, in FB, *Camera* (1991), pp. xxiv–xxv; Baldini to FB, FBV, S.83, VIII:2 (quote).
4. Miranda, *Bianchini* (2000), 251–2.
5. Miranda, *Bianchini* (2000), 179–80, 192–3, 196–7, 199–200, 259.
6. Miranda, *Bianchini* (2000), 181–5.
7. FB to Farnese's brother, n.d., in BPV, U.22, fos 79–83, on Farnese collecting.
8. MV 109; Miranda, *Bianchini* (2000), 97–8, 264–5; Bozzoli, *Cenni* (1873), 9; FB to Duchessa di Parma, 4 December 1726, in FBV, U.22, fos 179–81.
9. FB, *Palazzo* (1738), 6, 16 (quote).

10. Suetonius, *Caesars* (2014), ii. 353–61.

11. FB, *Palazzo* (1738), 8–10 (quote); Huelsen, *Mitteilungen*, 10 (1895), 252–3. The short quote from *Palazzo* does not occupy three pages; the book has a Latin translation facing the original Italian.

12. Lumisden, *Remarks* (1812), 157–8 (quote), 159–62.

13. FB, *Palazzo* (1738), tab. 17, and perhaps other plates; Miranda, *Bianchini* (2000), 74; *DBI*, s.v. "Nicoletti."

14. Panvinio, *De ludis* (1600), 8–9; Ferrary, *Panvinio* (1996), 29, 214; Huelsen, *Saggio* (1969), 101.

15. FB, *Palazzo* (1738), 36, 40, 44, 46; Lanciani, *Storia* (2000), 78, 86, 88; Miranda, *Bianchini* (2000), 116–22.

16. FB, *Palazzo* (1738), 50 (quote), 52, 54, 58; Cecamore, *Palatino* (2002), tab. V, "L'insieme del complesso domiziano," and pp. 58, 91, 189–92.

17. FB, *Palazzo* (1738), 80, 82, 86, 88; FB to Grand Duchess Violante, 29 December 1725, in FBV, U.22, fo. 161^r.

18. FB, *Palazzo* (1738), 108, 110, 116, 118, 120, 122, 124, 126, 128, 130, 132, 140, 148.

19. FB, *Palazzo* (1738), 154, 156, 158; Claridge, *Rome* (2010), 146–51.

20. FB, *Palazzo* (1738), 114.

21. FB, *Palazzo* (1738), 184–6 (quote), 188, 196, 232.

22. FB, *Palazzo* (1738), 194, 200, 204, 206, 216.

23. FB, *Palazzo* (1738), 218, 220, 226, 248, 272, 288.

24. FB, *Palazzo* (1738), 66.

25. MV 104; Valerio, *Diario*, iv. 568 (gives 23 August); FB to Grand Duchess of Tuscany, 19 January 1726, in FBV, U.22, fos 177–8, and *HP* 5–6 = *NP* 24; Bozzoli, *Cenni* (1873), 9. "Martyr der Antiquität": J. J. Volkmann, *Historisch-Kritische Nachrichten von Italien* (1770), quoted in Schröter, in Gaehtgens, *Winckelmann* (1986), 103 n. 19.

26. FB, *Palazzo* (1738), 266n., 300; Huelsen, *Mitteilungen*, 10 (1895), 23–4, 264, 273–5.

27. Huelsen, *Mitteilungen*, 10 (1895), 252–3, 270; Maffei to G. Bianchini, 8 March 1736, in *Lettere di vari illustri* (1841–3), ii. 67; Miranda, *Bianchini* (2000), 99–101.

28. e.g., Coarelli, *Rome* (2007), 134, 137; Oechslin, in *Piranesi* (1979), 110–11; Millon, in Scott and Scott, *Eius virtutis* (1993), 481, 483, 485; Pinto, *Speaking Ruins* (2012), 38–42; Engelberg, in KS 139, 144–5, 151–2.

29. Quoted from Davia by Miranda, *Bianchini* (2000), 267.

30. Polignac, *Eutopia*, 2/1 (1993), 46, 53, 61; FB, *Camera* (1727), pp. iii–vii; Kockel, in KS 117–22, and Polignac, in KS 169.

31. Kammerer-Grothaus, *Babesch*, 86/1 (2011), 97, and *Mélanges*, 91/1 (1979), 323–4, 334–5; FB to the Principessa of Tuscany, 1 and 29 December 1725, in FBV, U.22, fos 161–2.

32. FB, *Camera* (1727), pp. i–ii.

33. FB, *Camera* (1727), 17, 24–5; Fehl, in Magnusson et al., *Ultra* (1997), 91.

34. FB, *Camera* (1727), 5 (2nd quote), 8 (first), 15–16.

35. Kolendo, *Klio*, 71/2 (1989), 422, 429–30, and in FB, *Camera* (1991), p. xxiv; Treggiari, *Papers of the British School at Rome*, 4 (1975), 58 (55 jobs), 60–3.

36. FB, *Camera* (1727), 19–66, 86–7; Kolendo, in FB, *Camera* (1991), pp xxiv–xxvi; Treggiari, *Papers of the British School at Rome*, 4 (1975), 50–6, 71.

37. Kockel, in KS 115–22, 129n.; Gori, *Monumentum* (1727); Ghezzi, *Camere* (1731).

38. Battaglia, *Dialoghi di storia dell'arte*, 2 (1966), 58; Polignac, *Eutopia*, 2/1 (1993), 51–4; Oechslin, in *Piranesi* (1979), 108–10; Kolendo, *Klio*, 71/2 (1989), 426; Basso and Buonopane, in CR 292, 300–8.

39. "Benedict III," in Davis, *Liber* (1995), 168, 172 (quotes); Kelly, *Popes* (1988), 105–10.

40. "Hadrian II," in Davis, *Liber* (1995), 282, and editorial material, pp. 251–3, 282–6.

41. Duchesne, *Étude* (1877), 217.

42. AB i, fo. a3ᵛ, §5, and fo. a4ʳ, §§8–9 (quote).

43. AB i, fos b1ʳ–b2ᵛ, §§14–16, 20–1; b3, §§22–3; b4–c1ʳ, §26; c3ʳ, §§31–2; c4ᵛ, §35; d1, §38; d3–d4, §§48–51.

44. AB i, publisher's introduction; fo. e2ʳ, §60 (quote); AB iii, re popes St Anastasius I (399–40) and St Damasus I (366–84), resp.

45. AB i, fos e2ʳ, §58, and e2ᵛ, §60.

46. AB ii, pp. lxv–lxix.

47. AB ii, pp. x–xiv, violating the priority later (p. lxxxiii) reasserted; "Monsignore Bianchini fonda la sua Cronologia," FBV, S.83.2, fos 459–61.

48. Baldini, in *Vite*, v (1751), 116, 118, 124; Sölch, *Bianchini* (2007), 337–44.

49. MV 19.

50. AB ii, pp. lxxi–lxxiv, lxxvi (1st quote), lxviii, lxxxi (2nd quote); Duchesne, *Étude* (1877), 34–5, and *Liber* (1886–92), i, pp. xxvii–xxviii.

51. AB ii, p. lxxvii (quote), lxxxvii–cxl (the synopsis).

52. AB ii, pp. cix–cxxi, cxxii–cxl (the itinerary), cxli–cxlv (the cemeteries), cxlvii (quote).

53. AB ii, p. cxlviii.

54. AB ii, pp. clxi–ccxx, esp. cxlvii (1st quote), cxlvi, cxlviii (2nd quote).

55. MV 95.

56. Maffei to FB, 8 July 1724, in *Lettere di vari illustri* (1841–3), ii. 55, referring especially to the chronological parts (Maffei, in *IU* (1747), fo. a4ʳ); Istituto di Bologna to FB, February 1724, in FBV, U.15, fo. 72 (3rd quote); and Chiarlo, in Hübner-Wojciechowski, *Eredità* (1992), 182 (4th quote). For Muratori, see Nicolini, *Uomini* (1992), 117–19, and Duchesne, *Étude* (1877), 120; Muratori, *Scriptores*, iii.1 (1723), 55–91; Gibbon, *Decline*, ch. 49, n. 32, ed. Bury (1909–14), v. 275.

57. AB ii, iv.

58. AB iii, pp. ii–vi.

59. AB iii, pp. xxvii–xxi, xxxix–li; FBC ccclviii: 9.

60. AB iv, fo. **3, and pp. 78–9; Baldini, *Vite*, v (1751), 115.

61. Duchesne, *Étude* (1877), 85–6, 113; Spagnolo, *Accademia di agricultura, scienze e lettere, Verona, Memorie*, 74/2 (1898), 94–6.

62. Duchesne, *Étude* (1877), 119 (quote). Duchesne relies on Bianchini often in his *Liber* (1886), i, e.g., pp. lxxiv, lxxxix, xcv, cxlv, clxix–clxx, cxcv–cxcvi, ccii, 127, 333.

63. Duchesne, *Liber* (1886–92), i, Préface; FB to Marco Cornelio Bentivoglio d'Aragona, 14 June 1727, quoted in Miranda, *Bianchini* (2000), 94.

Chapter 8

1. Quincy, *Mémoires* (1741), 23–6, 32–3; Stoye, *Marsigli's Europe* (1994), 8–10, 15–20, 23–6; Marsigli, *Osservazioni* (1681), 4–6, 11.

2. Quincy, *Mémoires* (1741), 107–35, 148–53; Stoye, *Marsigli's Europe* (1994), 20–3, 31–6.

3. Quincy, *Mémoires* (1741), 168, 177–84, 202–12; Stoye, *Marsigli's Europe* (1994), 42–52, 202–15.

4. Quincy, *Mémoires* (1741), 188–9, 195, 215–16; Stoye, *Marsigli's Europe* (1994), 134–6, 149–50.

5. Stoye, *Marsigli's Europe* (1994), 229–34, 338–51.

6. Longhena, *Atti e memorie della Accademia di agricoltura, scienze e lettere di Verona*, 21 (1942–3), 179–87, 196–7, 201–12, 217–20; Stoye, *Marsigli's Europe* (1994), 221–8. The Commachio incident occurred in 1708.

7. Bolletti, *Origine* (1751), 13–20; Heilbron, *Historical Studies in the Physical Sciences*, 22/1 (1991), 61–5.

8. Quincy, *Mémoires* (1741), 191–2; FB to Marsigli, 6 July 1711, in Fantuzzi, *Memorie* (1770), 327–8; Stoye, *Marsigli's Europe* (1994), 255–6, 285, 316 n. 8; Dal Prete, *Nuncius*, 20/1 (2005), 110–11, 115; Uglietti 87, 173.

9. Bolletti, *Origine* (1751), 26–31; Biagi Maino, in Biagi Maino, *L'immagine* (2005), 55–60.

10. Takahashi, *Zeitschrift für Kunstgeschichte*, 82 (2019), 185–7; HS, 79–80, 94.

11. Johns, *Zeitschrift für Kunstgeschichte*, 55 (1992), 583–4; Bedini, in Merrill, *Proceedings* (1980), i, pp. xix–xx.

12. HS 167–70.

13. Anon., *Giornale de' letterati d'Italia*, 4 (1710), 434; HS 197.

14. Braccesi, *Giornale di astronomia*, 4 (1978), 342–9; Heilbron, in Bursill-Hall, *Boscovich* (1993), 390–6.

15. Manfredi to Marsigli, 30 August 1711, in Fantuzzi, *Memorie* (1770), 319–20.

16. Bedini, in Merrill, *Proceedings* (1980), i, pp. xxiii–xxv.

17. FB, in Baldi, *Mem.* (1724), 140–5.

18. MV 103–4; FB, in Baldi, *Memorie* (1724), 133–5, 138–9; Albani, in Baldi, *Memorie* (1724), "A chi legge;" FB to ?, 28 June and 12 July 1724, in FBV, U.22, fos 67–9; NP 7–8.

19. Maffei to FB, 12 June and 8 July 1724, in *Lettere di vari illustri* (1841–3), ii. 52–3 (quote), 55; Maffei, in IU (1747), a3[v].

20. Carvalho, *Astronomia* (1985), 40–2, 47–51, 106 n. 42; Tirapicos, *Mediterranean Archaeology and Archaeometry*, 16/4 (2016), 505. The result for the difference in longitude between Lisbon and Rome from observations of Jupiter's moons was quite good; FB, RSL, RBC, 14:273–4, and Carbone, RSL, RBC, 12:481.

21. Finocchiaro, in CR 334–6.

22. Castro, *Portugal*, i. 57–61; Carvalho, *Astronomia* (1985), 38–9; Donato, *Accademie romane* (2000), 75; Feist, *Sonne* (2013), 128, 132–7.

23. Tirapicos, *Ciência* (2017), 97–100; FB to João V, 20 November 1724, in FBV, U.20, fos 153, 154$^\text{v}$, 155$^\text{r}$.

24. Tirapicos, *Ciência* (2017), 150–2; Sousa Leitão, in *Estrelas* (2009), 49–64.

25. The measured χ must be diminished by a factor of cosδ. Cf. Hadley, *PT* 36/410 (1729), 161.

26. *NP* 124–33; cf. FB, *Observationes selectae* (1737), 139.

27. *NP* 134; *HP* 76, with "splendid" (*nobile*) substituted for *HP*'s "direct."

28. *NP* 17–18.

29. FB, *De aureis* (1717), 7 (2nd quote), 8–13, 21–2, 28 (1st quote).

30. *NP* 18, 133 (quote); *HP* 76 (quote).

31. *NP* 133–41; *NP* 45, 132 (*HP* 18, 75–6), for the discrepancy.

32. *NP* 21, 153–5; anon., *PT* 8 (1673), 5178, 5180; Van Helden, *Isis*, 65/1 (1974), 46–8.

33. *NP* 23; *HP* 5; Middlehurst and Burley, *Catalog* (1966), *passim*; Monaco, *Physis*, 21 (1983), 413–31.

34. *NP* 22–9; Feist, *Fragmenta*, 5 (2011), 310–16, 321–3, and in Feist and Rath, *Facetten* (2012), 305–6, 314–15.

35. *NP* 106–7; *HS* 255.

36. Melchior Briga, SJ, Florence, to FB, 3 September 1726, in *NP* 149.

37. FB, *Observationes selectae* (1737), 253–4; Manfredi, in FB, *Observationes selectae* (1737), p. v, and to Tomasso Deranio, 1729, quoted by Spagnolo, *Accademia di agricultura, scienze e lettere, Verona, Atti e memorie*, 74/2 (1898), 106–7.

38. Anon., *Present State*, 4 (1729), 242 (1st quote), 244; Maraldi to FB, 24 December 1726, in *Lettere di vari illustri* (1841–3), ii. 89 (2nd quote); Federici, *Elogi* (1818–19), iii. 17.

39. Fay, in *NP* (1996), 32, and Beaumont, *NP* 5.

40. Almeida, *Recreação* (1786–1800), vi (1795), 141–4; Hadley, *PT* 36 (1729), 159–60, had reservations about Bianchini's measurements.

41. Feist, *Sonne* (2013), 198, 210 (Maraldi); Dal Prete, *Astronomia*, 2 (2003), 16–17, and *Nuncius*, 20/1 (2005), 119, 134–43; Flammarion, *Terres* (1877), 204–11.

42. e.g., Dal Prete, *Astronomia*, 2 (2003); Feist, in Feist and Rath, *Facetten* (2012), 318–20; Tinazzi, *Fondazione Giorgio Ronchi, Atti*, 59 (2000), 443–53.

43. *NP* 87–101, quote on 99.

44. *NP* 87; *HP* 46.

45. Feist, in *Sonne* (2013), 157–60, 163–8.

46. *NP* 56–75.

47. Museu nacional dos coches, *Embaixaa* (1996), 19–20, 47; Tirapicos, *Ciência* (2017), 147–9; Barchiesi, *Estudos italianos em Portugal*, 23 (1964), 149, 151; MV 113–15; FBC ccccxxx: ii, fos 55–82 (Italian time).

48. Feist, *Sonne* (2013), 89–91.

49. FB, *Relazione delle cose* (1882), 8–10, 21–8.

50. FB, *Relazione delle cose* (1882), 28 (1st quote), 15, 16 (2nd quote), 17–19, 20 (FB's height), 21 (3rd quote).

51. Cf. Feist, *Sonne* (2013), 183, 186–7, 200–1, 235, 241–3.

52. International Astronomical Union, *Gazeteer*, s.v. "Bianchini;" Cattermole, *Venus* (1994), 2.

53. NP 37, 62, 74; Carvalho, *Astronomia* (1985), 13–15, 23–5.

54. NP 14 (quote); FB, *Observationes selectae* (1737), 254.

55. *Opuscula varia* (1754), ii. 81; NP 110.

56. FBC cccliv: v, fos 119, 122ᵛ, fo. 123ʳ, 87–92.

57. Manfredi to FB, 2 August 1727, quoted in Feist, in Feist and Rath, *Facetten* (2012), 322.

58. FB, *Observationes selectae* (1737), 261–6; Manfredi, in FB, *Observationes selectae*, 265–6; Salvago to FB, 12 October 1716 (quote), 31 July 1717, 31 December 1718 (bellissima e unica observazione), in Balestrieri, "Delizie."

59. Fontenelle, *HAS* (1729), 114.

60. NP 99 (Galileo), 100 (Cassini).

61. FBC cccliv: v, fo. 123.

62. FB, *Nuova raccolta*, 41 (1785), 4–5, 8–9 (quote), 21.

63. Marsigli to FB, 24 December 1726, in Rodolico, *Dictionary*, ix (1974), 135, s.v. "Marsili."

64. NP 19.

65. Gaetano Cenni, in FB, *Opuscula varia* (1754), i, p. iii.

66. FB, in G. Bianchini, *Demonstratio* (1752–4), i:1, pp. cclxxi–cccxxvi.

Chapter 9

1. Platania, *Ultimi Sobieski* (1990), 197, 200–3, 209–10, 221–2.

2. Azon, *Parentalia* (1736), 7–10.

3. Azon, *Parentalia* (1736), 23–4.

4. Platania, *Ultimi Sobieski* (1990), 249–52.

5. Valesio, *Diario*, iii. 745 (1706, noting FB's design); Olszewski, *Römisches Jahrbuch der Bibliotheca Hertziana*, 32 (1997–8), 555.

6. Cucco, *Albani* (2001), 31 n. 4; FB, in FBC ccccxxx: v, 10 June 1727, fos 152–3.

7. DBI, s.v. "Innocent XI;" the best-selling novel Monaldi and Sorti, *Imprimatur* (2002), 615–47, presents archival documents supporting its assertions about Innocent.

8. Corp, *Stuarts in Italy* (2011), 22, 25, 28–33.

9. *Funerali* (1766), pp. iv (2nd quote), xii (1st quote), xxviii–xxxi.

10. Wogan to Swift, 27 February 1732, in Swift, *Miscellaneous Pieces* (1789), 89–91.

11. O'Callaghan, *History* (1870), 314–15; Flood, *Life* (1922), 113; DNB, s.v. "Wogan."

12. Wogan, in Gilbert, *Narratives* (1894), 31–4; Miller, *Wife* (1965), 149, 152–3.

13. Flood, *Life* (1922), 15; Wogan to Swift, 27 February 1732/3, in Swift, *Correspondence* (1999–2014), iii. 591, and 519 n. 12.

14. Wharton, in Wogan, "Manuscript," 36, 40, Galway/Kilfenora Diocesan Archives; Melville, *Life* (1913), 76–7, 170–81, 193–4, 203–4. "Essay" = "trial, challenge."

15. Wogan to Swift, 27 February 1732/3, in Swift, *Correspondence* (1999–2014), iii. 590–2, and reply, iii. 514–17.

16. Wogan to Swift, in Swift, *Miscellaneous Pieces* (1789), 73–86; Law, Royal Society of Antiquaries of Ireland, *Journal*, 7 (1937), 253–64.

17. Swift to Wogan, 1735/6, in Swift, *Correspondenc* (1999–2014), iv. 272–3.

18. Weber, in Jamme and Poncet, *Offices* (2005), 568–70, 575, 580–6; Nicéron, in *Mémoires*, 29 (1734), 80–1.

19. FB, *Oratio* (1724), pp. ii, vii–x, xvi–xvii; Posterla, *Roma sacra* (1725), 694 (quote); MV 111–12; FBV, T.91, fo. 111.

20. Fiorani, *Concilio* (1978), 29–34, 37–44, 71–2.

21. Marzocchi, *Civiltà veronese*, 3rd ser. (1993), 47–8.

22. Clement to Barbarigo, 8 October 1720, in FBV, U.15, fo. 219; FB to Paolucci, 5 November 1720, in FB, U.22, fos 60–4; MV 99; Bonanni, *Gerarchia* (1720), 473–4, and ill. 133, for the *veste pavonezza*.

23. G. F. Barbarigo to FB, 11 July and 10 December 1728, in FB, U.15, fos 227, 233. Gregorio Barbarigo was beatified in 1761 and canonized in 1960.

24. MV 102; Favaretto, in KS 34–5.

25. MV 105–8; FB to Bentivoglio, 14 June 1727, in Sölch, *Bianchini* (2007), 369–70.

26. MV 34–6, 117–18.

27. Valesio, *Diario*, v. 26–7; Maffei, *Verona* (1771), pt 2, 167.

28. Maffei, *Verona* (1771), pt 2, 167–8; MV 119.

29. *HS* 148 (Leibniz); Newton to J. B. Mencke, 1724, in Newton, *Correspondence* (1959–77), vi. 255; Montesquieu, *Œuvres* (1949), i. 781.

30. Lanza, *Lettere italiane*, 10 (1958), 39–48, and Heilbron, in Biale and Westman, *Thinking* (2008), 261–7 (Vico); Winckelmann, *Kleine Schriften* (2002), 90, 122, and Schröter, in Gaehtgens, *Winckelmann* (1986), 87, 106 n. 46. In modern times the high authority Arnaldo Momigliano, *Contributo* (1955), 85–7, shares their high opinion of *Istoria universale*.

31. Lenglet du Fresnoy, *Méthode* (1729), ii. 69.

32. Giuliari, in FB, *Dei viaggi* (1877), dedication, and in FB, *Lettera* (1883), 5–8, *Relazione sulla fortezza di Guastalla* (1885), 5–6, and *Carte* (1871), 6, 8–10; Giuliari, *Capitolare* (1888), 259–60, 301, 336.

33. FB, "Descrizione di Roma sacra e profana," in FBC ccclv: ii; Uglietti, 61; FB to the electress of Bavaria, 5 and 17 May 1725, re visit of the Grand Dichess Violante of Tuscany, in FBV, U.22, fos 165, 167.

34. *DBI*, s.v. "Ottoboni;" Corp, *Stuarts in Italy* (2011), 27.

35. Lotz, *Römische Jahrbuch für Kunstgeschichte*, 12 (1969), 66, 69, 83–4; Johns, *Papal Art* (1993), 185–9.

36. Folkes, "Journal," in Bodleian Library, MS Eng. misc. c.444; Nollet, "Journal," Bibliothèque Municipale, Soissons, MS 150, fo. 42^r (6 September 1749).

37. *HS* 246–64.

38. Gierowski, in Kanceff and Lewanski, *Viaggiatori* (1988), 192–5, 200–2; Catamo and Lucarini, *Cielo* (2002), 84.

39. Lambertini to FB, 17 January 1724, FBV, U.19, fo. 23; Heilbron, in Messbarger et al., *Benedict XIV* (2016), 179–80.

40. Sölch, in CR 311–15, and *Bianchini* (2007), 89–128, 160–75 (the seventeen sketches), 149–53.

41. Sölch, *Bianchini* (2007), 154–8; Uglietti, 137–46; Miranda, *Bianchini* (2000), 31–5, 96, 98.

42. Marzocchi, *Civiltà veronese*, 3rd ser. (1993), 48 (quote), 50–4, from FBC ccccxxxvii; Giuliari, *Capitolare* (1888), 39–46, 49, 110; Spagnolo, *Manoscritti* (1996), 349–438. The books included nine incunabula.

43. Correspondence between Maffei and FB, 1716 to 1727, in *Lettere di vari illustri* (1841–3), ii. 43–4, 49, 51–2, 60; Huelsen, *Bulletino della Commissione archeologica comunale di Roma*, 18 (1890), 261 n. 2; Rusconi, in Messbarger et al., *Benedict XIV* (2016), 278–9.

44. Sölch, *Bianchini* (2007), 283, 287, 290–1, 300–8.

45. Cf. Huelsen, *Bulletino della Commissione archeologica comunale di Roma*, 18 (1890), 263–77, esp. 265.

46. Dixon, *Accademia* (2006), 78–81.

47. Michel Giuseppe Morei, custodian of the Arcadia, in Boscovich, *Vite*, v (1751), 128–9; Federici, *Elogi* (1818–19), iii. 42–3: "Noverat ille quidem nostro quidquid patet orbe | Noverat immenso, quidquid & orbe | Haec tamen haud fuerat studiorum meta suorum | Maius adeptus & est a probitate decus."

WORKS CITED

Bianchini's writings have come down to us in daunting abundance. Consequently, no general survey of them exists. An impression of the whole can be gained, however, by inspecting the catalogue of FBC in Spagnolo, *Manoscritti* (1996), 349–438; notices of parts of FBV in Baldini, in CR 75–99; sources of Bianchini's correspondence in Viola, CR 121–61; and the good, but incomplete, list of his publications in Bozzoli, *Cenni* (1873). Another useful bibliographical introduction is Arecco, CR 163–84, a study of the work of Salvatore Rotta, a modern specialist on Bianchini.

Main Manuscripts Consulted

Beinecke Library, Yale: General MSS File 113
Biblioteca Capitolare, Verona: FB Papers (FBC)
Biblioteca Civica, Verona, MS 2833: FB, "Trattati di matematico e di fisica composti e dettati del Sig.r Dottore Geminiano Montanari . . . scritti . . . negli anni 1682, 1683"
Biblioteca Vallicelliana, Rome: FB Papers (FBV)
Bibliothèque Municipale, Soissons, MS 150: J. A. Nollet, "Journal du voyage de Piedmont et d'Italie"
Bodleian Library, Oxford: MS Eng. misc. c. 444: Folkes, "Journal"
Bodleian Library, Oxford: Carte MSS
British Library, London (BL): MS 1572/868(2) FB, "Studium bibliothecae" [1690]
British Library, London (BL): Add. MSS 20549–20550 (Gualterio [Gualtieri] Papers)
British Library, London (BL): Add. MSS 31260–31261 (Nairne Papers)
Cambridge, Kings College, Keynes MS 130.5
Cambridge University Library: RGO 2/10 (Halley Papers); RGO 1/42, 1/69D
Galway/Kilfenora Diocesan Archives (Wogan MS)
Oxford University Archives, WPbeta/21/6
RSL, items from RBC series, Letter Books, and Journal Books

Printed Works

Ademollo, Agostino, *Gli spettacoli dell'antica Roma descrizione istorica* (Florence: the author, 1837).

Affò, Ireneo, *Istoria della città, e ducato, di Guastalla*, 4 vols (Guastalla: Regio-Ducale Stamperia, 1785–7).

Agnellus Ravennatus, *Liber pontificalis. Bischofsbuch*, ed. and trans. Claudia Nauerth, 2 vols (Freiburg: Herder, 1996).

Albani, Gio. Francesco, *Discorso detto nella Reale Accademia della Maestà di Cristina Regina di Svezia in lode di Giacomo II. Re della Gran Bretagna* (Rome: Tinassi, 1687).

Almeida, Teodoro de, *Recreação filosófica, ou diálogo sobre a filosófia natural*, 10 vols (Lisbon: Regia Officina Typografica, 1786–1800).

Altieri Biagi, Maria Luisa, and Bruno Basile (eds), *Scienziati del seicento* (Milan and Naples: Ricciardi, 1980).

Anastasius Bibliothecarius, *Liber pontificalis to the Pontificate of Gregory I*, trans. Louise Roper Loomis (New York: Columbia University Press, 1916).

Andreoli, Aldo, "Il Bacchini e il Muratori," *Benedictiana*, 6 (1952), 59–84.

Andreoli, Aldo, *Nel mondo di Lodovico Antonio Muratori* (Bologna: Il Mulino, 1972).

Anon., "A Discovery of Two New Planets about Saturn," *PT* 8 (1673), 5178–85.

Anon., "Preface," in D'Elci, *Present State* (1706), pp. iii–xlv.

Anon., [Zeno, Apostolo]. "Relazione della linea meridiane orizzontale, e della elissi polare fabricata in Roma l'anno 1702," *Giornale de' letterati d'Italia*, 4 (1710), 64–87.

Anon., *Hannibal not at our Gates: Or, an Enquiry into the Grounds of our Present Fears of Popery and the Pre-der* (London: s.n., 1714).

Anon., *At a . . . Meeting of the Vice Chancellor . . . [re] a Letter . . . Containing Treasonable Matters* (Oxford: s.n., 1714).

Anon., "Journal du séjour de S.M.R. à Rome en 1717," Bodleian Library, Oxford, Carte MSS ccviii, fos 338–59.

Anon., *Remarks upon a Jacobite Pamphlet Privately Handed about, Entitled Female Fortitude Exemplify'd* (London: J. Roberts, [1722]).

Anon., "Rome," The Present State of the Republick of Letters, 2 (December 1728), 453–4.

Anon., "New Observations of the Planet Venus," *Present State of the Republic of Letters*, 4 (July 1729), 242–8.

Anon., "[Relazione che porta li più singolari evenimenti accaduti nella fuga della principessa reale Clementina Sobieski,]" in Gilbert, *Narratives* (1894), 145–58.

Anon. "Narrative of the Seizure, Escape and Marriage of the Princess Clementina Sobieski," in Gilbert, *Narratives* (1894), 159–89.

Antonelli, Roberta, "Il viaggio lucchese di Clementina Sobieska," in Gaetano Platania (ed.), *Da est ad ovest, da ovest ad est: Viaggiatori per le strade del mondo* (Viterbo: Sette Città, 2006), 225–62.

Aquino, Carlo d', *Sacra exequialia in funere Jacobi II* (Rome: Typis Barberinis, 1702).

Arecco, Davide, "Il linguaggio di uno scienziato all'alba dell'Illuminismo: Note sugli studi bianchiniani di Salvatore Rotta," in CR 163–84.

Augustine, St, *The City of God*, trans. Marcus Dods, in *Great Books of the Western World*, 18 (1952), 129–618.

Azon, Filippo d', *Parentalia in anniversario funere Mariae Clementinae* (Rome: Congregatio de Propaganda Fede, 1736).

Bacchini, Benedetto (ed.), *Agnelli . . . Ravennatis, Liber pontificalis*, rev. L. A. Muratori, in L. A. Muratori, *Scriptores*, ii (Milan: Societas Palatina, 1723), 1–220.

Bagnani, Gilbert, *Parentalia Mariae Clementinae . . . Iussu Clementis XII* (Rome: Salvioni, 1736).

Baldi, Bernardini, et al., *Memorie concernenti la Città di Urbino dedicate alla Sagra Real Maestà di Giacomo III. Re della Gran Bretagna, etc* (Rome: Salvioni, 1724).

Baldini, Gian Francesco, "Vita di Monsignor Francesco Bianchini Veronese detto Selvaggio Afrodisio," *Le vite degli arcadi illustri*, v (Rome: A. de' Rossi, 1751), 115–29.

Baldini, Ugo, "Due raccolte romane di lettere di Eustachio Manfredi," in Renzo Cremante and Walter Tega (eds), *Scienza e letteratura nella cultura italiana del settecento* (Bologna: Il Mulino, 1984), 529–44.

Baldini, Ugo, "La rete di comunicazione astronomica di Francesco Bianchini: Un'analisi del fondo vallicelliano," in CR 75–99.

Baldovin, John F., "Hippolytus and the *Apostolic Tradition*: Recent Research and Commentary," *Theological Studies*, 64 (2003), 520–42.

Balestrieri, Riccardo, "Delizie in villa: Paris Maria Salvago: Epistolario con Francesco Bianchini," http://uranialigustica.altervista.org/salvago/lettere/bianchini_lettere.htm (13 December 2016).

Barbieri, Francesco, and Franca Cattelani Degani, "Tre lettere di Geminiano Montanari a Gian Domenico Cassini," *Nuncius*, 12/2 (1997), 433–41.

Barchiesi, Roberto, "I Bianchini e la Corte di Lisbona," *Estudos italianos em Portugal*, 23 (1964), 147–59.

Baroni, Pier Giovanni, *Un conformista del secolo diciottesimo: Il cardinale Pietro Ottoboni* (Bologna: Ponte nuovo, 1969).

Barret-Kriegel, Blandine, *Jean Mabillon* (Paris: PUF, 1988).

Barret-Kriegel, Blandine, *La Défaite de l'érudition* (Paris: PUF, 1988).

Bartoli, Pietro Santo, et al., *Terme di Tito* ([Rome]: n.p., n.d.)

Bartoli, Pietro Santo, et al., *Veteres arcus Augustorum triumphis insignes ex reliquiis quae Romae adhuc supersunt* (Rome: Ad Templum Sanctae Mariae de Pace, 1690).

Basso, Patrizia, and Alfredo Buonopane, "Bianchini fra archeologia ed epigrafia: Il Sepolcro degli schiavi e dei liberti di Livia," in CR 285–308.

Battaglia, Roberta, "Da Francesco Bianchini a Giovan Battista Piranesi: L'illustrazione delle Camere Sepocrali dei Liberti di Livia," *Dialoghi di storia dell'arte*, 2 (1996), 58–81.

Bayer, Johannes. *Uranometria: Omnium asterismorum continens schemata*. Augsburg: C. Mangus, 1603.

Béchu, Philippe, et al., *Le Cardinal Armand Gaston de Rohan, 1674–1749* (Paris: Archives Nationales, 2013).

Beck, Herbert, et al., *Antikensammlungen im 18. Jahrhundert* (Berlin: Mann, 1981).

Becker, Maria, "Iam nova progenies caelo demittitur alto. Ein Beitrag zur Vergil-Erklärung (ECL. 4/7)," *Hermes*, 131/4 (2003), 456–63.

Bedini, Silvio A., "The Vatican's Astronomical Paintings and the Institute of the Sciences of Bologna," in Russell B. Merrill (ed.), *Proceedings of the 11th Lunar and Planetary Science Conference*, 3 vols (New York: n.p., 1980), i, pp. xiii–xxxiii.

Bellori, Giovanni Pietro, "Carlo Maratti," in Bellori, *Le vite de' pittore, scultori et architetti moderni*, 3 vols (Pisa: Capurro, 1821), iii. 136–237.

Bély, Lucien, *Espions et ambassadeurs au temps de Louis XIV* (Paris: Fayard, 1990).

Benedetti, Silvano, "L'Accademia degli Aletofili di Verona," in *Accademie e cultura: Aspetti storici tra sei e settecento* (Florence: Olschki, 1979), 223–6.

Berkel, Klaas van, "'Cornelius Meijer invenit et fecit:' On the Representation of Science in Late Seventeenth-Century Rome," in Pamela H. Smith and Paula Findlen (eds), *Merchants and Marvels: Commerce, Science, and Art in Early Modern Europe* (New York: Routledge, 2002), 277–94.

Bertelli, Sergio, "La crisi dello scetticismo e il rapporto erudizione-scienza agli inizi del secolo xviii," *Società*, 11/3 (1955), 435–56.

Bertelli, Sergio, *Erudizione e storia in Lodovico Antonio Muratori* (Naples: Nella sede dell'Istituto, 1960).

Berveglieri, Roberto, "Tecnologia idraulica olandese in Italia nel secolo xvii: Cornelius Janszoon Meijer a Venezia," *Studi veneziani*, 10 (1985), 81–97.

Bevilacqua, Mario, "Cartografia e immagini urbane: Giovanni Battista Falda e Cornelis Meyer nella Roma di Innocenzo XI," in Richard Bösel et al. (eds), *Innocenzo XI Odescalchi: Papa, politico, committente* (Rome: Viella, 2014), 289–308.

Biagi Maino, Donatella, "I pittori per l'Istituto," in Donatella Biagi Maino (ed.), *L'immagine del settecento da Luigi Ferdinando Marsigli a Benedetto XIV* (Turin: Allemandi, 2005), 51–64.

Bianchini, Francesco, "Cometes anno 1684 mense junio julioque Romae observata," *Acta eruditorum* (1685), 189–90, 241–5.

Bianchini, Francesco, "Nova methodus cassiniana observandi parallaxes & distantias planetarum a terris," *Acta eruditorum* (1685), 470–8.

Bianchini, Francesco, "Eruditissimae virginis Helenae Corneliae imago inter astra refertur," in *Pompe funebri* (1686), 169–70.

Bianchini, Francesco, "De emblemate, nomine atque instituto Alethophilorum dissertatio," *Giornale de' letterati di Parma* (January 1687), 237. Cf. Gasperoni, *Scipione Maffei* (1954), 28–9.

Bianchini, Francesco, "Introduzione all'opera, e breve compendio della vita dell'autore," in Montanari, *Forze* (1694), fos a.6–b.12.

Bianchini, Francesco, *Tavola istorica degli avvenimenti più celebri de' principali per 1600 anni disposta secondo l'ordine de' tempi ed ornata con figure* (Rome: Stamperia della Regia Camera, 1695). (Partial copy in BL, *Tracts* 1572, 868 (1685–1724).)

Bianchini, Francesco, *La istoria universale provata con monumenti, e figurata con simboli degli antichi* (Rome: for the author, 1697); ed. Antonio Giuseppe Barbazza (Rome: A. De' Rossi, 1747); 5 vols (Venice: Battaggia, 1825–7). Cited as *IU*, *IU* (1747), and *IU* (1825).

Bianchini, Francesco, *De lapide antiati epistola ad . . . Franciscum Aquavivam* (Rome: A. de Rubeis, 1698).

Bianchini, Francesco, *Lettera ad un amico, in ragguaglio della legazione . . . alla maestà cattolica del rè Filippo V* (Rome: Olivieri, 1702).

Bianchini, Francesco, *Solutio problematis paschalis* (Rome: Camera Apostolica, 1703).

Bianchini, Francesco, *De kalendario et cyclo caesari, ac de paschali canone S. Hippolyti martyris dissertationes duae . . . Quibus inseritur descriptio, & explanatio basis, in Campo Martio nuper detectae sub columna Antonino Pio olim dicata: Hic accessit Enarratio per epistolam ad amicum De nummo et gnomone Clementino* (Rome: De Comitibus, 1703).

Bianchini, Francesco, *Considerazioni teoriche, e pratiche intorno al trasporto della colonna d'Antonino Pio collacata in Monte Citorio* (Rome: Camera apostolica, 1704).

Bianchini, Francesco, "Vita del cardinale Enrico Noris," in Giovan Mario Crescimbeni, *Le vite degli Arcadi illustri*, i (1708), 199–220.

Bianchini, Francesco, *De nobilissimo hospite Comitis de Trausnitz nomen professo et in villa pinciana Burghesiorum principum excepto die 27 Maii 1716. Epistola* (Rome: A. de Rubeis, 1716).

Bianchini, Francesco, *De aureis et argenteis cimeliis in arce perusina effossis* (n.p., 1717).

Bianchini, Francesco (ed.), *Vitae romanorum pontificum a B. Petro Apostolo ad Nicolaum I perductae cura Anastasii S.R.E. Bibliothecarii*, 4 vols (Rome: Salvioni, 1718–35). Cited as AB.

Bianchini, Francesco, *Il giorno natalizio della sac: Reale maestà di Giacomo III, Re d'Inghilterra &c celebrato nelle Colline d'Albano con cantata pastorale il dì 21 giugno 1720* (Rome: A. de' Rossi, 1720).

Bianchini, Francesco, "Spiegazione delle sculture contenute nelle lxxii tavole di marmo e bassorilievi collocati nel basamento esteriore del palazzo di Urbino," in Baldi, *Memorie* (1724), 79–132.

Bianchini, Francesco, "Notizie, e pruove della corografia del Ducato di Urbino, e della longitudine, e latitudine geografica della Città medesima, e delle vicine, che servono a stabilire quelle di tutta la Italia," in Baldi, *Memorie* (1724), 133–47.

Bianchini, Francesco, *Oratio de eligendo summo pontifice post obitum Innocentii XIII* (Rome: A. de' Rossi, [1724]).

Bianchini, Francesco, *Camera ed inscrizioni sepulcrali de' liberti, servi, ed ufficiali della casa di Augusto* (Rome: Salvioni, 1727).

Bianchini, Francesco, *Hesperi et phosphori nova phaenomena* (Rome: Salvioni, 1728). Cited as *HP*.

Bianchini, Francesco, *Astronomicae, ac geographicae observationes selectae*, ed. Eustachio Manfredi (Verona: D. Ramanzini, 1737).

Bianchini, Francesco, *Del palazzo de' cesari opera posthuma* (Verona: P. Berno, 1738).

Bianchini, Francesco, *De tribus generibus instrumentorum musicae veterum organicae dissertatio* (Rome: Bernabò & Lazzarini, 1742).

Bianchini, Francesco, "De globi farnesiani structura, figura et indicationibus," in G. Bianchini, *Demonstratio* (1752), i, pt 1, pp. cclxxx–cccxxvi.

Bianchini, Francesco, *Opuscula varia nunc in lucem edita*, 2 vols (Rome: Barbiellino, 1754).

Bianchini, Francesco, "Dissertazione ... da lui recitata nella radunanza dell'Accademia degli Aletofili [1687]," in A. Coglierà (ed.), *Nuova raccolta d'opuscoli scientifici e filologici*, 41 (1785), 3–37.

Bianchini, Francesco, *Carte da giuoco in servigio dell'istoria e della cronologia*, ed. G. B. C. Giuliari and N. Tommaso (Bologna: Romagnoli, 1871).

Bianchini, Francesco, *Dei viaggi di Monsignor Francesco Bianchini con alcune sue lettere*, ed. G. C. Giuliari (Verona: Tip. Vescovile, 1877). (*Nuova serie di aneddoti*, no. 19, 27 pp.)

Bianchini, Francesco, *Relazione delle cose più erudite e rare de' principi di Firenze e di Parma e nell'Istituto di Bologna mandata a S.M. Giovani V re di Portogallo*, ed. G. C. Giuliari (Verona: G. Civelli, 1882). (*Nuova serie di aneddoti*, no. 15, 28 pp.)

Bianchini, Francesco, *Lettera … agli eccellentissimi provveditori della Repubblica veneta intorno alla fortezza di Guastalla*, ed. G. C. Giuliari (Verona: P. Apollonio, 1883). (*Nuova serie di aneddoti*, no. 17, 12 pp.)

Bianchini, Francesco, *Relazione sulla fortezza di Guastalla: Presentata alla Repubblica veneta nel 1686*, ed. G. C. Giuliari (Verona: P. Apollonio, 1885). (*Nuova serie di aneddoti*, no. 36, 15 pp.)

Bianchini, Francesco, *Schizzi di carte celesti delineati da Francesco Bianchini sopra osservazioni proprie e di Geminiano Montanari*, ed. Francesco Porro (Genoa: Pagano, 1902).

Bianchini, Francesco, *Iter in britanniam* [1713], in Salvatore Rotta, *Francesco Bianchini in Inghilterra* (Brescia: Paideia, [1966]), 31–73. Now available in Italian translation: Mara Musante, *Viaggio in Inghhilterra* [1713] (Genoa: Città del Silenzio, 2020).

Bianchini, Francesco, *New Phenomena of Hesperus and Phosphorus, or rather Observations Concerning the Planet Venus*, trans. Sally Beaumont and Peter Fay (Berlin: Springer, 1996). Cited as *NP*.

Bianchini, Giuseppe, *Demonstratio historiae ecclesiasticae quadripartitae comprobatae monumentis pertinentibus ad fidem temporum et gestorum*, 2 vols in 4 parts (Rome: Apollinea, 1752–4).

Bibliotheca stoschiana sive catalogus selectissimoroum librorum quae collegerat Philippus Liber Baro de Stosch (Florence: n.p., 1759).

Biddle, Sheila, *Bolingbroke and Harley* (New York: Knopf, 1974).

Bignami Odier, Jeanne, and Anna Maria Partini, "Cristina di Svezia e le scienze occulte," *Physis*, 25 (1983), 251–78.

Blažeković, Zdravko, "Francesco Bianchini's Triplex Lyra in Eighteenth-Century Music Historiography," in Antonio Baldassarre (ed.), *Musik, Raum, Akkord, Bild* (Bern: Peter Lang, 2012), 581–95.

Bold, Edward, "Music at the Stuart Court at Urbino, 1717–1718," *Music & Letters*, 81/3 (2000), 351–63.

Bolingbroke, Henry St John, Viscount. *Letters and Papers*, ed. Gilbert Parke, 4 vols (London: G. G. and T. Robinson, 1798).

Bolletti, Giuseppe Gaetano, *Dell'origine e de' progressi dell'Istituto delle scienze di Bologna e di tutte le accademie ad esso unite* (Bologna: de la Volpe, 1751).

Boncompagni Ludovisi, Ugo, *Vita della Ven. Camilla Orsini-Borghese* (Rome: Libreria Salesiana, 1931).

Bonjour, Guillaume, *Calendarium romanum* (Rome: Buagni, 1701).

Bonanni, Filippo, *La gerarchia ecclesiastica considerata nelle vesti sagre, e civili usate da quelli, li quali la compongono* (Rome: G. Placho, 1720).

Borgi, Elena, and Giuseppe Ricuperati, "Appendice iconografica e bibliografica," *Rivista storica italiana*, 117 (2005), 944–73.

Boscovich, Roger, "[Selvaggio Afrodisio]," in Michel Giuseppe Morei (ed.), *Le vite degli Arcadi illustri*, v (Rome: A. de' Rossi, 1751), 128–9.

Bossuet, J. B., "Méditations pour le temps du jubilé (1696 et 1702)," in Bossuet, *Œuvres*, 43 vols (Versailles: Lebel, 1815–19), vi. 547–618.

Bots, Hans, and Françoise Waquet (eds), *Commercium litterarium, 1600–1700: La Communication dans la République des Lettres* (Amsterdam: APA-Holland University Press., 1994).

Bougainville, Jean Pierre, "Préface," in Nicolas Fréret, *Défense de la chronologie . . . contre le système chronologique de M. Newton* (Paris: Durand, 1758), pp. i–lv.

Boutier, Jean, et al. (eds), *Naples, Rome, Florence: Une histoire comparée des milieux intellectuels italiens (xviie-xviiie siècles)* (Rome: École française de Rome, 2005).

Boylan, Bonaventura, "A Journal of the Arrest and Escape of the Princess Sobieski," in Gilbert, *Narratives* (1894), 1–30.

Bozzoli, G. M., *Cenni biografici di Francesco Bianchini, estratti del Libro d'oro dei nostri tempi* (Borgomanero: Gerenia, 1873).

Braccesi, Alessandro, "Gli inizi della specola di Bologna," *Giornale di astronomia*, 4 (1978), 327–50.

Brent, Allen, "The Statue of Hippolytus: Discovery," in Brent, *Hippolytus and the Roman Church in the Third Century* (Leyden: Brill, 1995), 3–50.

Brodsky, Laurel, et al., "Prince Rupert's Drops," RSL, *Notes & Records*, 41 (1986), 1–26.

Bruce, Maurice, "The Duke of Mar in Exile, 1716–32," Royal Historical Society, *Transactions*, 20 (1937), 61–82.

Bruckmann, Patricia Carr, "'Men, Women and Poles': Samuel Richardson and the Romance of a Stuart Princess," *Eighteenth Century Life*, 27/3 (2003), 31–52.

Brugueres, Michele, "Il trionfo della virtù feminile," in *Pompe funebri* (1686), 13–24.

Buchwald, Jed, and Mordechai Feingold, *Newton and the Origin of Civilization* (Princeton: Princeton University Press, 2013).

Bulifon, Antonio, *Journal du voyage d'Italie de l'invincible & glorieux monarque Philippe V, roy d'Espagne et de Naples* (Naples: N. Bulifon, 1704).

Burke, John G., *Cosmic Debris: Meteorites in History* (Berkeley and Los Angeles: University of California Press, 1986).

Burlini Calapaj, Anna, "I rapporti tra Lamindo Pritanio e Bernardo Trevisan," in *Accademie e cultura: Aspetti storici tra sei e settecento* (Florence: Olschki, 1979), 73–94.

Caffiero, Marina, "Scienza e politica a Roma in un carteggio di Celestino Galiani," *Archivio della società romana di storia patria*, 101 (1978), 311–44.

Calvisius, Sethus, *Opus chronologicum, ad annum* MDCLXXXV *continuatum* (Frankfurt: C. Gensch, 1685).

Campori, Giuseppe, "Notizie e lettere inedite di Geminano Montanari," Deputazione di storia patria per le provincie modenesi e parmensi, *Atti e memorie*, 8 (1877), 65–96.

Caracciolo, Alfredo, *Domenico Passionei tra Roma e la Repubblica delle Lettere* (Rome: Storia e Letteratura, 1968).

Carbone, G. B., "Observationes astronomicae," *PT* 35 (1727–8), 471–6.

Carini, Isidoro, *L'Arcadia dal 1690 al 1890: Memorie storiche* (Rome: F. Cuggiani, 1891).

Carini, Isidoro, "Diciotto lettere inedite di Francesco Bianchini a Giovanni Ciampini," *Il muratori*, 1/4 (1892), 145–75.

Carli, Paol Francesco, "L'apparato dell'Accademia," in *Pompe funebri* (1686), 1–12.

Carvalho, Rómulo de, *Astronomia em Portugal no sécolo xviii* (Lisbon: Istituto de Cultura e Língua Portuguesa, 1985).

Casini, Paolo, *Newton e la coscienza europea* (Bologna: Il Mulino, 1983).

Casini, Paolo, "Bianchini e la questione copernicana: tra Leibniz e Newton," in CR 9–31.

Cassini, Giovanni Domenico, *La meridiana del tempio di S. Petronio* (Bologna: Erede di V. Benacci, 1695).

Cassini, Giovanni Domenico, "Remarques sur le calendrier du P. Bonjour," *Journal de Trévoux*, 2 (1702), 152–62.

Cassini, Giovanni Domenico, "Comparaison des premières observations de la comete du mois d'avril 1702 faites à Rome et à Berlin," *HAS* (1702), 121–30.

Castelli, Giulio, *Gli anni santi* (Bologna: Cappelli, 1949).

Castro, José de, *Portugal em Roma*, 2 vols (Lisbon: União gráfica, [1939]).

Catamo, Mario, and Cesare Lucarini, *Il cielo in basilica: La meridiana della Basilica di Santa Maria degli Angeli e dei Martiri a Roma* (Cosenza: ARPA, 2002).

Cattabiani, Alfredo, *Breve storia dei giubilei (1300–2000)* (Milan: Bompiani, 1999).

Cattelani Degani, Franca, and Mario Umberto Lugli, "Cinque lettere di Geminiano Montanari a Gian Domenico Cassini," *Nuncius*, 19/1 (2004), 205–23.

Cattermole, Peter, *Venus, the Geological Story* (Baltimore: Johns Hopkins University Press, 1994).

Cavazza, Marta, "La cometa del 1680–81: Astrologi e astronomi a confronto," *Studi e memorie per la storia dell'Università di Bologna*, 3 (1981), 409–66.

Cavazza, Marta, *Settecento inquieto: Alle origini dell'Istituto delle scienze di Bologna* (Bologna: Il Mulino, 1990).

Cavazza, Marta, "Il giovane Bianchini e la visione al buio," in CR 101–20.

Cecamore, Claudia, *Palatium: Topografia storica del Palatino tra iii sec. a.C. e il i sec. d.C.* (Rome: Bretschneider, 2002). (*Bulletino della Commissione archeologica comunale di Roma*, supp. 9.)

Celani, Enrico, "L'epistolario di Monsignor Francesco Bianchini, memoria e indice," *Archivio veneto*, 36 (1888), 155–87, 343–68.

Chevreau, Urbain, *Chevraeana, ou Diverses pensées d'histoire, de critique, d'érudition et de morale*, 2 vols (Amsterdam: T. Lombrail, 1700).

Chiarlo, Carlo R., "Francesco Bianchini e l'antiquaria italiana del settecento," in Joanna Hübner-Wojciechowski (ed.), *L'eredità classica in Italia e in Polonia nel settecento* (Breslau: Accademia polacca delle scienze, 1992), 162–86.

Chiarlo, Carlo R., "Considerazioni sull'apparato illustrativo de' La storia universale: Le immagini relative ai monumenti classici," in CR 245–58.

Chirico, Teresa, "L'inedita serenata alla regina Maria Casimira di Polonia: Pietro Ottoboni committente di cantate e serenate (1689–1708)," in Nicolò Maccavino (ed.), *La serenata tra seicento e settecento*, 2 vols (Reggio Calabria: Laruffa, 2007), ii. 397–449.

Ciampini, G. G., *Opera*, 3 vols (Rome: C. Giannini, 1747).

Ciancio, Luca, and Gian Paolo Romagnani (eds), *Unità del sapere, molteplicità dei saperi: Francesco Bianchini (1662–1729) tra natura, storia e religione* (Verona: QuiEdit, 2010). Cited as CR.

Cicero, Marcus Tullius, *Brutus: On the Nature of the Gods: On Divination. On Duties*, trans. H. M. Poteat (Chicago: University of Chicago Press, 1950).

Cipriani, Antonio, "Contributo per una storia politica dell'Arcadia settecentesca," *Atti e memorie dell'Arcadia*, 5/2–3 (1971), 101–66.

Claridge, Amanda, *Rome: An Oxford Archaeological Guide* (Oxford University Press, 2010).

Clarke de Dromantin, Patrick, *Les Réfugiés jacobites dans la France du xviiie siècle* (Pessac: Presses universitaires de Bordeaux, 2005).

Clement XI, *A Homily of Pope Clement XI as Delivered by him in Saint Peter's in the Vatican on the Feast of the Holy Apostles St Peter and St Paul, in the Year of our Lord 1717; upon the Occasion of the Pretender's being there Present* (London: T. Warner, 1717).

Clement XI, *Opera omnia* (Frankfurt: Weidemann, 1729).

Coarelli, Filippo, *Rome and Environs: An Archaeological Guide* (Berkeley and Los Angeles: University of California Press, 2007).

Cochrane, Eric W., "The Settecento Medievalists," *Journal for the History of Ideas*, 19 (1958), 35–61.

Cochrane, Eric W., "Muratori: The Vocation of an Historian," *Catholic Historical Review*, 51 (1965), 153–72.

Condren, John, "The Dynastic Triangle in International Relations: Modena, England, and France, 1678–85," *International History Review*, 37/4 (2015), 700–20.

Connors, Joseph, "The One-Room Apartment of Cornelis Meijer," in Nebahat Acioglu and Allison Sherman (eds), *Artistic Practices and Cultural Transfer in Early Modern Italy* (Aldershot: Ashgate, 2015), 45–64.

Cornaro, Elena Lucrezia, *Opera*, ed. Benedetto Bacchini (Rome: n.p., 1688).

Corp, Edward (ed.), *The Stuart Court in Rome: The Legacy of Exile* (Aldershot: Ashgate, 2003).

Corp, Edward, *The Jacobites at Urbino: An Exiled Court in Transition* (Basingstoke: Palgrave-Macmillan, 2009).

Corp, Edward, "The Location of the Stuart Court in Rome: The Palazzo del Re," in Monod et al., *Loyalty* (2010), 180–205.

Corp, Edward, *The Stuarts in Italy 1719–1766: A Royal Court in Permanent Exile* (Cambridge: Cambridge University Press, 2011).

Corp, Edward, *Sir David Nairne: The Life of a Scottish Jacobite at the Court of the Exiled Stuarts* (Oxford: Peter Lang, 2018).

Corp, Edward, "Clementina Sobieska at the Jacobite court," in Maher, *Irish* (2021), 121–45.

Crescimbeni, Giovan Maria (ed.), *Le vite degli Arcadi illustri*, 4 vols (Rome: A de' Rossi, 1708–27).

Crescimbeni, Giovan Maria (ed.), *Prose degli Arcadi*, 3 vols (Rome: A de' Rossi, 1718).

Crescimbeni, Giovan Maria, "Clemente XI," in Crescimbeni, *Notizie*, iii (1721), 361–77.

Crescimbeni, Giovan Maria (ed.), *Notizie istoriche degli Arcadi morti*, 3 vols (Rome: A. de' Rossi, 1720–1).

Croce, Benedetto, "Francesco Bianchini e G.B. Vico," in Croce, *Conversazioni critiche*, 2 vols (Bari: Laterza, 1924), ii. 101–9.

Cucco, Giuseppe, *Papa Albani e le arti a Urbino e Roma 1700–1721* (Venice: Marsilio, 2001).

Dal Prete, Ivano, "*Hesperi et phosphori nova phaenomena*: Le osservazioni venusiane di Francesco Bianchini," *Astronomia*, 2 (2003), 10–18.

Dal Prete, Ivano, "Scienza e società nella terraferma veneta: Il caso veronese 1680–1796," Ph.D. thesis, University of Verona, 2004.

Dal Prete, Ivano, "Francesco Bianchini e il pianeta Venere: Astronomia, cronologia e storia della terra tra Roma e Parigi all'inizio del xviii secolo," *Nuncius*, 20/1 (2005), 95–152.

Dal Prete, Ivano, "Francesco Bianchini e la cultura scientifica veronese tra sei e settecento," in CR 207–41.

Daugnon, F. F. de, *Gli italiani in Polonia dal ix secolo al xviii*, i (Crema: Plausi e Cattaneo, 1905).

Davillé, Louis, *Leibniz historien* (Paris: Alcan, 1909).

Davis, Raymond, *The Lives of the Ninth-Century Popes (Liber pontificalis)* (Liverpool: Liverpool University Press, 1995).

De Brosses, Charles, *Lettres familières sur l'Italie*, ed. Yvonne Bezard, 2 vols (Paris: Firmin-Didot, 1931).

[Defoe, Daniel], *Hannibal at the Gates: Or, the Progress of Jacobitism, with the Present Danger of the Pretender* (London: T. Baker, 1714).

Derenzini, Tullio, "Alcune lettere di Giovanni Alfonso Borelli ad Alessandro Marchetti," *Physis*, 1 (1959), 224–43.

De Sanctis, Francesco, *Storia della letteratura italiana*, 2 vols (Naples: A. Morano, 1870).

Desimoni, Cornelio, "Notizie di Paris Maria Salvago e del suo osservatorio astronomico in Carbonara," *Giornale ligustico di archeologia, storia e belle arti*, 2 (1875), 465–86; 3 (1876), 41–65.

Dix, T. Keith, "Virgil in the Grynean Grove: Two Riddles in the Third Eclogue," *Classical Philology*, 90/3 (1995), 256–62.

Dixon, Susan M., "Piranesi and Francesco Bianchini: *Capricci* in the Service of Pre-Scientific Archeology," *Art History*, 22/2 (1999), 184–213.

Dixon, Susan M., "Francesco Bianchini's Images and his Legacy in the Mid-Eighteenth Century: From Capricci to Playing Cards to Proscenium and back," in KS 83–106.

Dixon, Susan M., *Between the Real and the Ideal: The Accademia degli Arcadi and its Garden in Eighteenth-Century Rome* (Newark: University of Delaware Press, 2006).

Donato, Maria Pio, *Accademie romane: Una storia sociale (1671–1824)* (Naples: Edizioni scientifiche italiane, 2000).

Donato, Maria Pio, "L'onere della prova: Il Sant'Uffizio, l'atomismo e i medici romani," *Nuncius*, 18/1 (2003), 69–87.

Donato, Maria Pio, "Le due accademie dei concili a Roma," in Boutier et al., *Naples* (2005), 243–55.

Duchesne, Louis, *Étude sur le Liber pontificalis* (Paris: E. Thorin, 1877).

Duchesne, Louis, *Le Liber pontificalis: Texte, introduction et commentaire*, 2 vols (Paris: E. Thorin, 1886–92).

Duke, Dennis W., "Analysis of the Farnese Globe," *Journal for the History of Astronomy*, 37 (2006), 87–100.

Dupront, Alphonse, *L. A. Muratori et la société européenne des pré-lumières* (Florence: Olschki, 1976).

Elci, Orazio, Conte d', *The Present State of the Court of Rome: Or, the Lives of the Present Pope Clement XI and of the Present College of Cardinals* (London: "the Booksellers," 1706; London: G. Strahan, 1721).

Engelberg, Meinrad von, "Ricavare l'idea del tutto: Francesco Bianchinis 'Del palazzo de' Cesari,'" in KS 135–63.

Eschinardi, Francesco, *Raguagli ... dati ad un'amico in Parigi [G. D. Cassini] sopra alcuni pensieri sperimentali proposti nell'Accademia Fisicomatematica di Roma* (Rome: Tinassi, 1680).

Eschinardi, Francesco, *Lettere ... nelle quale si contengono alcuni discorsi fisicomatematici* (Rome: Tinassi, 1681).

Eschinardi, Francesco, *De impetu tractatus duplex* (Rome: A. Bernabò, 1684).

Eschinardi, Francesco, *Cursus philosophicus* (Rome: Komarek, 1689).

Fabiani, Ferdinando, *Il merito applaudito ... [del] Monsig. Giovanni Ciampini* (Fermo: G. F. Bolis, 1694).

Faccini, Annamaria, "Francesco Bianchini e le sue relazioni con l'Università di Oxford," in Carlo Albarello and Giuseppe Zivelonghi (eds), *Per Alberto Piazzi* (Verona: Biblioteca capitolare, 1998).

Fantuzzi, Giovanni, *Memorie della vita del generale Co: Luigi Ferdinando Marsigli* (Bologna: dalla Volpe, 1770).

Favaretto, Irene, "Ogni genere d'erudite anticaglie: Francesco Bianchini e l'ambiente veronese," in KS 27–39.

Fèa, Carlo, *Storia delle acque antiche sorgenti in Roma perdutae, ii. Dei condotti antico-moderni delle acque, Vergine, Felice e Paola, e loro autori* (Rome: RCA, 1832).

Federici, Luigi, "Franceso Bianchini," in Federici, *Elogi istorici de' più illustri ecclesiastici veronesi*, 3 vols (Verona: Ramanzini, 1818–19), iii. 1–48.

Fehl, Maria Raina, "Archeologists at Work in 1726: The Columbarium of the Household of Livia Augusta," in Börje Magnusson et al. (eds), *Ultra terminum vagari: Scritti in onore di Carl Nylander* (Rome: Quasar, 1997), 89–112.

Feist, Ulrike, "Francesco Bianchini's Discovery of the Venus Markings," *Fragmenta*, 5 (2011), 309–33.

Feist, Ulrike, "Mondkrater und Venusflekten: Francesco Bianchini in der Nachfolge Galileo Galileis," in Ulrike Feist and Markus Rath (eds), *Et in imagine ego: Facetten von Bildakt und Verkörperung: Festgabe für Horst Bredekamp* (Berlin: Akademie Verlag, 2012), 301–23.

Feist, Ulrike, *Sonne, Mond und Venus: Visualisierungen astronomischen Wissens im frühneuzeitlichen Rome* (Berlin: Akademie Verlag, 2013).

Ferraris, Paola, "Il Bosco Parrasio dell'Arcadia (1721–1726)," in Sandra Vasco Rocca and Gabriele Borghini (eds), *Giovanni V di Portogallo (1707–1750) e la cultura romana del suo tempo* (Rome: Argos, 1995), 137–52.

Ferrary, Jean-Louis, *Onofrio Panvinio et les antiquités romaines* (Rome: École française de Rome, 1996).

Ferrone, Vincenzo, "Celestino Galiani e la diffusione del Newtonianismo," *Giornale critico della filosofia italiana*, 61 (1982), 1–33.

Ferrone, Vincenzo, *The Intellectual Roots of the Italian Enlightenment: Newtonian Science, Religion, and Politics in the Early Eighteenth Century*, trans. Sue Brotherton (Atlantic Heights: Humanities Press, 1995).

Ferroni, Giuseppe, *Dialogo fisico-astronomico contro il sistema copernicano* (Bologna: G. Longi, 1680).

Fieldhouse, H. N., "A Note on the Negotiations for the Peace of Utrecht," *American Historical Review*, 40 (1935), 274–8.

Fieldhouse, H. N., "Oxford, Bolingbroke, and the Pretender's Place of Residence, 1711–14," *English Historical Review*, 52 (1937), 289–96.

Fieldhouse, H. N., "Bolingbroke's Share in the Jacobite Intrigues of 1710–1714," *English Historical Review*, 52 (1937), 443–59.

Finocchiaro, Giuseppe, "Indugiatore e umorista: Francesco Bianchini e le accademie di papa Albani," in CR 323–37.

Fiorani, Luigi, *Il concilio romano del 1725* (Rome: Storia e letteratura, 1978).

Flammarion, Camille, *Les Terres du ciel, description astronomique, physique, climatologique, géographique des planètes* (Paris: Didier, 1877).

Flamsteed, John, "An Abstract of a Letter . . . Giving an Account of the Eclipses of Jupiter's Satellites," *PT* 15/177 (1685), 1215–25.

Flamsteed, John, *Correspondence*, ed. Eric G. Forbes et al., 3 vols (Bristol: Institute of Physics, 1995–2002).

Flood, John M., *The Life of Chevalier Charles Wogan: An Irish Soldier of Fortune* (Dublin: Talbot; London: T. Fisher Unwin, 1922).

Folkes, Martin, "Journal while in Venice and Rome [1733–35]," Bodleian Library, Oxford, MS Eng. misc.c.444.

Fontanini, G., *La vita della . . . serva di Dio . . . C. Orsini Borghese . . . di poi Suor Maria Vittoria* (Rome: Francesco Gonzaga, 1717).

Fontenelle, Bernard le Bovier de, "Eloge de M. Bianchini," *HAS* (1729), 110–14 = Fontenelle, *Eloges des académiciens*, 2 vols (The Hague: I. Vander Kloot, 1740), 374–99.

Foscolo, Ugo, *Opere*, ed. Franco Gavazzeni, ii (Milan: R. Ricciardi, 1981).

Fransen, Petronella, *Leibniz und di Friedenschlüsse von Utrecht und Rastatt-Baden* (Pumerend: J. Muusses, 1933).

Frati, Ludovico, "Maria Clementina Sobieski a Roma," *Nuova antologia*, 220 (August–September 1908), 420–30.

Frey, Linda S., and Marsha Frey (eds), *The Treaties of the War of the Spanish Succession: An Historical and Critical Dictionary* (Westport: Greenwood, 1995).

Fröhner, Wilhelm, *Notice de la sculpture antique du Musée impérial du Louvre* (Paris: De Mourgues, 1878).

Funerali di Giacomo III: Re della Gran Brettagna (Rome: Salvioni, 1766).

Galilei, Galileo, *Dialogue Concerning the Two Chief World Systems*, trans. Stillman Drake (Berkeley and Los Angeles: University of California Press, 1953).

Galilei, Galileo, *Two New Sciences*, trans. Stillman Drake (Madison: University of Wisconsin Press, 1974).

Gallo, Daniela, "Rome, mythe et réalité pour le citoyen de la République des Lettres," in Bots and Waquet, *Commercium* (1994), 191–205.

Gallo, Daniela, "Pour une histoire des antiquaires romains au xviiie siècle," in Boutier et al., *Naples* (2005), 257–75.

Galluzzi, Paolo, "Lettere di Giovani Alfonso Borelli ad Antonio Magliabechi," *Physis*, 12 (1970), 267–98.

Gascoigne, John, "'The Wisdom of the Egyptians' and the Secularization of History in the Age of Newton," in Gaukroger, *Uses* (1991), 171–212.

Gasperoni, Gaetano, *Scipione Maffei e Verona settecentesca* (Verona: Valdonega, 1955).

Gaukroger, Stephen (ed.), *The Uses of Antiquity. The Scientific Revolution and the Classical Tradition* (Dordrecht: Kluwer, 1991).

Gaydon, Richard, "Narrative," in Gilbert, *Narratives* (1894), 109–44.

Genet-Rouffiac, Nathalie, "The Irish Jacobite Exile in France, 1692–1715," in Toby Barnard and Jane Fenlon (eds), *The Dukes of Ormonde, 1610–1745* (Woodbridge: Boydell and Brewer, 2000).

Genet-Rouffiac, Nathalie, "The Irish Jacobite Regiments and the French Army," in Monod et al., *Loyalty* (2010), 206–28.

Geneva, Ann, *Astrology in the Seventeenth Century Mind* (Manchester: Manchester University Press, 1995).

Ghezzi, Pier Leone, *Camere sepolcrali de liberti e liberte di Livia Augusta e d'altri cesari* (Rome: F. de' Rossi, 1731).

Gibbon, Edward, *The Decline and Fall of the Roman Empire*, ed. J. B. Bury, 7 vols (London: Methuen, 1909–14).

Gierowski, Józef A., "Viaggi in Italia del principe Federico Augusto," in Kanceff and Lewanski, *Viaggiatori* (1988), 191–202.

Gilbert, John T. (ed.), *Narratives of the Detention, Liberation and Marriage of Maria Clementina Stuart* (Dublin: J. Dolland, 1894).

Giorgetti Vichi, Anna Maria, *Gli arcadi dal 1690 al 1800: Onomasticon* (Rome: Arcadia, 1977).

Gittins, Estelle, "Princess Clementina's Marriage Certificate," in Maher, *Irish* (2021), 167–76.

Giuliari, G. B. Carlo, *La Capitolare Biblioteca di Verona* (Verona: n.p., 1888).

Giustino, Sandra, "Il carteggio fra i Cassini e Eustachio Manfredi (1699–1737)," *Bollettino di storia delle scienze matematiche*, 21 (2001), 5–180.

Golinelli, Paolo, "Alle origini della storiografia scientifica in Italia," *Atti e memorie della Deputazione di storia patria per le antiche provincie modenesi*, 11 (1976), 143–72.

Goméz Lopéz, Susana, *Le passioni degli atomi: Montanari e Rossetti: Una polemica tra galileiani* (Florence: Olschki, 1997).

Goméz Lopéz, Susana, "The Royal Society and Post-Galilean Science in Italy," RSL, *Notes & Records*, 51/1 (1997), 35–44.

Gori, Antonio Francesco, *Monumentum sive columbarium libertorum et servorum Liviae Augustae et Caesarum* (Florence: Typis Regiae Celsitudinis, 1727).

Grafton, Anthony, *Defenders of the Text: The Traditions of Scholarship in an Age of Science, 1450–1800* (Cambridge, MA: Harvard University Press, 1991).

Grafton, Anthony, "Morhof and History," in Waquet, *Mapping* (2000), 155–77.

Graziosi, Maria Teresa Acquaro, *L'Arcadia: Trecento anni di storia* (Rome: Palombi, 1991).

Greco, Gianpasquale, "Un monumento 'a se stesso:' La tomba di Carlo Maratti in Santa Maria degli Angeli, Roma," *Ricerche di storia dell'arte*, 118 (2016), 89–94.

Gregg, Edward, "The Jacobite Career of John, Earl of Mar," in Eveline Cruickshanks (ed.), *Ideology and Conspiracy: Aspects of Jacobitism, 1689–1759* (Edinburgh: J. Donald, 1982), 179–200.

Gregg, Edward, "'Power, Friends and Alliances:' The Search for the Pretender's Bride," *Studies in History and Politics*, 4 (1985), 35–54.

Gregg, Edward, "The Financial Vicissitudes of James III in Rome," in Corp, *Stuart Court* (2003), 65–83.

Grendler, Paul F., *The Jesuits and the Italian Universities, 1548–1773* (Washington: Catholic University of America, 2017).

Gross, Hanns, *Rome in the Age of Enlightenment* (Cambridge: Cambridge University Press, 1990).

Gualandi, Andrea, *Teorie delle comete: Da Galileo a Newton* (Milan: FrancoAngeli, 2009).

Guerrini, Anna, "The Tory Newtonians: Gregory, Pitcairne, and their Circle," *Journal of British Studies*, 25 (1986), 288–311.

Gyllenborg, Carl, *Letters which Passed between Count Gyllenborg, the Barons Gortz, Sparre and Others, Relating to the Design of Raising Rebellion in His Majesty's Dominions, to be Supported by a Force from Sweden* (London: S. Buckley, 1717).

Hadley, John, "An Account of a Book Entituled *Hesperi et phosphori nova phenomena*," *PT* 36/410 (1729), 158–63.

Haile, Martin, *James Francis Edward, "The Old Cavalier"* (London: Dent, 1907).

Hall, A. Rupert, "Further Newtonian Correspondence," RSL, *Notes & Records*, 37 (1982–3), 7–34.

Halley, Edmond, "An Account of the Measure of the Thickness of Gold upon Gilt-Wire, together with a Demonstration of the Exceeding Minuteness of the Atoms or Constituent Parts of Gold," *PT* 17/194 (1691), 540–2.

Hamel du Breuil, Jean de, "Le Mariage du Prétendant (1719)," *Revue d'histoire diplomatique*, 9 (1895), 53–96.

Harrison, John, *The Library of Isaac Newton* (Cambridge: Cambridge University Press, 2008).

Hauksbee, Francis, *Esperienze fisico-meccaniche sopra vari soggetti*, trans. Thomas Dereham (Florence: Stamperia Reale, 1716).

Hearne, Thomas, *Remarks and Collections*, ed. C. E. Dobbs et al., 11 vols (Oxford: Oxford Historical Society, 1885–1921).

Heilbron, J. L., *Electricity in the Seventeenth and Eighteenth Centuries: A Study of Early Modern Science* (Berkeley and Los Angeles: University of California Press, 1979; New York: Dover, 1999).

Heilbron, J. L., "The Contributions of Bologna to Galvanism," *Historical Studies in the Physical Sciences*, 22/1 (1991), 57–84.

Heilbron, J. L., "Meridiane and Meridians in Early Modern Science," in Piers Bursill-Hall (ed.), *R. J. Boscovich vita e attività scientifica: His Life and Scientific Work* (Rome: Enciclopedia italiana, 1993), 385–406.

Heilbron, J. L., *The Sun in the Church: Cathedrals as Solar Observatories* (Cambridge, MA: Harvard University Press, 1999). Cited as HS.

Heilbron, J. L., "Censorship of Astronomy in Italy after Galileo," in Ernan McMullin (ed.), *The Church and Galileo* (Notre Dame: Notre Dame University Press, 2005), 279–322.

Heilbron, J. L., "Bianchini as an Astronomer," in KS 57–82.

Heilbron, J. L., "Francesco Bianchini, Historian: In Memory of Amos Funkenstein," in David Biale and Robert Westman (eds), *Thinking Impossibilities: The Intellectual Legacy of Amos Funkenstein* (Toronto: University of Toronto Press, 2008), 227–78.

Heilbron, J. L., "Bianchini and Natural Philosophy," in CR 33–73.

Heilbron, J. L., "Boscovich in Britain," in Theodore Arabatzis et al. (eds), *Relocating the History of Science* (Cham, Switzerland: Springer, 2015), 99–116.

Heilbron, J. L., "Benedict XIV and the Natural Sciences," in Rebecca Messbarger et al. (eds), *Benedict XIV and the Enlightenment* (Toronto: University of Toronto Press, 2016), 177–205.

Heinz, Werner, *Römische Thermen* (Munich: Hirmer, 1983).

Hill, B. W., "Oxford, Bolingbroke, and the Peace of Utrecht," *Historical Journal*, 16 (1973), 241–63.

Hippolytus, St, *Opera*, ed. J. A. Fabricius, 2 vols (Hamburg: Liebezeit, 1716–18).

Historical Manuscripts Commission, *Calendar of the Stuart Papers*, 7 vols (London: HMSO, 1902–23). Cited as SP.

Holmes, Geoffrey, *British Politics in the Age of Anne* (London: Hambledon, 1987).

Hoogewerff, G. J., "Cornelis Jansz. Meijer, Amsterdamsch ingenieur in Italië (1629–1701)," *Oud-Holland*, 38 (1920), 83–103.

Huelsen, Christian, "Il 'Museo ecclesiastico' di Clemente XI. Albani," *Bulletino della Commissione archeologica comunale di Roma*, 18 (1890), 260–77.

Huelsen, Christian, *Saggio di bibliografia ragionata delle piante iconografiche e prospettiche di Roma dal 1551 ad 1748* (Rome: Bardi, 1969).

Huelsen, Christian, "Untersuchungen zur Topographie des Palatins: Die Ausgrabungen in den farnesischen Gärten 1720–1730," *Mitteilungen des deutschen archaeologischen Instituts, Römische Abteilungen*, 10 (1895), 3–37, 252–76.

International Astronomical Union, *Gazeteer of Planetary Nomenclature* (USGS Astrogeology Science Center, n.d.).

Jankovic, Vladimir, "The Politics of Sky Battles in Early Hanoverian Britain," *Journal of British Studies*, 41/4 (2002), 429–59.

Jervis, Simon, "Multum in parvo," *Furniture History*, 21 (1985), 1–10.

Johns, Christopher M. S., "Art and Science in Eighteenth-Century Bologna: Donato Creti's Astronomical Landscape Paintings," *Zeitschrift für Kunstgeschichte*, 55 (1992), 578–89.

Johns, Christopher M. S., *Papal Art and Cultural Politics: Rome in the Age of Clement XI* (Cambridge: Cambridge University Press, 1993).

Johns, Christopher M. S., "Papa Albani and Francesco Bianchini: Intellectual and Visual Culture in Early Eighteenth-Century Rome," in KS 41–55.

Justi, Carl, "Philipp von Stosch und seine Zeit," *Zeitschrift für bildende Kunst*, 7 (1872), 293–308, 333–46.

Kahn, David, *The Codebreakers* (London: Weidenfeld and Nicolson, 1967).

Kammerer Grothaus, Heike, "Camere sepolcrali de' liberti e liberte di Livia Augusta ed altri caesari," *Mélanges de l'École française de Rome, Antiquité*, 91 (1979), 315–47.

Kammerer Grothaus, Heike, "Momentum augusti: Das sogenannte Columbarium der freigelassenen des Augustus," *Babesch: Bulletin antieke beschaving*, 86/1 (2011), 95–110.

Kanceff, Emanuele, and Richard Lewanski (eds), *Viaggiatori polacchi in Italia* (Geneva: Slatkine, 1988).

Karmon, David, "Restoring the Ancient Water Supply System in Renaissance Rome: The Popes, the Civic Administration, and the Acqua Virgine," *Waters of Rome*, 3 (2005), 13 pp.

Karmon, David, "Michelangelo's 'Minimalism' in the Design of Santa Maria degli Angeli," *Annali di architettura*, 20 (2008), 141–52.

Kelly, J. N. D., *The Oxford Dictionary of Popes* (Oxford: Oxford University Press, 1988).

Kircher, Athanasius, *Ars magna lucis et umbrae* (Rome: H. Scheus, 1646).

Klempt, Adalbert, *Die Säkularisierung der universalhistorischen Auffassung: Zum Wandel des Geschichtsdenkens im 16. und 17. Jahrhundert* (Göttingen: Musterschmidt, 1960).

Kockel, Valentine, "Ichnographia—Orthographia—Scaenographia: Abbildungsmodi antiker Architektur am Beispiel des 'Colombarium der Liberti der Livia,'" in KS 107–33.

Kockel, Valentine, and Brigitte Sölch (eds), *Francesco Bianchini (1662–1729) und die europäische gelehrte Welt um 1700* (Berlin: Akademie Verlag, 2005). Abbreviated KS.

Kolendo, Jerzy, "Les Recherches sur l'esclavage au début du xviiie siècle (à propos du livre de F. Bianchini)," *Klio*, 71/2 (1989), 420–31.

Kolendo, Jerzy, "Nota di lettura," in Francesco Bianchini, *Camera ed inscrizioni sepulcrali de' liberti, servi, ed ufficiali della casa di Augusto* (Naples: Jorene, 1991), pp. xii–xxxii.

Komaszynski, Michel, "Il viaggio triomfale di una Regina di Polonia in Italia," in Kanceff and Lewanski, *Viaggiatori* (1988), 153–63.

Laderchi, Giacomo, *Lettera . . . in risposta ad alcune difficoltà, e dubbiezze motivate contro gl'atti de' medesimi Santi [Cresci e Compagni] dati in luce dal P. Giacomo Laderchi* (Florence: J. Guiducci, 1711).

Le Mothe Le Vayer, François de, *Discours pour montrer que les doutes de la philosophie sceptique sont de grand usage dans les sciences* (Paris: Billaine, 1669).

La Mothe Le Vayer, François, *Œuvres*, 15 vols (Paris: Billaine, 1669).

Lanciani, Rodolfo, *Storia degli scavi di Roma*, vi [1700–1878] (Rome: Quasar, 2000).

Lanza, Franco, "L'*Istoria universale* del Bianchini e la *Scienza nuova*," *Lettere italiane*, 10 (1958), 339–48.

Law, Hugh A., "Sir Charles Wogan," Royal Society of Antiquaries of Ireland, *Journal*, 7 (1937), 253–64.

Le Clerc, Jean, "[Review of Bianchini, *De kalendario*]," *Bibliothèque choisie*, 27 (1713), 177–96.

Legg, Leopold George Wickham, *Matthew Prior: A Study of his Public Career and Correspondence* (Cambridge: Cambridge University Press, 1921).

Legrelle, Arsène, *La Diplomatie française et la succession d'Espagne*, 2nd edn, 6 vols (Braine-la-comte: Zech, 1895–9).

Lenglet du Fresnoy, Pierre Nicolas, *Méthode pour étudier l'histoire, avec un catalogue des principaux historiens, & des remarques sur la bonté de leurs ouvrages*, 3 vols (Paris: P. Gandouin, 1729).

Leonio, Vincenzo, "Vita di monsig. Gio. Giustino Ciampini, Romano," in G. M. Crescimbeni (ed.), *Le vite degli Arcadi illustri*, ii (Rome: A. de' Rossi, 1710), 195–253.

Lequien, Michel, *Defense du texte hébreu et de la version vulgate* (Paris: A. Auroy, 1690).

Lettere di vari illustri italiani del secolo xviii e xix a' loro amici, e de' massimi scienziati e letterati nazionali e stranieri al celebre abate Lazzaro Spallanzani, 2 vols (Reggio nell'Emilia: Torreggiani, 1841–3).

Lewis, Lesley, *Connoisseurs and Agents in Eighteenth-Century Rome* (London: Chatto & Windus, 1961).

Liebenwein, Wolfgang, "Der Porticus Clemens' XI. und sein Statuenschmuck. Antikenrezeption und Kapitolsidee im frühen 18. Jahrhundert," in Beck et al., *Antikensammlungen* (1981), 73–118.

Lippincott, Kristen, "A Chapter in the 'Nachleben' of the Farnese Atlas: Martin Folkes's Globe," *Journal of the Warburg and Courtauld Institutes*, 74 (2011), 281–99.

Liverani, Paolo, "Il 'Museo Ecclesiastico' e dintorni," in KS 207–34.

Lo Bianco, Anne (ed.), *Pier Leone Ghezzi: Settecento alla moda* (Venice: Marsilio, 1999).

Longhena, Mario, "Lettere inedite di E. Manfredi a L.F. Marsigli: Gli inizi dell'Istituto della scienza e della specola astronomica," *Atti e memorie della Accademia di agricoltura, scienze e lettere di Verona*, 21 (1942–3), 177–247.

Loret, Mattia, "I lavori artistici durante il pontificato di Clemente XI," *Archivi*, 3 (1936), 55–59.

Lotz, Wolfgang, "Die Spanische Treppe: Architektur als Mittel der Diplomatie," *Römische Jahrbuch für Kunstgeschichte*, 12 (1969), 39–94.

Lumisden, Andrew, *Remarks on the Antiquities of Rome and its Environs* (London: W. Bulmer et al., 1812).

Luttrell, Narcissus, *A Brief Historical Relation of State Affairs from September 1678 to April 1714*, 6 vols (Oxford: Oxford University Press, 1857).

Maffei, Scipione, *La Verona illustrata ridotta in compendio principalmente per uso de' forestieri* [1731], 2 parts (Verona: Moroni, 1771).

Maffei, Scipione, "Vita di Monsignor Francesco Bianchini," in Bianchini, *IU* (1747), a2^{v}–a5^{r}.

Maffei, Scipione, *Epistolario*, ed. Celestino Garibotto, 2 vols (Milan: A. Giuffrì, 1955).

Maher, Richard K., "Poems from the Prison Yard: Poetic Correspondence between Charles Wogan and William Turnstall," *History Ireland*, 25/2 (2017), 20–3.

Maher, Richard K., "Service and Exile: Sir Charles Wogan, 1715–1719," in Maher, *Irish* (2021), 73–94.

Maher, Richard K. (ed.), *The Irish to the Rescue: The Tercentenary of the Polish Princess Clementina's Escape* (Oxford: Peter Lang, 2021).

Mamiani, Maurizio, "La 'nuova scienza' del 'Giornale de' letterati' di Benedetto Bacchini (Parma, Modena 1686–1697)," in Renzo Cremante and Walter Tega (eds), *Scienza e letteratura nella cultura italiana del settecento* (Bologna: Il Mulino, 1984), 373–9.

Manfredi, Eustachio, "Praefatio," in Francesco Bianchini, Observationes (1737), pp. i–xiii.

Manilius, Marcus, *Astronomicon*, ed. Richard Bentley (London: H. Woodfall, 1739).

Manuel, Frank E., *Isaac Newton Historian* (Cambridge, MA: Harvard University Press, 1963).

Marchi, Gian Paolo, "Un confronto ineludibile: Scipione Maffei e Ludovico Antonio Muratori," in Romagnani, *Scipione Maffei* (1998), 363–97.

Marini, Gaetano, *Iscrizioni antiche delle ville e de' palazzi Albani* (Rome: Giunchi, 1785).

Markuszewska, Aneta, "Political Allusions in Music Dedicated to James and Maria Clementina in 1719," in Maher, *Irish* (2021), 147–65.

Marsigli, Luigi Ferdinando, *Osservazioni intorno al Bosforo tracio, overo Canale di Constantinopoli* (Rome: Tinassi, 1681).

Martianay, Jean, *Défense du texte hébreu et de la chronologie de la vulgate contre le livre de l'Antiquité des tems retablie* (Paris: L. Rouland, 1689).

Marzocchi, Roberto, "Contributo alla figura e alla personalità di Francesco Bianchini 'erudito bibliotecario,'" *Civiltà veronese*, 3rd ser., 1/3 (1999), 39–55.

Masini, Antonio, *Bologna perlustrata*, 3rd edn (Bologna: Benacci, 1666).

Mastai Ferretti, Paolino, *Notizie storiche delle accademie d'Europa* (Rome: Lazzazrini, 1792).

Matitti, Flavia, "Il cardinale Pietro Ottoboni mecenate delle arti: Cronache e documenti (1689–1740)," *Storia dell'arte*, 84 (1995), 156–243.

Matitti, Flavia, "Le anticihità di casa Ottoboni," *Storia dell'arte*, 90 (1997), 201–49.

Maugain, Gabriel, *Étude sur l'évolution intellectuelle de l'Italie de 1657 à 1750 environ* (Paris: Hachette, 1909).

Mazzoleni, Alessandro, *Vita di monsignor Francesco Bianchini, Veronese* (Verona: Targa, 1735). Cited as MV.

Meijer, Cornelis, *L'arte di restituire à Roma la tralasciata navigatione del suo Tevere . . . [e] d'alcun' altre propositioni proficue per lo stato ecclesiastico* (Rome: Tinassi, 1685).

Meijer, Cornelis, *Nuovi ritrovamenti . . . per eccitare l'ingegno de' virtuosi* (Rome: Komarek, 1689, 1696).

Melville, Lewis, *The Life and Writings of Philip, Duke of Wharton* (London: John Lane, 1913).

Menniti Ippolito, Antonio, *Fortuna e sfortune di una famiglia veneziana nel seicento: Gli Ottoboni al tempo dell'aggregazione al patriziato* (Venice: Istituto di scienze, lettere ed arti, 1996).

Merolla, Riccardo, "Lo stato della chiesa," in *Letteratura italiana: Storia e geografia: L'età moderna*, 2 vols (Turin: Einaudi, 1988), ii. 1019–1109.

Merrifield, Mary Philadelphia, *The Art of Fresco Painting as Practised by the Old Italian and Spanish Masters* (London: for the author, 1846).

Messbarger, Rebecca, et al. (eds), *Benedict XIV and the Enlightenment: Art, Science, and Spirituality* (Toronto: Toronto University Press, 2016).

Metzler, Josef, "L'Accademia dei Concili nel Collegio Urbano di 'Propaganda Fede' (1671–1756)," *Euntes docete*, 36 (1983), 233–46.

Middlehurst, B. M., and J. M. Burley, *Chronological Catalog of Reported Lunar Events from 1540 to 1966*, NASA Technical Report Tr R-277, Goddard Space Flight Center (April 1966).

Middleton, W. E. Knowles, *The Experimenters: A Study of the Accademia del Cimento* (Baltimore: Johns Hopkins University Press, 1971).

Middleton, W. E. Knowles, "Science in Rome, 1675–1700, and the Accademia Fisico-matematica of Giovanni Giustino Ciampini," *British Journal for the History of Science*, 8 (1975), 138–54.

Miller, Peggy, *A Wife for the Pretender* (London: George Allen and Unwin, 1965).

Millon, Henry A., "Reconstruction of the Palatine in the 18th Century," in Ann R. Scott and Russell T. Scott (eds), *Eius virtutis studiosi: Classical and Postclassical Studies in Memory of Frank Edward Brown (1908–1988)* (Washington: National Gallery of Art, 1993), 479–93.

Miranda, Silvana, *Francesco Bianchini e lo scavo farnesiano del Palatino (1700–1729)* (Florence: Nuova Italia, 2000).

Molari, Luisa, and Pier Gabriele Molari, *Il trionfo del'ingegneria nel fregio del palazzo ducale d'Urbino* (Pisa: ETS, 2006).

Momigliano, Arnaldo Dante, *Contributo alla storia degli studi classici e del mondo antico* (Rome: Storia e letteratura, 1955).

Momigliano, Arnaldo Dante, *Terzo contributo alla storia degli studi classici e del mondo antico*, 2 vols (Rome: Storia e letteratura, 1966).

Monaco, Giuseppe, "Un parere di Francesco Bianchini sui telescopi di Giuseppe Campani," *Physis*, 21 (1983), 413–31.

Monaldi, Rita, and Francesco Sorti, *Imprimatur* (Edinburgh: Polygon, 2009).

Monod, Paul, Murray Pittock, and Daniel Szechi (eds), *Loyalty and Identity: Jacobites at Home and Abroad* (Basingstoke: Palgrave, 2010).

Montalto, Lina, "Un ateneo internazionale vagheggiato in Roma sulla fine del secolo xvii," *Studi romani*, 10 (1962), 660–73.

Montanari, Geminiano, *Cometes Bononiae observatus anno 1664 & 1665: Astronomophysica disseratio* (Bologna: G. B. Ferroni, 1665).

Montanari, Geminiano, *Pensieri fisico-matematici sopra alcune esperienze fatte in Bologna . . . intorno diversi effetti de' liquidi in cannuccie di vetro, & altri vasi* (Bologna: Manolessi, 1667).

Montanari, Geminiano, *Speculazioni fisiche . . . sopra gli effetti di que' vetri temprati, che rotti in una parte si risolvono tutti in polvere* (Bologna: Manolessi, 1671).

Montanari, Geminiano, "Sopra la sparizione d'alcune stelle, et altre novità celesti: Discorso astronomico," in Accademia de' Gelati, Bologna, *Prose* (Bologna: Manolessi, 1671), 369–92.

Montanari, Geminiano, "Discorso del vacuo recitato nell' Accademia della Traccia [28 Nov 1675]," in Montanari, *Forze* (1694), 271–312.

Montanari, Geminiano, *La fiamma volante: Gran meteora veduta sopra l'Italia la sera de 31. Marzo mdclxxvi: Speculazioni fisiche, et astronomiche* (Bologna: Manolessi, 1676).

Montanari, Geminiano, *Manueletto dei bombisti* (Venice: A. Poletti, 1680, 1682, 1684).

Montanari, Geminiano, *Copia di due lettere . . . sopra i moti, e le apparenze delle due comete ultimamente apparse sul fin di Novembre 1680* (Venice: Poletti, 1681).

Montanari, Geminiano, "Della natura et uso degli atomi o sia corpuscoli appresso i moderni: Trattato fisico-matematico [1682–3]," in Altieri Biagi and Basile, *Scienziati* (1980), 537–52.

Montanari, Geminiano, *La zecca in consulta di stato* [1683], in Augusto Graziani (ed.), *Economisti del cinque e seicento* (Bari: Laterza, 1913), 237–79.

Montanari, Geminiano, *L'astrologia convinta di falso col mezzo di nuove esperienze, e ragioni fisico-astronomiche* (Venice: F. Nicolini, 1685).

Montanari, Geminiano, *Le forze d'Eolo: Dialogo fisico-matematico sopra gli effetti del vortice . . . che il giorno 29 luglio 1686 ha scorso e flagellato molto ville . . .. Opera postuma* (Parma: A. Poletti, 1694).

Montanari, Tomaso, "La dispersione delle collezioni di Cristina di Svezia: Gli Azzolini, gli Ottoboni e gli Odescalchi," *Storia dell'arte*, 90 (1997), 250–300.

Montesquieu, *Œuvres completes*, 2 vols (Paris: Gallimard, 1949).

Montfaucon, Bernard de, *The Travels of the Learned Father Montfaucon from Paris thro' Italy* (London: E. Curll et al., 1712).

Montfaucon, Bernard de, *Supplément au livre de l'antiquité*, 3 vols (Paris: Delaulne et al., 1724).

Morei, Michel Giuseppe, *Memorie istoriche dell' Adunanze degli Arcadi* (Rome: Rossi, 1761).

Morello, Nicoletta, "Tra diluvio e vulcani: Le concezioni geologiche di Francesco Bianchini e del suo tempo," in CR 185–206.

Morét, Stefan, "Einige Bemerkungen zu Agostino Masucci," in Sybille Ebert-Schifferer et al. (eds), *Maratti e la sua fortuna* (Rome: Campisano, 2016), 211–26.

Mori, Attilio, "Studi, trattive e proposte per la costruzione di una carta geografica della toscana nella seconda metà del secolo xviii," *Archivio storico italiano*, 35 (1705), 369–424.

Muratori, Ludovico Antonio, *Riflessioni sopra il buon gusto nelle scienze e nelle arti*, 2 pts (1708, 1715), in Muratori, *Opere* (1964), i. 221–85.

Muratori, Ludovico Antonio, *Rerum italicarum scriptores*, 25 vols (Milan: Societas Palatina, 1723–51).

Muratori, Ludovico Antonio, *Delle forze dell'intendimento umano* (Venice: Pasquale, 1748).

Muratori, Ludovico Antonio, *Epistolario*, ed. Matteo Càmpori et al., 14 vols (Modena: Società tipografica modenese, 1901–22).

Muratori, Ludovico Antonio, *Opere*, ed. Giorgio Falco and Fiorenzo Forti, 2 vols (Milan: Ricciardi, 1964).

Muratori, Ludovico Antonio, *Carteggi*, ed. Ennio Ferraglio et al., vii (Florence: Olschki, 2014).

Murtagh, Diarmuid, "Colonel Edward Wogan," *Irish Sword*, 2 (1954–56), 43–53.

Museu nacional dos coches, *Embaixada de D. Rodrigo Amas de Sá Almeida e Menezes, Marquês de Fontes enviada por D. João V ao Papa Clemente XI* (Lisbon: Museu nacional dos coches, 1996).

Nauerth, Claudia, "Einleitung," in Agnellus, *Liber* (1996), i. 9–75.

Needham, Joseph, and Wang Ling, *Science and Civilization in China*, iii. *Mathematics and the Sciences of the Heavens and Earth* (Cambridge: Cambridge University Press, 1970).

Negroni, Niccolò, "Ragionamento . . . in occasione d'interpretare la seconda risposta dell'Oracolo data da [Cardinal Pietro Ottoboni] . . . interrogato se l'Arcadia nel corso della nuova Olimpiade doveva essere felice," in Crescimbeni, *Prose* (1718), iii. 58–62.

Newton, Isaac, *Opticks* [1704, 1730] (New York: Dover, 1952).

Newton, Isaac, *Mathematical Principles of Natural Philosophy*, trans. Alexander Motte (1729), ed. Florian Cajori (Berkeley and Los Angeles: University of California Press, 1934).

Newton, Isaac, *The Correspondence*, ed. H. W. Turnbull et al., 7 vols (Cambridge: Cambridge University Press, 1959–77).

Newton, Isaac, *Philosophiae naturalis principia mathematica: The Third Edition (1726) with Variant Readings*, ed. Alexandre Koyré and I. B. Cohen, 2 vols (Cambridge, MA: Harvard University Press, 1972).

Nicéron, Jean Pierre, "François Bianchini," *Mémoires pour servir à l'histoire des hommes illustres dans la République des Lettres*, 29 (1734), 77–89.

Nicolini, Fausto, *Uomini di spada di chiesa di toga di studio ai tempi di Giambattista Vico* (Bologna: Il Mulino, 1992).

O'Callaghan, John Cornelius, *History of the Irish Brigades in the Service of France* (Glasgow: Cameron and Ferguson, 1870).

Oechslin, Werner, "Storia d'archeologia prima del Piranesi: Nota su Francesco Bianchini," in *Piranesi nei luoghi di Piranesi*: iii. *Archaeologia Piranesiana* (Rome: Multigrafica, 1979), 107–20.

Oldenburg, Henry, *Correspondence*, ed. A. Rupert Hall and Marie Boas Hall, 13 vols (Madison: University of Wisconsin Press, 1965–73 (vols i–ix); London: Mansell, 1975–7 (vols x–xi); London: Taylor and Francis, 1986 (vols xii–xiii)).

Olson, Roberta J. M., and Jay M. Pasachoff, *Fire in the Sky: Comets and Meteors* (Cambridge: Cambridge University Press, 1998).

Olszewski, Edward J., "The Painters in Cardinal Pietro Ottoboni's Court of the Cancelleria, 1689–1740," *Römisches Jahrbuch der Bibliotheca Hertziana*, 32 (1997–8), 535–66.

Olszewski, Edward J., *Cardinal Pietro Ottoboni (1667–1740) and the Vatican Tomb of Alexander VIII* (Philadelphia: American Philosophical Society, 2004).

Onorati, Francesco Maria, *Apologia . . . per la passonata fatta sopra il Tevere fuora di Porta del Poplo in difesa della strada Flaminia con la direttione del Signor Cornelio Meijer* (Rome: Bernabò, 1698).

Ormonde, James Butler, 2nd Duke of, *The Letter of the Duke of Or—nd to All True Lovers of the Church of England, and their Country* (n.p., n.d. [1714]). Copy in Bodleian Library, Oxford, G. Pamph. 1686: 36.

Ormonde, James Butler, 2nd Duke of, *Mémoires de la vie de Mylord Duc d'Ormond*, 2 vols (The Hague: Aux depens de la compagnie, 1737).

Palumbo, Genoveffa, *Giubileo giubilei* (Rome: RAI-ERI, 1999).

Panvinio, Onofrio, *De ludis circensibus libri II* (Venice: G. B. Ciotti, 1600).

Papa, E., "Politica ecclesiastica nel Regno di Napoli tra il 1708 e il 1710," *Gregorianum*, 36 (1955), 626–68; 37 (1956), 55–87.

Paragallo, Gaspare, *Istoria naturale del Monte Vesuvio* (Naples: G. Raillard, 1705).

Paravia, Pier-Alessandro, "Vita di M. Francesco Bianchini," in Francesco Bianchini, *Istoria universale*, 5 vols (Venice: G. Battagia, 1825–7), i, pp. xv–xliv.

Paschini, Pio, "Mons. Giovanni Ciampini e la Conferenza dei concili a 'Propaganda Fide'," Atti della Pontificia Accademia romana di archeologia, *Rendiconti*, 11 (1935), 95–106.

Pastor, Ludwig, *The History of the Popes*, 40 vols, vols i–ii (London: J. Hodges, 1891); vols iii–xl (London: Kegan Paul, 1894–1953).

Patin, Charles, *Introduzione alla storia della pratica delle medaglie* (Venice: G. G. Hertz, 1673).

Patin, Charles, *Lyceum patavinum, sive icones et vitae professorum Patavii, 1682, publice docentium* (Padua: P. M. Frambotti, 1682).

Patin, Charles, *Histoire des médailles, ou Introduction à la connoissance de cette science* (Paris: Cramoisi, 1695).

Perizonius, Jacobus, *Aegyptiarum originum et temporum antiquissimorum investigatio* (Leyden: J. vander Linden, 1711).

Petavius, Dionysius (Pétau, Denis), *Uranologion, sive systema variorum authorum qui de sphaera ac sideribus eorumque motibus graecè commentati sunt* (Paris: Cremoisy, 1630).

Petavius, Dionysius (Pétau, Denis), *Abrégé chronologique de l'histoire universelle sacrée et profane*, 3 vols (Paris: L. Billaine, 1682).

Petavius, Dionysius (Pétau, Denis), *Opus de doctrina temporum auctius . . . notis et emendationibus quampluribus* (Antwerp: Gallet, 1703).

Pezron, Paul, *L'Antiquité des tems rétablie et défendue contre les Juifs et les nouveaux chronologistes* (Paris: Martin et al., 1687).

Pezzini, Grazia Bernini, *Il fregio dell'arte della guerra nel Palazzo Ducale di Urbino: Catalogo dei rilievi* (Rome: Istituto poligrafico e Zecca dello stato, 1985).

Phipson, T. L., *Meteorites, Aerolites, and Falling Stars* (London: L. Reeve, 1867).

Pighetti, Clelia, *L'influsso scientifico di Robert Boyle nel tardo '600 italiano* (Milan: FrancoAngeli, 1988).

Pighetti, Clelia, *Il vuoto e la quiete: Scienza e mistica nel '600: Elena Cornaro e Carlo Rinaldini* (Milan: FrancoAngeli, 2005).

Pinto, John, *Speaking Ruins: Piranesi, Architects, and Antiquity in Eighteenth-Century Rome* (Ann Arbor: University of Michigan Press, 2012).

Piperno, Franco, "Crateo, Olinto, Archimede e l'Arcadia: Rime per alcuni spettacoli operistici romani (1710–1711)," in Nino Pirotta and Agostino Zijno (eds), *Händel e Scarlatti a Roma* (Florence: Olschki, 1987), 349–65.

Platania, Gaetano, "Una pagina inedita del soggiorno romano di Maria Casimira Sobieska," *Studia italo-polonica*, 3 (1987), 81–113.

Platania, Gaetano, "Viaggio in Italia e soggiorno romano di una dama polacca: Maria Casimira Sobieska," in Kanceff and Lewanski, *Viaggiatori* (1988), 165–81.

Platania, Gaetano, *Gli ultimi Sobieski a Roma: Fasti e miserie di una famiglia reale polacca tra sei e settecento (1699–1715)* (Rome: Vecchiarelli, 1990).

Platania, Gaetano, *La politica europea e il matrimonio inglese di una principessa polacca, Maria Clementina Sobieska* (Rome: Vecchiarelli, 1993). (Accademia polacca delle scienze, Conferenze, 101.)

Platania, Gaetano, "Maria Casimira Sobieska e Roma: Alcuni episodi del soggiorno romano di una regina polacca," in Istituto nazionale di studi romani (ed.), *Il viaggio* (Rome: Bulzoni, 1995), 3–48.

Platania, Gaetano (ed.), *L'Europa di Giovanni Sobieski: Cultura, politica, mercatura e società* (Viterbo: Sette Città, 2005).

Platania, Gaetano (ed.), *Da est ad ovest: Viaggiatori per le strade del mondo* (Viterbo: Sette Città, 2006).

Platania, Gaetano, *Polonia e Curia Romana: Corrispondenza tra Giovanni III Sobieski, re di Polonia con Carlo Barberini prottetore del regno, 1681–1696* (Viterbo: Sette citta, 2011).

Polidori, Pietro, *De vita e rebus gestis Clementis XI* (Urbino: A. Fantauzzi, 1727).

Polignac, François de, "La 'Fortune' du columbarium," *Eutopia*, 2/1 (1993), 41–63.

Polignac, François de, "Francesco Bianchini et les 'cardinaux antiquaries:' Archéologie, science et politique," in KS 165–78.

Polignac, François de, "La doppia chiave del S. Pietro: Il museo di storia ecclesiastica spiegato in una lettera di Bianchini al cardinale Gualterio," in CR 271–84.

Pometti, Francesco, "Studii sul pontificato di Clemente XI," *Archivio della Società romana di storia patria*, 21 (1898), 279–457; 22 (1899), 109–79; 23 (1900), 239–76, 449–515.

Le pompe funebri celebrate da' signori Academici infecondi di Roma per la morte dell'illustrissima Signora Elenena Lucrezia Cornaro Piscopia (Padua: Cadorino, 1686).

Ponzi, Giuseppe Dionisio, *Cometicae observationes habitae ab Academia physicomathematica Romana, anno 1680, et 1681* (Rome: Tinassi, 1685).

Popkin, Richard, *The History of Skepticism from Erasmus to Spinoza* (Berkeley and Los Angeles: University of California Press, 1979).

Popkin, Richard, *Isaac de la Peyrère, 1596–1676* (Leyden: Brill, 1987).

Porro, Francesco, *Schizzi di carte celeste deliniate da Francesco Bianchini sopra osservazioni proprie e di Geminiano Montanari* (Genoa: Pagano, 1902).

Porzio, Luc'Antonio, *Del sorgimento de' licori nelle fistole aperte d'ambidue gli estremi . . . Discorso* (Venice [for Naples]: n.p., 1667).

Posterla, Francesco, et al., *Roma sacra e moderna . . . aggiuntovi . . . un Diario istorico, che contiene tutto ciò che è accaduto di più memorabile in Roma dalla clausura delle porte sante del 1700 fino all'apertura delle medisime nell'anno 1724* (Rome: Mainardi, 1725).

Previdi, Elena, "Francesco Bianchini (1662–1729) e la sua dissertazione sugli strumenti musicali dell'antichità," *Fonti musicali italiane*, 12 (2007), 39–69.

Pucci, Giuseppe, "L'archeologia di Francesco Bianchini," in CR 259–70.

Quincy, L. D. C. H. D., *Mémoires sur la vie de Mr le Comte de Marsigli*, i (Zurich: C. Orell, 1741).

Quondam, Amedeo, "L'Instituzione arcadia: Sociologia e ideologia di un'accademia," *Quaderni storici*, 8/2 (1973), 389–438.

Raimondi, Ezio, "I padri Maurini e l'opera del Muratori," *Giornale storico della letteratura italiana*, 128 (1951), 429–71.

Réaumur, René Antoine Ferchault de, "Description d'une machine portative, propre à soûtenir des verres de très-grand foyers: Présentée à l'Académie par M. Bianchini," *HAS* (1713), 299–306.

Reboulet, Simon, *Histoire de Clément XI, Pape*, 2 vols (Avignon: Delorme and Girard, 1752).

Remmert, Volker R., "Picturing Jesuit Anti-Copernican Consensus: Astronomy and Biblical Exegesis in the Engraved Title-Page of Clavius' *Opera mathematica*," in John W. O'Malley et al. (eds), *The Jesuits II: Cultures, Sciences, and the Arts, 1540–1773* (Toronto: University of Toronto Press, 2016), 291–313.

Renaldo, John J., *Daniello Bartoli. A Letterato of the Seicento* (Naples: Istituto italiano per gli studi storici, 1979).

Reventlow, Henning Graf, "Richard Simon und seine Bedeutung für die kritische Erforschung der Bibel," in Schwaiger, *Kritik* (1980), 11–36.

Ribouillault, Denis, "Sundials on the Quirinal: Astronomy and the Early Modern Garden," in H. Fischer et al. (eds), *Gardens, Knowledge and the Sciences in the Early Modern Period* (Switzerland: Springer International, 2016), 103–34.

Ricuperati, Giuseppe, "Franceso Bianchini e l'idea di storia universale 'figurata,'" *Rivista storica italiana*, 117 (2005), 872–943.

Rinne, Katherine Wentworth, *The Waters of Rome: Aqueducts, Fountains, and the Birth of the Baroque City* (New Haven: Yale University Press, 2010).

Ripperdà, Juan Guillermo, baron de, *Mémoires*, trans. John Campbell (London: Stagg and Browne, 1740).

Robinet, André, *G. W. Leibniz: Iter italicum (mars 1689–mars 1690)* (Florence: Olschki, 1988).

Rodolico, Francesco, "Marsili (or Marsigli), Luigi Ferdinando," in *Dictionary of Scientific Biography*, ed. Charles Gillispie, ix (New York: Scribners, 1974), 134–6.

Romagnani, Gian Paolo (ed.), *Scipione Maffei nell'Europa del settecento* (Verona: Cierre, 1998).

Romagnani, Gian Paolo, *"Sotto la bandiera dell'istoria:" Eruditi e uomini di lettere nell' Italia del settecento* (Verona: Cierre, 1999).

Romanin, Samuele, *Storia documentata di Venezia*, vii (Venice: Naratovich, 1858).

Rosa, Mario, "Un 'médiateur' dans la République des Lettres: Le Bibliothécaire," in Bots and Waquet, *Commercium* (1994), 81–99.

Roncetti, Antonio, *Lettere inedite scientifico-letterarie di Lodovico Muratori [et al.]* (Milan: G. Silvestri, 1845).

Roscommon, Wentworth Dillon, 4th Earl of, *An Essay on Translated Verse* [1684] (London: H. Hills, 1709).

Rossi, Teodosio, *Horarium universale perpetuum* (Rome: Typis Vaticanis, 1637).

Rotta, Salvatore, *Francesco Bianchini in Inghilterra: Contributo alla storia del newtonianismo in Italia* (Brescia: Paideia, 1966).

Rotta, Salvatore, "Scienza e 'pubblica felicità' in Geminiano Montanari," *Miscellenea seicento*, 2 (1971), 64–210. Also in Rotta, *Montanari* (2021), 21–141.

Rotta, Salvatore, "L'Accademia fisicomatematica Ciampiniana: Un'iniziativa di Cristina?" in W. Di Palma et al. (eds), *Cristina di Svezia. Scienza ed alchimia nella Roma barocca* (Bari: Dedalo, 1990), 99–186. Also in Rotta, *Montanari* (2021), 179–236.

Rotta, Salvatore, *Geminiano Montanari e altri studi di storia della scienza nella prima età moderna*, ed. Davide Arecco et al. (Milan: Mimesis Edizioni, 2021).

Rowlands, Guy, "Louis XIV, Vittorio Amedeo II and French Military Failure in Italy, 1689–96," *English Historical Review*, 115 (2000), 534–69.

Rowlinson, John S., "John Freind: Physician, Chemist, Jacobite, and Friend of Voltaire's," RSL, *Notes & Records*, 61 (2007), 109–27.

Rusconi, Roberto, "Benedict XIV and the Holiness of the Popes in the First Half of the Eighteenth Century," in Messbarger et al., *Benedict XIV* (2016), 276–96.

Sabbatini, Renzo, "Una repubblica tra due re: La visita a Lucca del Pretendente Stuart nelle settimane dell'Atterbury plot," *Mediterranea, ricerche storiche*, 15 (2018), 95–124, 541–66.

Sanctorio, Giovanni. "Joannis Justini Ciampini vita," in Ciampini, *Opera* (1747), i, pp. xvii–xxxi.

Sankey, Margaret, *Jacobite Prisoners of the 1715 Rebellion* (Burlington: Ashgate, 2005).

Saunders, E. Stuart, "Abbé François Gaultier: Secret Envoy," in Frey and Frey, *Treaties* (1995), 177–80.

Scano, Gaetanina, "Dalle pagine di un diario: Visite, incontri e cortesie tra un pontefice e una regina," *Strenna dei romanisti*, 25 (1964), 451–55.

Schaefer, Bradley E., "The Latitude and Epoch for the Origin of the Astronomical Lore of Eudoxus," *Journal for the History of Astronomy*, 35 (2004), 161–223.

Schaefer, Bradley E., "The Epoch of the Constellations on the Farnese Atlas and their Origin in Hipparchus's Lost Catalogue," *Journal for the History of Astronomy*, 36 (2005), 167–96.

Schiavo, Armando, *Il palazzo della Cancelleria* (Rome: Staderini, 1964).

Schiavo, Armando, *La meridiana di S. Maria degli Angeli* (Rome: Istituto poligrafico e Zecca dello stato, 1993).

Schröter, Elisabeth, "Winckelmanns Projekt einer Beschreibung der Altertümer in den Villen und Palästen Roms," in T. W. Gaehtgens (ed.), *Johann Joachim Winckelmann, 1717–1768* (Hamburg: F. Meiner, 1986), 55–119.

Schwaiger, Georg, *Historische Kritik in der Theologie* (Göttingen: Vandenhoek & Ruprecht, 1980).

Seton, Walter, "Itinerary of King James III, October to December 1715," *Scottish Historical Review*, 21 (1923–4), 249–66.

Seward, Desmond, *The King over the Water: A Complete History of the Jacobites* (Edinburgh: Birlinn, 2021).

Sherburn, George Wiley, *The Early Career of Alexander Pope* (Oxford: Oxford University Press, 1934).

Shield, Alice, and Andrew Lang, *The King over the Water* (London: Longmans, Green, 1907).

Siebenhühner, Herbert, "S. Maria degli Angeli in Rom," *Münchner Jahrbuch der bildenden Kunst*, 6 (1955), 179–206.

Sigismondi, Costatino, "Misura del ritario accumulato della rotazione terrestre, ΔUT1, alla meridiana clementina," in Manuela Incerti (ed.), *Mensura caeli, Convegno Nazionale della Società Italiana di Archeoastronomia, VIII, Atti* (2010), 240–8.

Sigismondi, Costatino, "Lo gnomone Clementino: Astronomia meridiana in chiesa del '700 ad oggi," *Astronomia* (March–April 2011), 56–62.

Sigismondi, Costatino, "Lo gnomone Clementino: Astronomia meridiana in basilica," *Gerbertus*, 7 (2014), 3–80.

Sigismondi, Costatino, "Summer Solstice Decorations on the Meridian Line of S. Maria degli Angeli in Rome Are Forgotten Obliquity Meters," *Gerbertus*, 11 (2018), 21–26.

Sigismondi, Costatino, "Nuovi studi astronometri sulla meridiana della Madonna degli Angeli in Roma (2018–20) e sul'astronomia del xviii secolo," *Gerbertus*, 12 (2019), 1–6.

Simon, Richard, *Histoire critique du Vieux Testament* (Paris: Veuve Billaine, 1678).

Smith, James Edward, *A Sketch of a Tour on the Continent*, 2nd edn, 3 vols (London: Longman et al., 1807).

Sölch, Brigitte, "Das 'Museo Ecclesiastico:' Beginn einer neuen Sammlungsära im Vatikan," in KS 179–205.

Sölch, Brigitte, *Francesco Bianchini (1662–1729), und die Anfänge öffentlicher Museen in Rom* (Munich: Deutsche Kunstverlag, 2007).

Sölch, Brigitte, "Bianchini e l'inizio dei musei pubblici a Roma," in CR 309–21.

Soli Muratori, Gian-Francesco, *Vita del proposto Lodovico Antonio Muratori* (Arezzo: Bellotti, 1767). (Muratori, *Opere*, i.)

Soppelsa, Marialaura, *Genesi del metodo Galileiano e tramonto dell'aristotelismo nella scuola di Padova* (Padua: Atenore, 1974).

Sorbelli, Tommaso, "Benedetto Bacchini e la Repubblica letteraria del Muratori," *Benedictina*, 6 (1952), 85–98.

Sousa Leitão, Henrique de, "D. João V patrono do astrónomo Bianchini," in *Estrelas de papel—livros de astronomia dos sécolos xiv a xviii* (Lisbon: Biblioteca national, 2009), 49–64.

Spagnolo, Antonio, "Francesco Bianchini e le sue opera," *Atti e memorie dell'Accademia di agricultura, scienze e lettere di Verona*, 74/2 (1898), 89–122.

Spagnolo, Antonio, *I manoscritti della Biblioteca capitolare di Verona* (Verona: Mazziana, 1996).

Stanhope, Philip Henry, Earl of, *History of England, Comprising the Reign of Queen Anne until the Peace of Utrecht, 1701–1713*, 2 vols (London: J. Murray, 1872).

Stark, Carl Bernhard, *Handbuch der Archäologie der Kunst*, i. *Systematik und Geschichte* (Leipzig: Engelmann, 1880).

Stosch, Phlipp von, *Gemmae antiquae caelatae/Pierres antiques gravées* (Amsterdam: B. Picart, 1724).

Stoye, John, *Marsigli's Europe, 1680–1730: The Life and Times of Luigi Ferdinando Marsigli, Soldier and Virtuoso* (New Haven: Yale University Press, 1994).

Suetonius, Gaius, *Lives of the Caesars*, trans. J. C. Rolfe, ed. K. R. Bradley, 2 vols (Cambridge, MA: Harvard University Press, 2014).

Swift, Jonathan, *Miscellaneous Pieces, in Prose and Verse*, ed. John Nichols (London: C. Dilly, 1789).

Swift, Jonathan, *Correspondence*, ed. David Woolley, 4 vols (Frankfurt am Main and New York: Peter Lang, 1999–2014).

Szechi, Daniel, "The Image of the Court: Idealism, Politics, and the Evolution of the Stuart Court, 1689–1730," in Corp, *Stuart Court* (2003), 49–64.

Tabarroni, Giorgio, "La posizione degli equinozi sulla sfera dell'Atlante Farnese," *Coelum*, 24 (1956), 169–74.

Takahashi, Kenichi, "Il cannocchiale in Arcadia: Nuove proposte per le *Osservazioni astronomiche* di Donato Creti," *Zeitschrift für Kunstgeschichte*, 82 (2019), 179–96.

Targioni-Tozzetti, Giovanni, *Notizie degli aggradimenti delle scienze fisiche accaduti in Toscana nel corso di anni lx del secolo xvii*, 3 vols in 4 (Florence: G. Bouchard, 1780).

Tayler, Henrietta, "Introduction," in Tayler, *Jacobite Court* (1938), 3–48, 111–42.

Tayler, Henrietta, *The Jacobite Court at Rome in 1719, from Original Documents* (Edinburgh: Scottish Historical Society, 1938).

Thompson, James Westfall, *A History of Historical Writing*, 2 vols (New York: Macmillan, 1942).

Tinazzi, Massimo, "I disegni inediti dei manoscritti di Francesco Bianchini conservati presso la Biblioteca Capitolare di Verona," Fondazione Giorgio Ronchi, *Atti*, 59/3 (2004), 407–56.

Tinazzi, Massimo, "Le ricerche di Francesco Bianchini sul globo (Atlante) farnesiano," *Società italiana di archeoastronomia, V Congresso* (September 2005), 69–85.

Tiraboschi, Girolamo, "Montanari, Geminiano," in Girolamo Tiraboschi, *Biblioteca modenese*, iii (Modena: Società Tipografica, 1783), 254–79.

Tirapicos, Luís, "The Old and the New Rome: Francesco Bianchini's Astronomical Exchanges with the Court of Lisbon," *Mediterranean Archaeology and Archaeometry*, 16/4 (2016), 503–8.

Tirapicos, Luís, *Ciência e diplomacia na corte de D. João V: A acção de João Baptista Corbone, 1722–1750*, Ph. D. thesis: University of Lisbon, 2017, http://hdl.handle.net/10456l/35028.

Todi, Simonetta, "L' 'Astro-Theology' di William Derham nella Lombardia del settecento: La 'Confutazione' di Giovanni Cadonici," *Giornale critico della filosofia italiana*, 79 (2000), 401–30.

Toomer, G. J., *Ptolemy's Almagest* (London: Duckworth, 1984).

Torcy, Jean-Baptiste Colbert, marquis de, *Mémoires*, 2 vols (London: P. Vaillant, 1757).

Torrini, Maurizio, "Giuseppe Ferroni, gesuita e galileiano," *Physis*, 15/4 (1973), 411–23.

Treggiari, Susan, "Jobs in the Household of Livia," *Papers of the British School at Rome*, 43 (1975), 48–77.

Trompf, Garry W., "On Newtonian History," in Gaukroger, Uses (1991), 213–49.

Uglietti, Francesco, *Un erudito veronese alle soglie del settecento: Mons. Francesco Bianchini 1662–1729* (Verona: Biblioteca Capitolare, 1986). Cited as Uglietti.

Ussher, James, *The Annals of the World: Deduced from the Origin of Time* (London: J. Crook and G. Bedell, 1658).

Vacant, A., and E. Mangenot (eds), *Dictionnaire de théologie catholique*, 2nd edn, 15 vols in 30 (Paris: Letouzey, 1935–50).

Valerio, Vladimiro, "Historiographic and Numerical Notes on the Atlante Farnese and its Celestial Sphere," *Der Globusfreund*, 35–7 (1987), 97–124.

Valery, Antoine Claude Pasquin, *Correspondance inédite de Mabillon et Montfaucon avec l'Italie*, 3 vols (Paris: Guilbert, 1846–7).

Valesio, Francesco, *Diario di Roma*, ed. G. Scano, 6 vols (Milan: Longanesi, 1977–9).

Van Helden, Albert, "The Telescope in the Seventeenth Century," *Isis*, 65/1 (1974), 38–58.

Vasina, Augusto, *Lineamenti culturali dell'Emilia-Romagna: Antiquaria, erudizione, storiografia dal xiv al xvii secolo* (Ravenna: Longo, 1978).

Vecchi, Alberto, "La nuova accademia letteraria d'Italia," in *Accademie e cultura: Aspetti storici tra sei e settecento* (Florence: Olschki, 1979), 39–72.

Vignoles, Alphonse de, *Chronologie de l'histoire sainte et des histoires étrangères qui la concernent depuis la sortie d'Egypte jusqu'à la captivité de Babylone*, 2 vols (Berlin: A. Haude, 1738).

Viola, Corrado, "Per un inventario dei carteggi bianchiniani," in CR 121–61.

Virgil, [*Works*], trans. H. Rushton Fairclough (Cambridge, MA: Harvard University Press, 1916).

Vogel, Lise, *The Column of Antoninus Pius* (Cambridge, MA: Harvard Uuniversity Press, 1973).

Volpato, Giancarlo, "Francesco Bianchini bibliotecario e lettore 'per censura,'" *Atti e memorie dell'Accademia di agricoltura, scienze e lettere di Verona*, 181 (2004–5), 451–516.

Waquet, Françoise, *Le Modèle français et l'Italie savante (1660–1750)* (Rome: École française de Rome, 1989).

Waquet, Françoise (ed.), *Mapping the World of Learning: The Polyhistor of Daniel Georg Morhof* (Wiesbaden: Harrassowitz, 2000).

Waquet, Françoise, "Ludovico Antonio Muratori: Le 'pio letterato' à l'épreuve des faits," in *Die Europäische Gelehrtenrepublik im Zeitalter des Konfessionalismus* (Wolfenbüttel: Herzog August Bibliothek, 2001), 87–103.

Waquet, Françoise, "De la 'Repubblica letteraria' au 'pio letterato,'" in Boutier et al., *Naples* (2005), 637–50.

Weber, Christoph, "Il referendariato di ambedue le Signature: Una forma speciale del 'servizio pubblico' delle Corte Roma e dello Stato Ponteficio," in Armand Jamme and Olivier Poncet (eds), *Offices et papauté (xive–xviie siècle): Charges, hommes, destins* (Rome: École française de Rome, 2005), 565–86.

Weitlauff, Manfred, "Die Mauriner und ihr historisch-kritisches Werk," in Schwaiger, *Kritik* (1980), 153–709.

Westfall, Richard S., *Never at Rest: A Biography of Isaac Newton* (Cambridge: Cambridge University Press, 1980).

Whiston, William, *Praelectiones physico-mathematicae*, 2nd edn, 2 vols (London: B. Motte, 1726).

Winckelmann, Johann Joachim, *Nachrichten von den neuesten Herculanischen Entdeckungen*, ed. Marianne Gross et al. (Mainz: Von Zabern, 1997).

Winckelmann, Johann Joachim, *Kleine Schriften* (Berlin: De Gruyter, 2002).

Winton, Calhoun, "Steele, Swift, and the Queen's Physician," in Henry Knight Miller et al. (eds), *The Augustan Milieu* (Oxford: Oxford University Press, 1970), 138–54.

Wogan, Charles, *The Preston Prisoners to the Ladies about Court and Town: By Way of Comfort* (London: J. Roberts, 1716), 1 p.

Wogan, Charles, *Female Fortitude, Exemplify'd, in an Impartial Narrative, of the Seizure, Escape and Marriage of the Princess Clementina Sobiesky, as it was Particularly Set down by Mr Charles Wogan* (London: n.p., 1722).

Wogan, Charles, "Mémoires sur l'entreprise d'Innsbruck en 1719," in Gilbert, *Narratives* (1894), 31–108.

Yates, Frances, "Queen Elizabeth as Astraea," *Journal of the Warburg and Courtauld Institutes*, 10 (1947), 27–82.

Zaccagnini, Guido, "Gli ultimi due anni di Cristina di Svezia a Rome (1687–1689)," *Rivista abbruzzese*, 14 (1899), 1–7, 145–55, 249–56, 357–61.

LIST OF FIGURE CREDITS AND PLATES

Figures

31. © British Library Board/Bridgeman Images
32. © British Library Board/Bridgeman Images
33. © British Library Board/Bridgeman Images
34. © British Library Board/Bridgeman Images
35. British Library/Alamy Stock Photo
36. Image courtesy of Internet Archive
37. Drawn by oup
38. Drawn by oup
39. © The Governing Body of Christ Church, Oxford
40. © The Governing Body of Christ Church, Oxford
41. Public domain, via Wikimedia Commons

Plates

01. Courtesy of Vatican Museums/Public domain
02. (a) Public domain, from G.D. Cassini, *La meridiana del tempio di S. Petronio* (Bologna: Benacci, 1695), fig. 9 after p. 75 (b) Peter Barritt/Alamo Stock Photo, from J.L. Heilbron, "Fisica e astronomia nel Settecento," in William R. Shea (ed.), *Storia delle scienze: Le scienze fisiche e astronomiche* (Turin: Einaudi, 1992), 318–443, on 349.
03. Catamo and Lucarini, *Il cielo in basilica* (2002), 50, 53, 56 / photo by Fabio Gallo
04. Jean-Pol Grandmont/Wikimedia Commons (CC BY-SA 3.0)
05. Photo courtesy of Stefano Gattei
06. Catamo and Lucarini, *Il cielo in basilica* (2002), folding plate after p. 37 / photo by Fabio Gallo
07. Catamo and Lucarini, *Il cielo in basilica* (2002), folding plate after p. 37 / photo by Fabio Gallo
08. Catamo and Lucarini, *Il cielo in basilica* (2002), 33 / photo by Fabio Gallo
09. By permission of the Ministry of Culture, National Archaeological Museum of Naples / photo by Giorgio Albano
10. National Portrait Gallery of Scotland (CC BY-NC 2.0), on permanent loan from the National Library of Scotland
11. National Galleries Scotland, purchased 1918 (CC BY-NC 2.0)
12. National Galleries Scotland, purchased 1977 with assistance from the Art Fund, the Pilgrim Trust and private donors
13. (a) Catamo and Lucarini, *Il cielo in basilica* (2002), folding plate after p. 37 / photo by Fabio Gallo (b) photo courtesy of Stefano Gattei
14. Catamo and Lucarini, *Il cielo in basilica* (2002), 83 / photo by Fabio Gallo
15. Webpicture. Wikimedia Commons/Didier Descouens (CC BY-SA 4.0)
16. Photo courtesy of Stefano Gattei

INDEX

Entries of institutions if localized will be found under the place, thus, "Rome, Santa Maria degli Angeli;" names of titled people under their highest dignity, thus, Alexander VIII Ottoboni, pope. The following short forms are used: FB, Francesco Bianchini; HRE, Holy Roman Emperor; SJ, Jesuit.